ROBOT DYNAMICS AND CONTROL

ROBOT DYNAMICS AND CONTROL

MARK W. SPONG

University of Illinois at Urbana-Champaign

M. VIDYASAGAR

University of Waterloo

JOHN WILEY & SONS

New York • Chichester • Brisbane • Toronto • Singapore

Library of Congress Cataloging in Publication Data:

Spong, Mark W.
Robot dynamics and control / Mark W. Spong, M. Vidyasagar.

1. Robots—Dynamics. 2. Robots—Control. I. Vidyasagar, M.
(Mathukumalli) II. Title.
TJ211.4.S66 1989 88-22724
629.8'92—dc19
ISBN 0-471-61243-X

Printed in the United States of America

10 9 8 7 6 5 4 3 2 1

To Matthew and John (MWS)
To Shakunthala and Aparna (MV)

PREFACE

The present age has been given many labels, among them The Nuclear Age, The Space Age, The Computer Age, and The Age of Automation. Modern science and technology is the common thread that binds these various labels together. Progress in science and technology began in prehistory when man first fashioned crude tools from stone and wood and learned that he could begin to exercise a degree of control over his environment. With the advent of the so-called Scientific Method of Investigation more than three hundred years ago, scientific and technological advancements began to accelerate until, at the present time, scientific knowledge is increasing at a phenomenal rate. This unprecedented explosion of knowledge and technological invention strains the fabric of our political, educational, religious, and legal institutions as they struggle to assimilate the changes brought about by rapid progress.

In the field of manufacturing, due in large part to recent advances in computer technology, mankind is at the threshold of a second Industrial Revolution, a revolution in Automation. The word automation, short for "Automatic Motivation," was coined in the 1940s at the Ford Motor Company, and was used to describe the collective operation of many interconnected machines. Automation is defined by the Oxford

English dictionary[1] as the "automatic control of manufacturing product through successive stages; use of automatic equipment to save mental and manual labor[2]." To the lay person, however, automation means the process of replacing human workers by machines. Thus it is often true that the word automation evokes fear and mistrust. Being "replaced by a machine" is no longer an idle threat to a significant portion of modern society. Yet historically, every advancement in technology that has led to an increase in productivity has been good in the long run. In the nineteenth century most people in the United States made their livings on farms. Farm automation took away most of those jobs so that at the present time less than two per cent of the population of the U.S. is engaged in farming. Yet there are more jobs now than ever before due to advances in technology and changing lifestyles. The same will likely be true of automation in the long run. While automation eliminates some jobs in welding, machining, assembly line work, etc., it also creates other jobs, such as robot installation and maintenance, software development and so forth. At the same time, the increased productivity brought about by automation enables a society to compete successfully in a global economy with the result that the general standard of living is raised.

Increasingly, automation has come to mean the introduction of robots into the manufacturing environment. At the present time the robot represents the highest form of automation. Although technically just machines, robots are viewed by most people in a much different light than other machines. This special, almost romantic, position occupied by robots is due partly to the Hollywood view of a robot as a "mechanical being," often with evil intent, or as a lovable companion such as R2D2 of the motion picture *Star Wars*. The word *robot* itself was introduced into our vocabulary by the Czech playwright Karel Capek in his 1920 play *Rossum's Universal Robots*; the word *robota* being the Czech word for work.

Even without the Hollywood imagery, however, humans would probably have a special attraction to robots compared to other machines for the simple reason that robots are inherently anthropomorphic devices. Indeed, the components of a robot are analogous to the human brain, senses, arms, hands, and legs. The ultimate robot would be one that could see, hear, feel, speak, move about, manipulate objects, and even think. Although our present level of technology is not able to produce such a machine, this is one of the ultimate goals of robotics research.

There have been many predictions made about the future of the robotics industry, such as that it will grow explosively into a two-

---

[1] Oxford University Press, 1971.

[2] The same dictionary defines "automatic" to mean "working of itself, without direct human actuation."

billion-dollar-a-year industry by 1990. Most of these predictions have failed to materialize for a variety of reasons. Many applications, such as assembly, have proved to be extremely difficult to automate. The present level of robotics technology, in such areas as machine vision, tactile sensing, artificial intelligence, etc., while spectacular, is still primitive compared to the marvelous adaptability and dexterity of humans. One gains a sense of awe and respect for human eyes, hands, and brains when attempting to program a robot to perform even simple tasks. There are also economic reasons why robots have not lived up to their expectations to date. A successful application of robotics requires much more than simply installing a robot on the factory floor. Most often it requires a rethinking and redesign of the entire process that is to be automated. The new field of applications engineer is beginning to address the issues of how best to automate manufacturing processes with robots.

Another very important reason for the slow growth of the robotics industry deals with the interdisciplinary nature of robotics itself. The field of robotics combines aspects of electrical, mechanical, and industrial engineering with computer science, mathematics, and economics. There is at present a critical shortage of trained people with the cross-disciplinary knowledge necessary to integrate successfully the various technologies involved in robotic applications. It is the task of the universities to provide such cross-disciplinary education.

The present text grew out of a set of lecture notes developed by the first author during 1983–84 in the School of Electrical Engineering at Cornell University. These notes have been extensively revised through subsequent courses taught in the Department of General Engineering and the Department of Electrical and Computer Engineering at the University of Illinois at Urbana–Champaign, and in the Department of Electrical Engineering at the University of Waterloo, Ontario, Canada.

The book has been written primarily for electrical engineering students who may have little or no exposure to the subjects of kinematics and dynamics of mechanical systems, but who have some knowledge of linear algebra and feedback control systems. The text can also be used in a mechanical engineering curriculum, however, and has been taught to students from nearly all other engineering disciplines, including industrial engineering, aerospace engineering, agricultural engineering, and computer science.

Recently a number of excellent textbooks have appeared that provide comprehensive treatments of robot kinematics, dynamics, trajectory planning, computer interfacing, artificial intelligence, sensing and vision, and applications. With the exception of Asada and Slotine (Chapter Two, Reference [2]), however, the area of robot control, particularly advanced methods, has received comparatively little attention to date. It is our primary intent in the present text, therefore, to provide a self-contained introduction to robot kinematics and dynamics followed by a more comprehensive treatment of robot control.

This textbook can be used for courses at two levels. Chapters One through Nine, can be used in a first course in robotics at the senior level with only a first course in feedback control systems as a prerequisite. It is very helpful in such a course to have access to a laboratory where the students can learn a specific programming language and can test some of the material on real systems. The instructor may also wish to supplement the material with outside reading on computer interfacing, sensors, and vision. At the graduate level, the entire book can be covered in one semester. The latter chapters will be most useful to students who have studied at least the state space theory of dynamical systems and who have access to computational facilities to design and simulate the advanced control algorithms presented. After covering the chapters on advanced control a graduate student should be prepared to tackle the research literature on robot dynamics and control.

The text is organized as follows. After an introductory chapter defining the basic terminology of robotics and outlining the text, Chapter Two gives some background on rotations and homogeneous transformations sufficient to follow the subsequent development. Chapter Three discusses the forward kinematics problem using the Denavit–Hartenberg convention for assigning coordinate frames to the links of a manipulator. Chapter Four discusses the problem of inverse kinematics from a geometric viewpoint that is sufficient to handle the most common robot configurations the student is likely to encounter. Chapter Five discusses velocity relationships and derives the manipulator Jacobian in the so-called cross product form and includes a discussion of singularities. Chapter Six discusses dynamics.

We have found that many students at the senior level have little exposure to Lagrangian dynamics, so we have made Chapter Six self-contained by including a derivation of the Euler–Lagrange equations from the principle of virtual work. We also discuss the recursive Newton–Euler formulation of manipulator dynamics.

Chapter Seven begins our discussion of robot control by treating the simplest approach, namely, independent joint control. We also discuss, in this chapter, the most basic approaches for trajectory interpolation. A robot is much more than just a series of rigid mechanical linkages. Thus we include material on actuator dynamics and drive-train dynamics and show how both significantly impact the manipulator control problem. We also introduce the idea of feedforward control and computed torque, which paves the way to the more advanced nonlinear control theory to follow.

Chapter Eight discusses robot control in the context of multivariable systems. Here we discuss the method of inverse dynamics and introduce the reader to the idea of robust control. Chapter Nine discusses force control, chiefly hybrid position/force control and impedance control, which are the two most common approaches to force control to date. Chapter Ten introduces the idea of feedback linearization of nonlinear systems. This is a recent and quite advanced concept in control theory. We have attempted to give a self-contained introduction to the subject while minimizing the mathematical background required of the reader. Thus, although we introduce, for example, the notion of Lie brackets of vector fields, we do so primarily as a notational convenience. All calculations are performed in local coordinates in IR^n so that we avoid the need to introduce rigorous definitions of more advanced concepts such as differentiable manifolds, tangent bundles, etc. We hope that the loss of mathematical rigor is compensated for by an increase in readability; in any case, the level of the treatment is sufficient to handle most cases of interest in robot control. Chapter Eleven discusses variable structure and adaptive control of robots. Again, we give specific applications of these ideas to the robot control problem rather than attempting to cover the most general theory. While these techniques are important to date primarily from a theoretical standpoint, they hold the promise of future practicality and also point the way to the research literature on robot control.

We had two main goals in writing this text. The first was to present the material as rigorously as possible at a senior/first-year graduate level by giving mathematical justification and proofs wherever possible. The second was to present the theory in such a way that, having read it, the reader is able actually to compute examples for himself. Thus most of the problems fall into two groups. The fist group asks the reader to fill in the gaps of some of the derivations and proofs, whereas the second group asks the reader to take a specific example and work out the various equations on his or her own.

ACKNOWLEDGMENTS

The first author would like to acknowledge the help and support of many individuals in the preparation of this manuscript. First and foremost are my colleagues in the Decision and Control Group in the Coordinated Science Laboratory at the University of Illinois, Tamer Basar, Jessy Grizzle, Petar Kokotovic, P.R. Kumar, Juraj Medanic, Bill Perkins, Kameshwar Poolla, and Mac Van Valkenburg, who provide a truly outstanding environment for scholarship and research. Next, my students Robert Anderson, Scott Bortoff, Fathi Ghorbel, John Hung, and William Scheid provided helpful comments and found numerous mistakes in early drafts of the manuscript. The simulation results in this book were performed using the TUTSIM, PC-Matlab, and PC-Simnon packages in the Robotics Laboratory in the Department of General Engineering at the University of Illinois. I would like to thank my former chairman, Jerry Dobrovolny, and my current chairman, Thomas Conry, for their support which made this facility possible and for their continued support and encouragement. The second author would like to thank his students, Y. C. Chen, Chris Ma, A. Sankar, and David Wang for educating him in robotics.

The initial discussions which eventually led to this book began during the summer of 1984 while both authors were in the Control Technology Branch at General Electric's Corporate Research and Development Headquarters in Schenectady, New York. We would like to thank John Cassidy and Larry Sweet for their support during that period, which made this collaboration possible. We would also like to thank Bruce Krogh and Ken Loparo for their critical reviews of the manuscript, and Lila Acosta for proofreading the manuscript. We would also like to thank Riccardo Marino, Romeo Ortega, Hebertt Sira-Ramirez, and Jean-Jacques Slotine, for numerous helpful discussions on the robot control problem.

Finally, we would like to thank Christina Mediate, Joe Ford, and Michael Jung at John Wiley for their patience and guidance in the preparation of this text.

MARK W. SPONG
M. VIDYASAGAR

CONTENTS

CHAPTER ONE

INTRODUCTION

1.1 INTRODUCTION

Robotics is a relatively new field of modern technology that crosses traditional engineering boundaries. Understanding the complexity of robots and their applications requires knowledge of electrical engineering, mechanical engineering, industrial engineering, computer science, economics, and mathematics. New disciplines of engineering, such as manufacturing engineering, applications engineering, and knowledge engineering, are beginning to emerge to deal with the complexity of the field of robotics and the larger area of factory automation. Within a few years it is possible that **robotics engineering** will stand on its own as a distinct engineering discipline.

In this text we explore the **kinematics, dynamics,** and **control** of robotic manipulators. In doing so we omit many other areas such as locomotion, machine vision, artificial intelligence, computer architectures, programming languages, computer-aided design, sensing, grasping, and manipulation, which collectively make up the discipline known as robotics. While the subject areas that we omit are important to the science of robotics, a firm understanding of the kinematics, dynamics, and control of manipulators is basic to the understanding and application of these other areas, so that a first course in robotics should begin with these three subjects.

1.2 ROBOTICS

The term **robot** was first introduced into our vocabulary by the Czech playwright Karel Capek in his 1920 play *Rossum's Universal Robots*, the word *robota* being the Czech word for work. Since then the term has been applied to a great variety of mechanical devices, such as teleoperators, underwater vehicles, autonomous land rovers, etc. Virtually anything that operates with some degree of autonomy, usually under computer control, has at some point been called a robot. In this text the term robot will mean a computer controlled industrial manipulator of the type shown in Figure 1-1. This type of robot is essentially a mechanical arm operating under computer control. Such devices, though far from the robots of science fiction, are nevertheless ex-

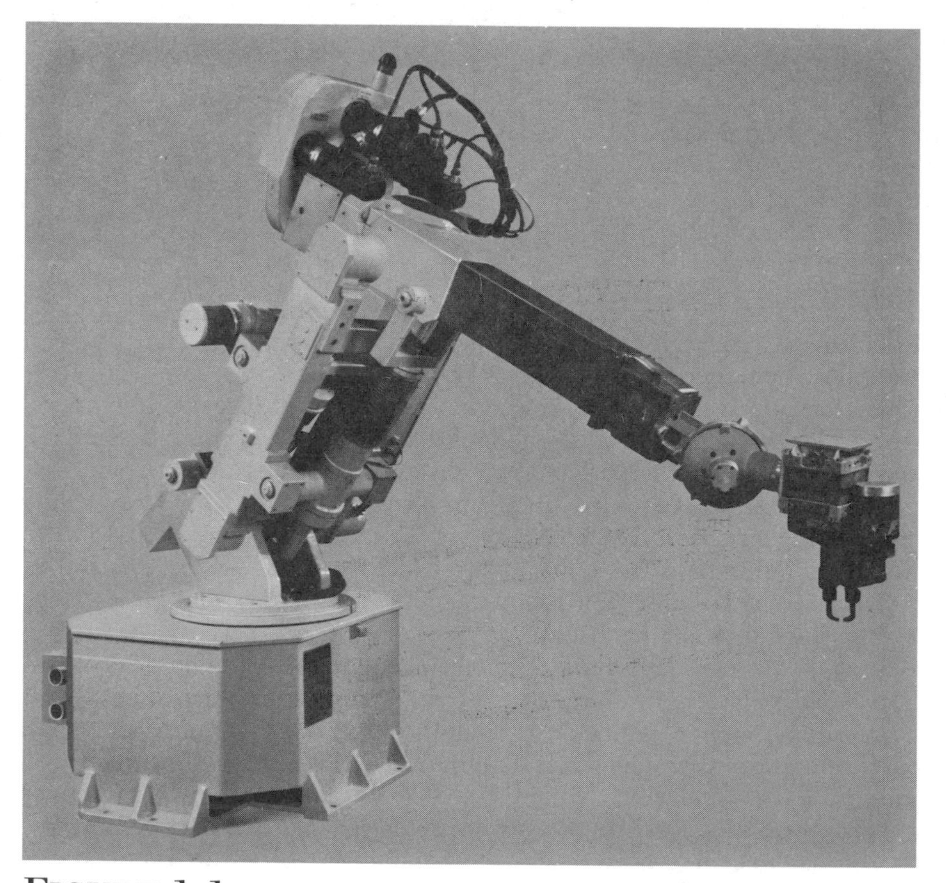

FIGURE 1-1
The Cincinnati Milacron T^3 industrial manipulator. Photo courtesy of Cincinnati Milacron.

tremely complex electro-mechanical systems whose analytical description requires advanced methods, and which present many challenging and interesting research problems.

An official definition of such a robot comes from the **Robot Institute of America** (RIA): *A robot is a reprogrammable multifunctional manipulator designed to move material, parts, tools, or specialized devices through variable programmed motions for the performance of a variety of tasks.*

The key element in the above definition is the reprogrammability of robots. It is the computer brain that gives the robot its utility and adaptability. The so-called robotics revolution is, in fact, part of the larger computer revolution.

Even this restricted version of a robot has several features that make it attractive in an industrial environment. Among the advantages often cited in favor of the introduction of robots are decreased labor costs, increased precision and productivity, increased flexibility compared with specialized machines, and more humane working conditions as dull, repetitive, or hazardous jobs are performed by robots.

The robot, as we have defined it, was born out of the marriage of two earlier technologies: that of **teleoperators** and **numerically controlled milling machines**. Teleoperators, or master–slave devices, were developed during the second world war to handle radioactive materials. Computer numerical control (CNC) was developed because of the high precision required in the machining of certain items, such as components of high performance aircraft. The first robots essentially combined the mechanical linkages of the teleoperator with the autonomy and programmability of CNC machines. Several milestones on the road to present day robot technology are listed below[17],[21]:

1947— the first servoed electric powered teleoperator is developed

1948— a teleoperator is developed incorporating force feedback

1949— research on numerically controlled milling machines is initiated

1954— George Devol designs the first **programmable robot**

1956— Joseph Engelberger, a Columbia University physics student, buys the rights to Devol's robot and founds the Unimation Company

1961— the first **Unimate** robot is installed in a Trenton, New Jersey plant of General Motors (to tend a die casting machine)

1961— the first robot incorporating force feedback information is developed

1963— the first robot vision system is developed

1971— the **Stanford Arm** is developed at Stanford University

1973—the first robot programming language (WAVE) is developed at Stanford

1974—Cincinnati Milacron introduces the T^3 robot with computer control

1975—Unimation Inc. registers its first financial profit

1976—the Remote Center Compliance (RCC) device for part insertion in assembly is developed at Draper Labs in Boston

1978—Unimation introduces the PUMA robot, based on designs from a General Motors study

1979—the SCARA robot design is introduced in Japan

1981—the first **direct-drive** robot is developed at Carnegie–Mellon University

The first successful applications of robot manipulators generally involved some sort of material transfer, such as injection molding or stamping where the robot merely attended a press to unload and either transfer or stack the finished part. These first robots were capable of being programmed to execute a sequence of movements, such as moving to a location A, closing a gripper, moving to a location B, etc., but had no external sensory capability. More complex applications, such as welding, grinding, deburring, and assembly require not only more complex motion but also some form of external sensing such as vision, tactile, or force sensing, due to the increased interaction of the robot with its environment. For a comprehensive discussion of robotic applications the reader is referred to the references at the end of this chapter, especially [6] and [8].

It should be pointed out that the important applications of robots are by no means limited to those industrial jobs where the robot is directly replacing a human worker. There are many other applications of robotics in areas where the use of humans is impractical or undesirable. Among these are undersea and planetary exploration, satellite retrieval and repair, the defusing of explosive devices, and work in radioactive environments. Finally, prostheses, such as artificial limbs, are themselves robotic devices requiring methods of analysis and design similar to those of industrial manipulators.

1.3 COMPONENTS AND STRUCTURE OF ROBOTS

Robot Manipulators are composed of **links** connected by **joints** into an **open kinematic chain**. Joints are typically rotary (revolute) or linear (prismatic). A **revolute** joint is like a hinge and allows relative rotation between two links. A **prismatic** joint allows a linear relative motion between two links. We use the convention (R) for representing revolute joints and (P) for prismatic joints as shown in Figure 1-2. Each joint represents the interconnection between two links, say, ℓ_i and ℓ_{i+1}. We denote the axis of rotation of a revolute joint, or the axis along which a prismatic joint slides, by z_i if the joint is the interconnection of links i and $i+1$. The **joint variables**, denoted by θ_i for a revolute joint and d_i for a prismatic joint, represent the relative displacement between adjacent links. We will make this precise in Chapter Three.

The joints of a manipulator may be electrically, hydraulically, or pneumatically actuated. The number of joints determines the **degrees-of-freedom** (DOF) of the manipulator. Typically, a manipulator should possess at least six independent DOF: three for **positioning** and three for **orientation**. With fewer than six DOF the arm cannot reach every point in its work environment with arbitrary orientation. Certain applications such as reaching around or behind obstacles require more than six DOF. The difficulty of controlling a manipulator increases rapidly with the number of links. A manipulator having more than six links is referred to as a **kinematically redundant** manipulator.

The **workspace** of a manipulator is the total volume swept out by the end-effector as the manipulator executes all possible motions. The workspace is constrained by the geometry of the manipulator as well as mechanical constraints on the joints. For example, a revolute joint may be limited to less than a full 360° of motion. The workspace is

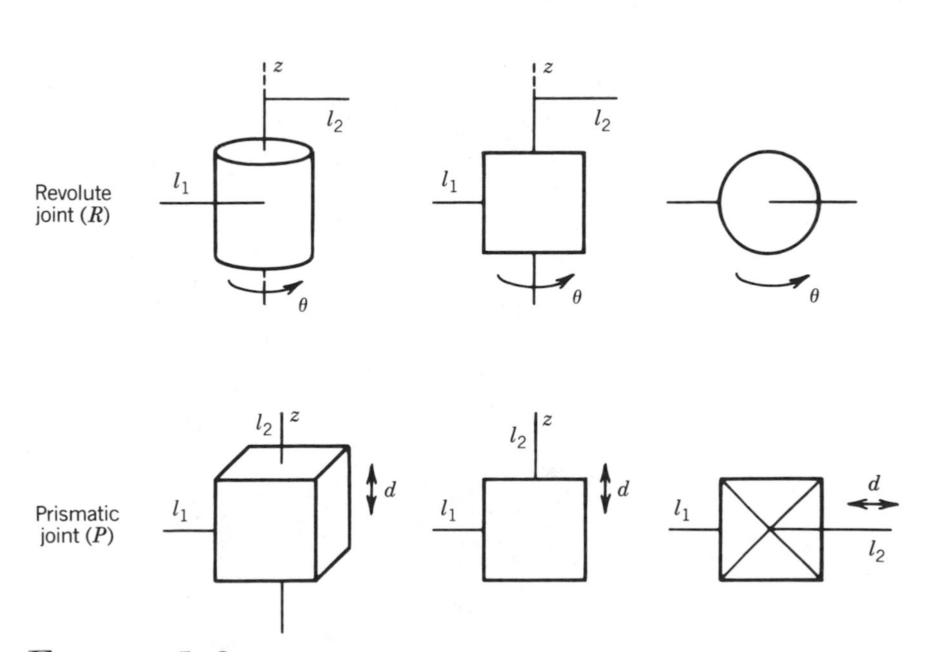

FIGURE 1-2
Symbolic representation of robot joints

often broken down into a **reachable workspace** and a **dextrous workspace**. The reachable workspace is the entire set of points reachable by the manipulator, whereas the dextrous workspace consists of those points that the manipulator can reach with an arbitrary orientation of the end-effector. Obviously the dextrous workspace is a subset of the reachable workspace.

A robot manipulator should be viewed as more than just a series of mechanical linkages. The mechanical arm is just one component to an overall **Robotic System**, shown in Figure 1-3, which consists of the **arm**, **external power source**, **end-of-arm tooling**, **external and internal sensors**, **servo**, **computer interface**, and **control computer**. Even the programmed software should be considered as an integral part of the overall system, since the manner in which the robot is programmed and controlled can have a major impact on its performance and subsequent range of applications.

1.3.1 Accuracy and Repeatability

The **accuracy** of a manipulator is a measure of how close the manipulator can come to a given point within its workspace. **Repeatability** is a measure of how close a manipulator can return to a previously taught point. Most present day manipulators are highly repeatable but not very accurate. The primary method of sensing positioning errors in most cases is with position encoders located at the joints, either on the shaft of the motor that actuates the joint or on the joint itself. There is

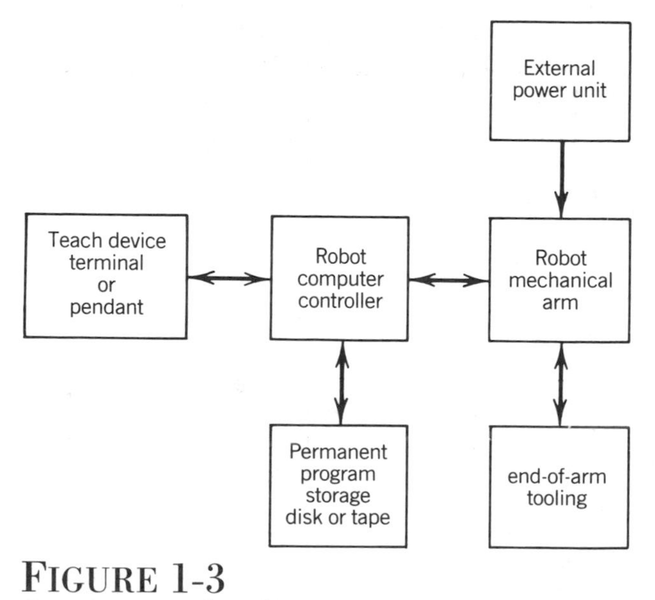

FIGURE 1-3
Components of a robotic system

typically no direct measurement of the end-effector position and orientation. One must rely on the assumed geometry of the manipulator and its rigidity to infer (i.e., to calculate) the end-effector position from the measured joint angles. Accuracy is affected therefore by computational errors, machining accuracy in the construction of the manipulator, flexibility effects such as the bending of the links under gravitational and other loads, gear backlash, and a host of other static and dynamic effects. It is primarily for this reason that robots are designed with extremely high rigidity. Without high rigidity, accuracy can only be improved by some sort of direct sensing of the end-effector position, such as with vision.

Once a point is taught to the manipulator, however, say with a teach pendant, the above effects are taken into account and the proper encoder values necessary to return to the given point are stored by the controlling computer. Repeatability therefore is affected primarily by the controller resolution. **Controller resolution** means the smallest increment of motion that the controller can sense. The resolution is computed as the total distance traveled by the tip divided by 2^n, where n is the number of bits of encoder accuracy. In this context linear axes, that is, prismatic joints, typically have higher resolution than revolute joints, since the straight line distance traversed by the tip of a linear axis between two points is less than the corresponding arclength traced by the tip of a rotational link.

In addition, as we will see in later chapters, rotational axes usually result in a large amount of kinematic and dynamic coupling among the links with a resultant accumulation of errors and a more difficult control problem. One may wonder then what the advantages of revolute joints are in manipulator design. The answer lies primarily in the increased dexterity and compactness of revolute joint designs. For example, Figure 1-4 shows that for the same range of motion, a rotational link can be made much smaller than a link with linear motion. Thus manipulators made from revolute joints occupy a smaller working volume than manipulators with linear axes. This increases the ability of the manipulator to work in the same space with other robots, machines, and people. At the same time revolute joint manipulators are better able to maneuver around obstacles and have a wider range of possible applications.

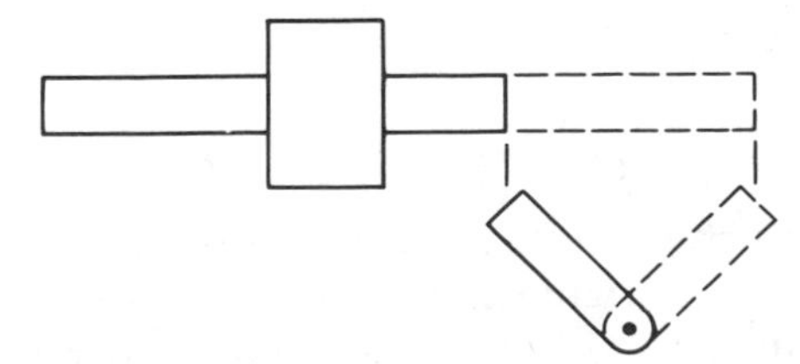

FIGURE 1-4
Linear vs. rotational link motion

1.4 COMMON KINEMATIC ARRANGEMENTS

Although in principle a manipulator is a general purpose device, in practice manipulators are usually designed with at least a broad class of applications in mind, such as welding, materials handling, and assembly. These applications largely dictate the choice of various design parameters of the manipulator, including its kinematic structure. For example, assembly of circuit boards is naturally performed by a SCARA type manipulator (Figure 1-13), while a spherical manipulator (Figure 1-10) may be better suited for tending a punch press.

Robot manipulators can be classified by several criteria, such as their geometry, or kinematic structure, the type of application for which they are designed, the manner in which they are controlled, etc. In this text we will mainly classify manipulators according to their geometry. Most industrial manipulators at the present time have six or fewer degrees-of-freedom. These manipulators are usually classified kinematically on the basis of the arm or first three joints, with the wrist being described separately. The majority of these manipulators fall into one of five geometric types: **articulated (RRR)**, **spherical (RRP)**, **SCARA (RRP)**, **cylindrical (RPP)**, or **cartesian (PPP)**.

1.4.1 ARTICULATED CONFIGURATION(RRR)

The articulated manipulator is also called a **revolute**, or **anthropomorphic** manipulator. Two common revolute designs are the **elbow** type manipulator, such as the PUMA, shown in Figure 1-5, and the **parallelogram linkage** such as the Cincinnati Milacron T^3 735, shown in Figure 1-6. In these arrangements joint axis z_3 is parallel to z_2 and both z_2 and z_3 are perpendicular to z_1. The structure and terminology associated with the elbow manipulator are shown in Figure 1-7. Its workspace is shown in Figure 1-8. This configuration provides for relatively large freedom of movement in a compact space. The parallelogram linkage, although less dextrous typically than the elbow manipulator configuration, nevertheless has several advantages that make it an attractive and popular design. The most notable feature of the parallelogram linkage configuration is that the actuator for joint 3 is located on link 1. Since the weight of the motor is born by link 1, links two and three can be made more lightweight and the motors themselves can be less powerful. Also the dynamics of the parallelogram manipulator are simpler than those of the elbow manipulator, thus making it easier to control.

1.4.2 SPHERICAL CONFIGURATION(RRP)

By replacing the third or elbow joint in the revolute configuration by a prismatic joint one obtains the spherical configuration shown in Figure 1-9. The term **spherical configuration** derives from the fact that the

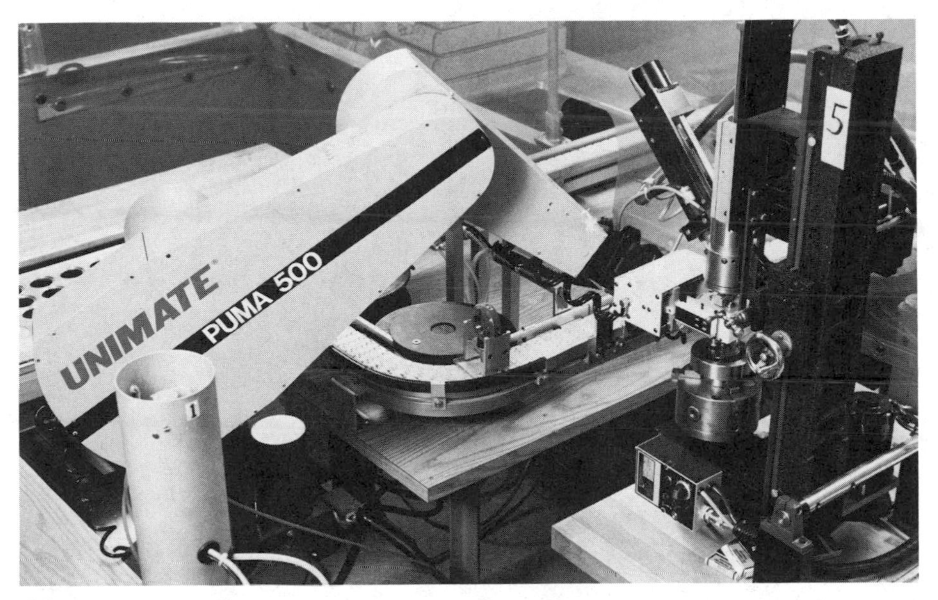

FIGURE 1-5
The Unimation PUMA (Programmable Universal Manipulator for Assembly) Photo courtesy of Westinghouse Automation Division/Unimation Incorporated.

spherical coordinates defining the position of the end-effector with respect to a frame whose origin lies at the intersection of the axes z_1 and z_2 are the same as the first three joint variables. A common manipulator with this configuration is the Stanford manipulator (Figure 1-10). The workspace of a spherical manipulator is shown in Figure 1-11.

1.4.3 SCARA CONFIGURATION (RRP)

The so-called **SCARA** (for **S**elective **C**ompliant **A**rticulated **R**obot for **A**ssembly) shown in Figure 1-12 is a recent and increasingly popular configuration, which, as its name suggests, is tailored for assembly operations. Although the SCARA has an RRP structure, it is quite different from the spherical configuration in both appearance and in its range of applications. Unlike the (spherical) Stanford design, which has z_0, z_1, z_2 mutually perpendicular, the SCARA has z_0, z_1, z_2 parallel. Figure 1-13 shows the AdeptOne, a manipulator of this type. The SCARA manipulator workspace is shown in Figure 1-14.

1.4.4 CYLINDRICAL CONFIGURATION (RPP)

The cylindrical configuration is shown in Figure 1-15. The first joint is revolute and produces a rotation about the base, while the second and third joints are prismatic. As the name suggests, the joint variables are

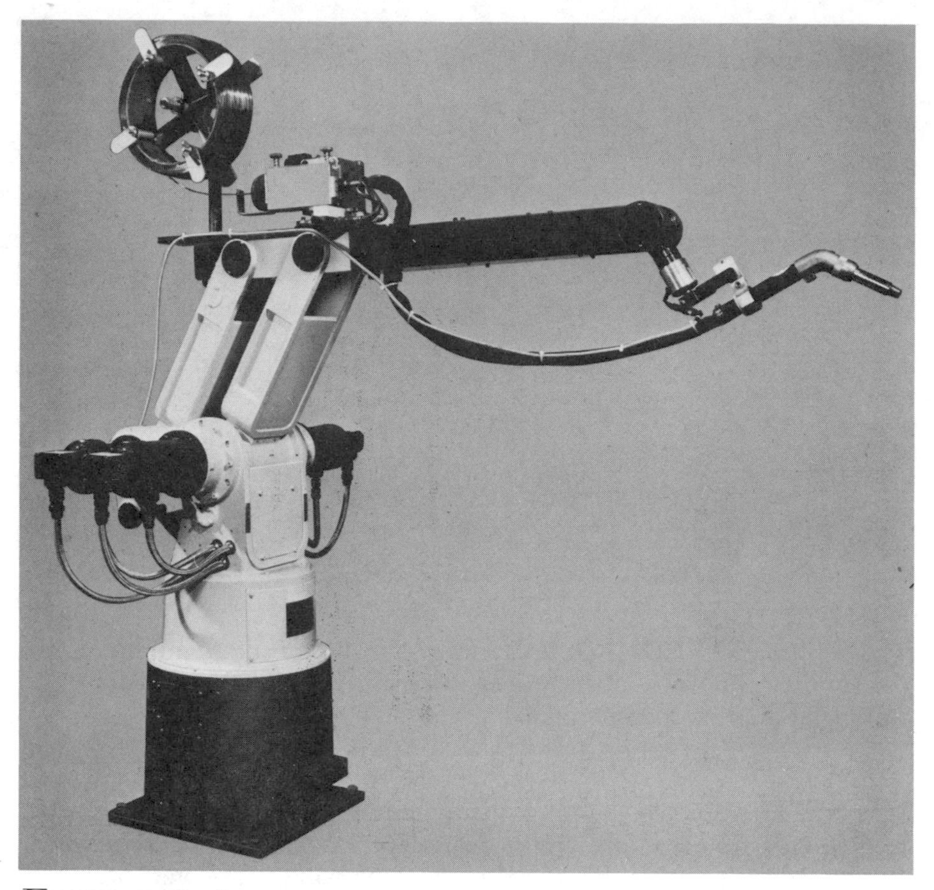

FIGURE 1-6
The Cincinnati Milacron T^3 735 robot. Courtesy of Cincinnati Milacron.

the cylindrical coordinates of the end-effector with respect to the base. A cylindrical robot, the GMF M-100, is shown in Figure 1-16, with its workspace shown in Figure 1-17.

1.4.5 Cartesian Configuration (PPP)

A manipulator whose first three joints are prismatic is known as a cartesian manipulator, shown in Figure 1-18. For the cartesian manipulator the joint variables are the cartesian coordinates of the end-effector with respect to the base. As might be expected the kinematic description of this manipulator is the simplest of all configurations. Cartesian configurations are useful for table-top assembly applications and, as

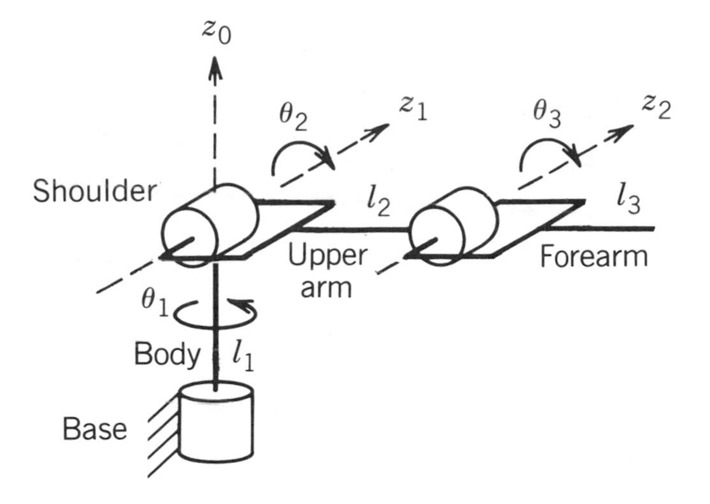

FIGURE 1-7
Structure of the elbow manipulator

gantry robots, for transfer of material or cargo. An example of a cartesian robot, the Cincinnati Milacron T^3 gantry robot, is shown in Figure 1-19. The workspace of a cartesian manipulator is shown in Figure 1-20.

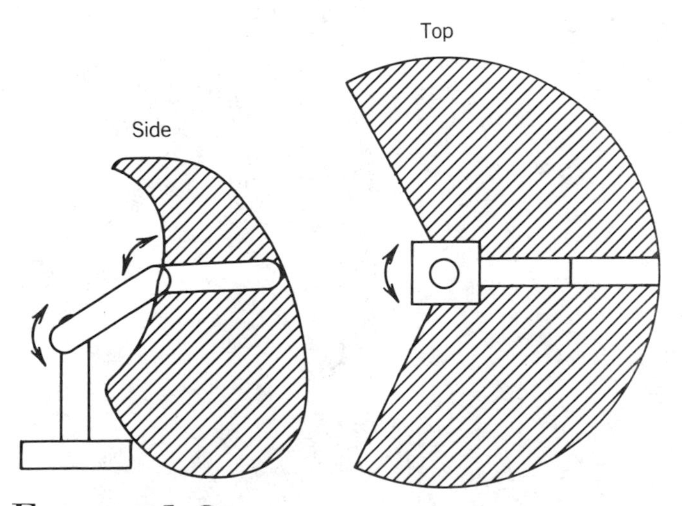

FIGURE 1-8
Workspace of the elbow manipulator

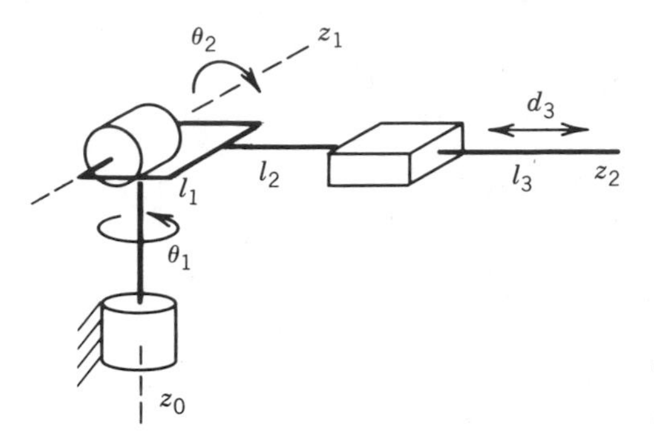

FIGURE 1-9
The spherical manipulator configuration

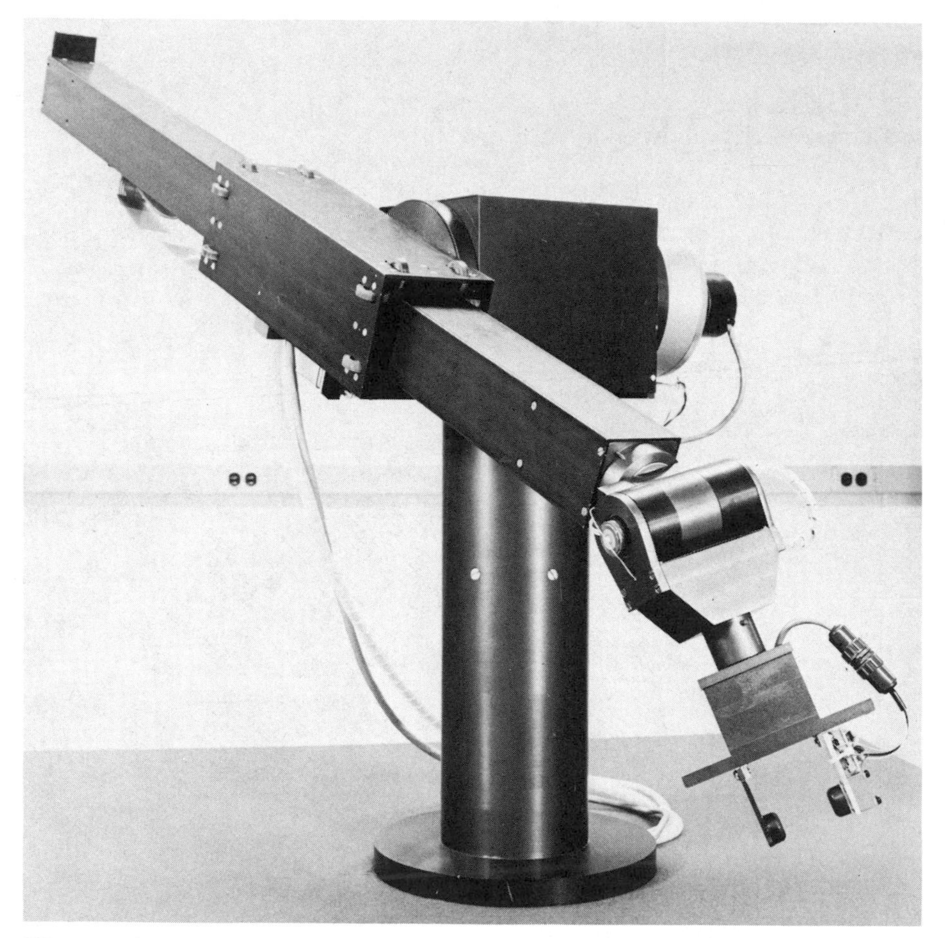

FIGURE 1-10
The Stanford manipulator

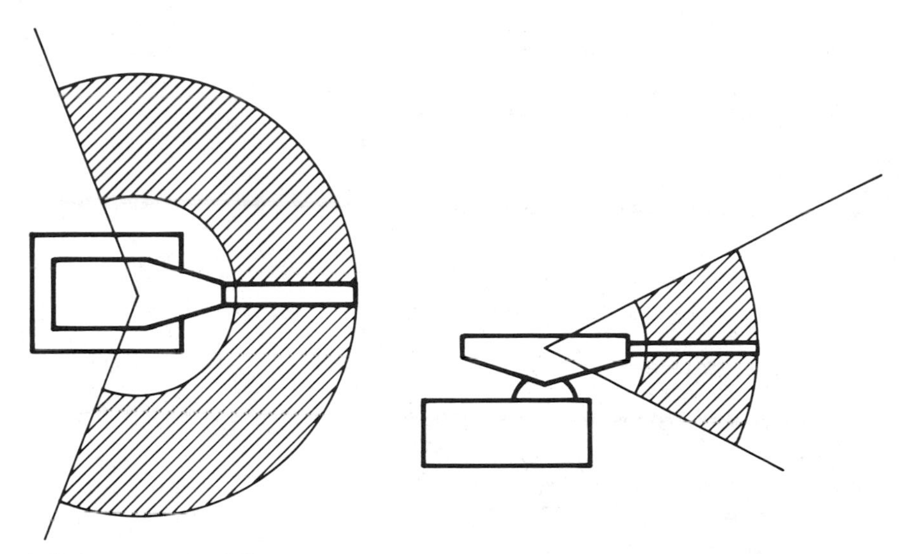

FIGURE 1-11
Workspace of the spherical manipulator

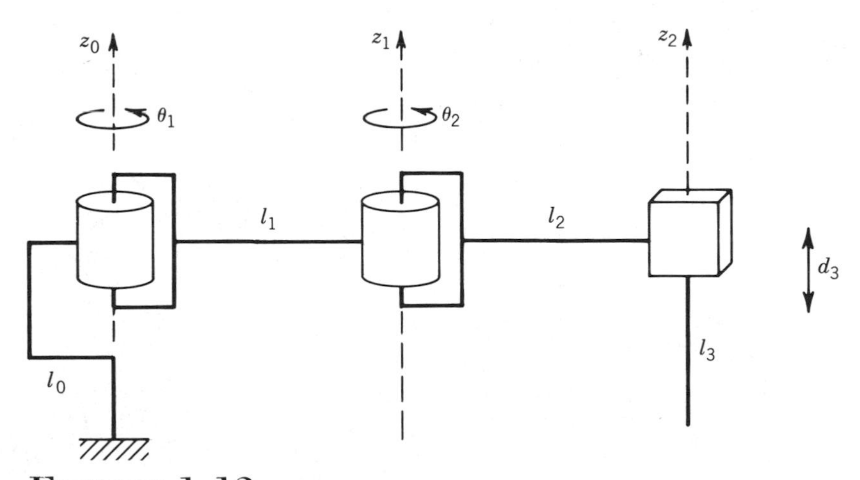

FIGURE 1-12
The SCARA (Selective Compliant Articulated Robot for Assembly)

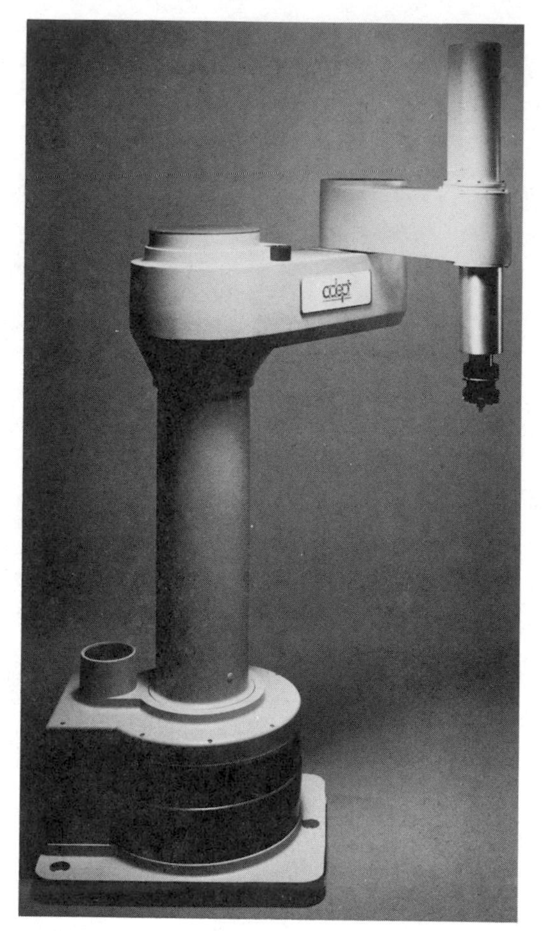

FIGURE 1-13
The AdeptOne robot. Photo courtesy of Adept Technologies.

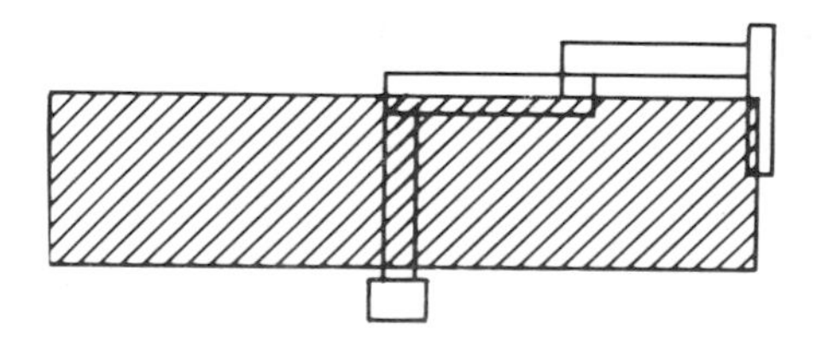

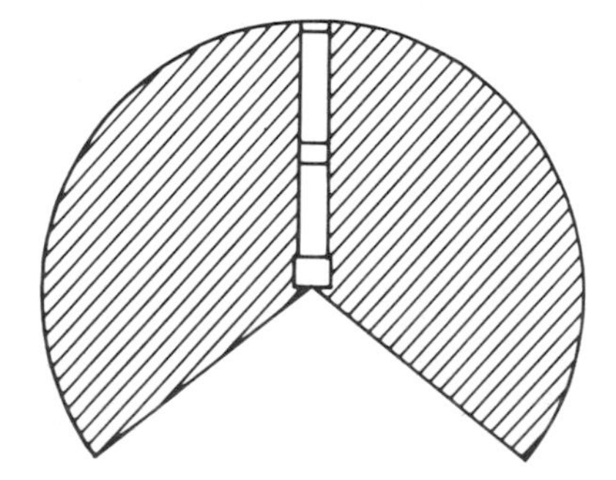

FIGURE 1-14
The workspace of the SCARA manipulator

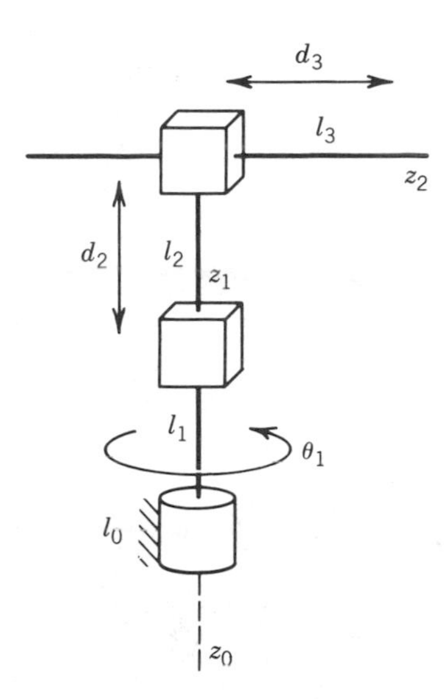

FIGURE 1-15
The cylindrical manipulator configuration

1.4.6 OTHER METHODS OF CLASSIFYING ROBOTS

Other common ways to classify robots are by their **power source**, **application area**, and **method of control**.

(i) *Power Source*

Typically, robots are either electrically, hydraulically, or pneumatically powered. Hydraulic actuators are unrivaled in their speed of response and torque producing capability. Therefore hydraulic robots are used primarily for lifting heavy loads. The drawbacks of hydraulic robots are that they tend to leak hydraulic fluid, require much more peripheral equipment, such as pumps, which also requires more maintenance, and they are noisy. Robots driven by DC- or AC-servo motors are increasingly popular since they are cheaper, cleaner and quieter. Pneumatic robots are inexpensive and simple but cannot be controlled precisely. As a result, pneumatic robots are limited in their range of applications and popularity.

(ii) *Application Area*

The largest projected area of future application of robots is in assembly. Therefore, robots are often classified by application into **assembly** and **non-assembly** robots. Assembly robots tend to be small, electrically driven and either revolute or SCARA in design. The main non-assembly application areas to date have been in welding, spray-painting, material handling, and machine loading and unloading.

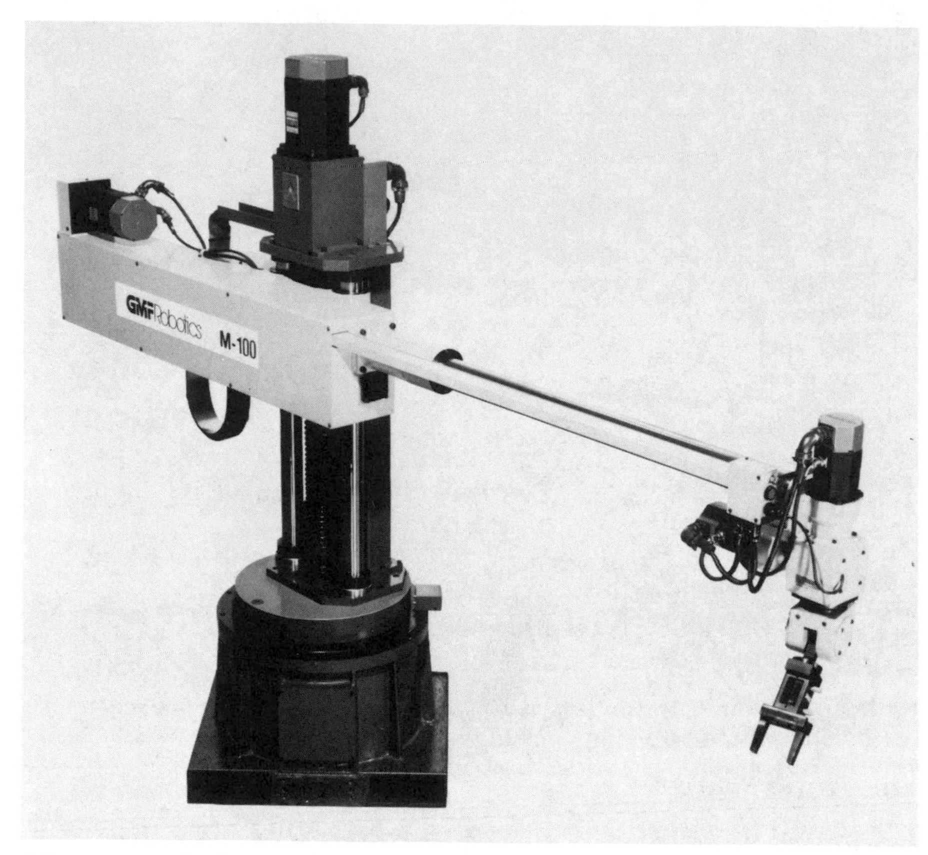

FIGURE 1-16
The GMF M-100 robot. Photo courtesy of GMF Robotics.

(iii) *Method of Control*

Robots are classified by control method into **servo** and **non-servo** robots. The earliest robots were non-servo robots. These robots are essentially open-loop devices whose movement is limited to predetermined mechanical stops, and are useful primarily for materials transfer. In fact, according to the definition given previously, fixed stop robots hardly qualify as robots. Servo robots use closed-loop computer control to determine their motion and are thus capable of being truly multifunctional, reprogrammable devices.

Servo controlled robots are further classified according to the method that the controller uses to guide the end-effector. The simplest type of robot in this class is the **point-to-point** robot. A point-to-point robot can be taught a discrete set of points but there is no control on the path of the end-effector in between taught points. Such robots are

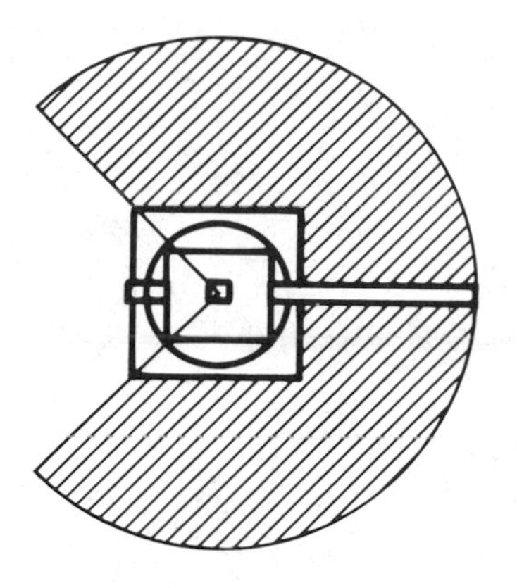

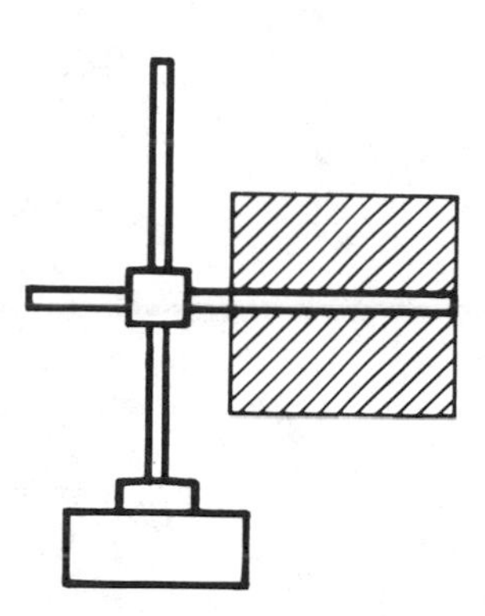

FIGURE 1-17
Workspace of the cylindrical manipulator

usually taught a series of points with a teach pendant. The points are then stored and played back. Point-to-point robots are severely limited in their range of applications. In **continuous path** robots, on the other hand, the entire path of the end-effector can be controlled. For example, the robot end-effector can be taught to follow a straight line between two points or even to follow a contour such as a welding seam. In addition, the velocity and/or acceleration of the end-effector can often be controlled. These are the most advanced robots and require the most sophisticated computer controllers and software development.

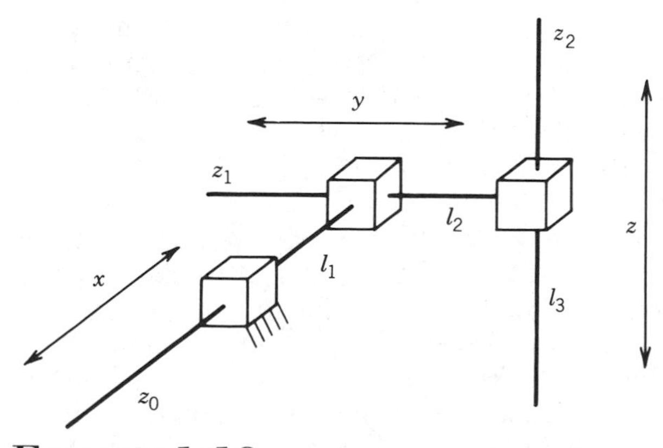

FIGURE 1-18
The cartesian manipulator configuration.

FIGURE 1-19
A Gantry robot, the Cincinnati Milacron T^3 886. Photo courtesy of Cincinnati Milacron.

1.4.7 WRISTS AND END-EFFECTORS

The **wrist** of a manipulator refers to the joints in the kinematic chain between the arm and hand. The wrist joints are nearly always all revolute. It is increasingly common to design manipulators with **spherical wrists**, by which we mean wrists whose joint axes intersect at a common point. The Cincinnati Milacron T^3 of Figure 1-1 and the Stanford manipulator of Figure 1-10 both have three degree-of-freedom spherical wrists. The spherical wrist is represented symbolically in Figure 1-21. The spherical wrist greatly simplifies the kinematic analysis, effectively allowing one to decouple the positioning and orientation of an object to as great an extent as possible. Typically therefore, the manipulator will possess three positional degrees-of-freedom, which are produced by three or more joints in the arm. The number of orientational degrees-of-freedom will then depend on the degrees-of-freedom of the wrist. It is common to find wrists having one, two, or three degrees-of-freedom depending of the application. For example, the AdeptOne (Figure 1-13) has four degrees-of-freedom. The wrist has only a roll about the final z-axis. The Cincinnati Milacron T^3 735 has

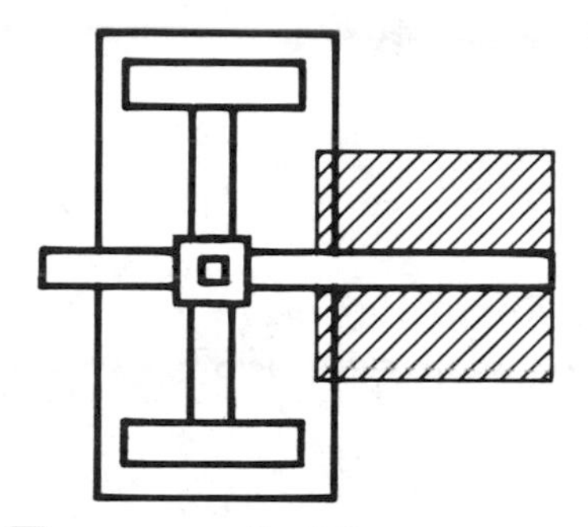
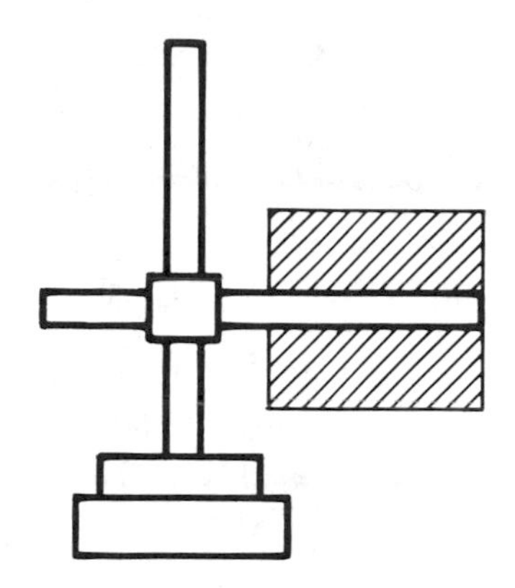

FIGURE 1-20
Workspace of the cartesian manipulator.

five degrees-of-freedom. The wrist has pitch and roll but no yaw motion. The PUMA has a full three degrees-of-freedom spherical wrist and hence the manipulator possesses six degrees-of-freedom.

It has been said that a robot is only as good as its **hand** or **end-effector**. The arm and wrist assemblies of a robot are used primarily for positioning the end-effector and any tool it may carry. It is the end-effector or tool that actually performs the work. The simplest type of end-effectors are grippers, such as shown in Figures 1-22 and 1-23, which usually are capable of only two actions, **opening** and **closing**.

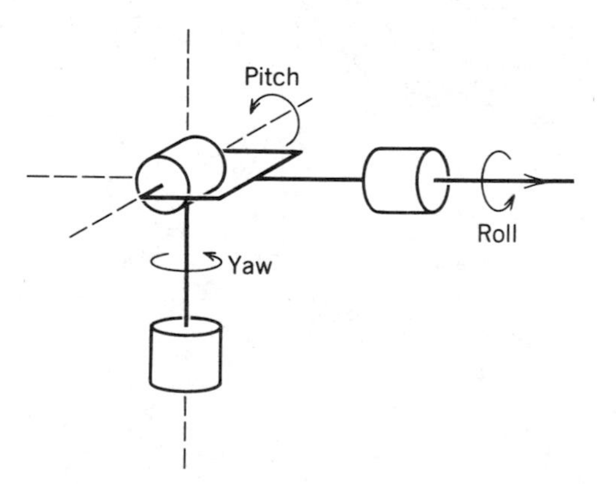

FIGURE 1-21
Structure of a spherical wrist.

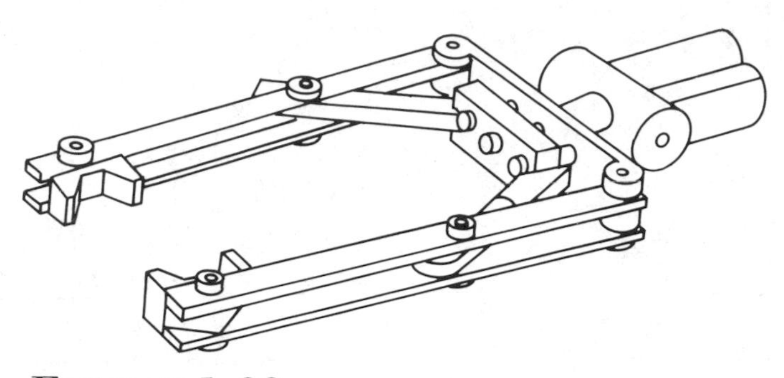

FIGURE 1-22
A parallel jaw gripper. (From: *A Robot Engineering Textbook*, by Mohsen Shahinpoor. Copyright 1987, Harper & Row Publishers, Inc.)

While this is adequate for materials transfer, some parts handling, or gripping simple tools, it is not adequate for other tasks such as welding, assembly, grinding, etc. A great deal of research is therefore being devoted to the design of special purpose end-effectors as well as tools that can be rapidly changed as the task dictates. There is also much research being devoted to the development of anthropomorphic hands. Such hands are being developed both for prosthetic use and for use in manufacturing. Since we are concerned with the analysis and control of the manipulator itself and not in the particular application or end-effector, we will not discuss end-effector design or the study of grasping and manipulation.

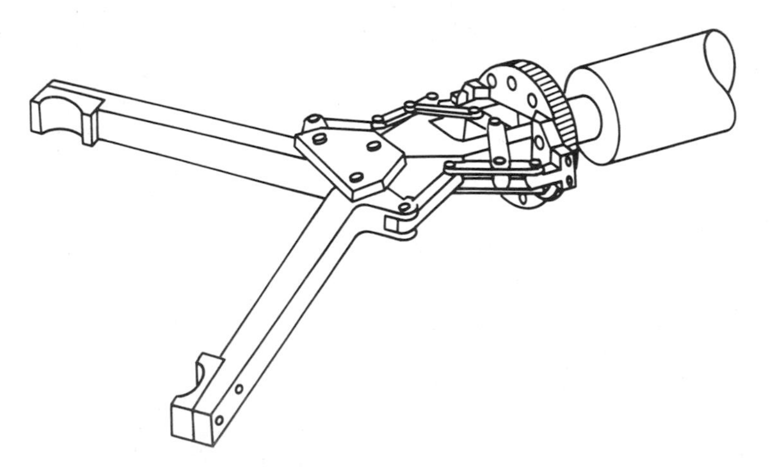

FIGURE 1-23
A two-fingered gripper. (From: *A Robot Engineering Handbook*, by Mohsen Shahinpoor. Copyright 1987, Harper & Row Publishers, Inc.)

1.5 OUTLINE OF THE TEXT

A typical application involving an industrial manipulator is shown in Figure 1-24. The manipulator, which possesses six degrees-of-freedom, is shown with a grinding tool, which it must use to remove a certain amount of metal from a surface. In the present text we are concerned with the following question: *What are the basic issues to be resolved and what must we learn in order to be able to program a robot to perform tasks such as the above?*

The ability to answer this question for a full six degree-of-freedom manipulator represents the goal of the present text. The answer itself is too complicated to be presented at this point. We can, however, use a simple two-link planar mechanism to illustrate the major issues involved and to preview the topics covered in this text.

Accordingly, consider a two-link robot as shown in Figure 1-25. Attached to the end of the manipulator is a tool of some sort, such as a grinding or cutting wheel. Suppose we wish to move the manipulator from its **home** position A to position B from which point the robot is to follow the contour of the surface S to C at constant velocity while maintaining a prescribed force F normal to the surface. In so doing the robot will cut or grind the surface according to a predetermined specification.

1.5.1 Problem 1: Forward Kinematics

The first problem encountered is to describe both the position of the tool and the locations A and B (and most likely the entire surface S) with respect to a common coordinate system. In Chapter Two we give some background on representations of coordinate systems and transformations among various coordinate systems.

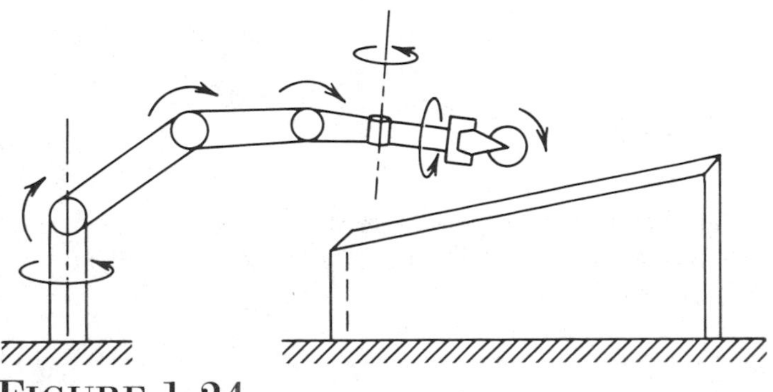

FIGURE 1-24
A 6–DOF robot with grinding tool.

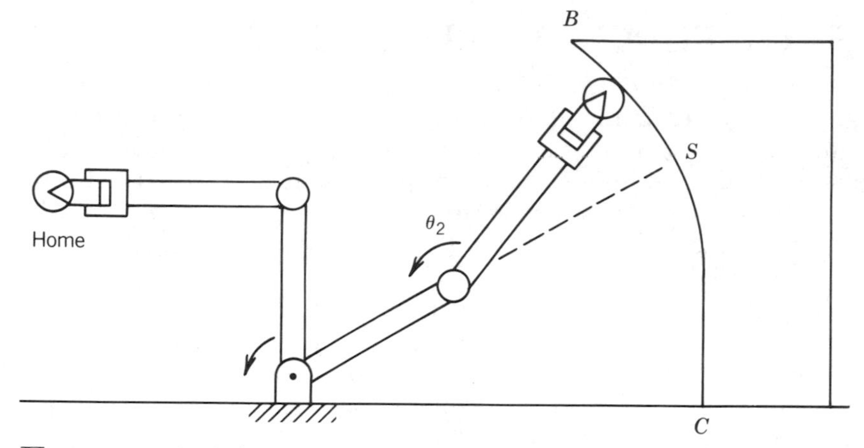

FIGURE 1-25
Two-link planar robot example.

Typically, the manipulator will be able to sense its own position in some manner using internal sensors (position encoders) located at joints 1 and 2, which can measure directly the joint angles θ_1 and θ_2. We also need therefore to express the positions A and B in terms of these joint angles. This leads to the **forward kinematics problem** studied in Chapter Three, which is to determine the position and orientation of the end-effector or tool in terms of the joint variables.

It is customary to establish a fixed coordinate system, called the **world** or **base** frame to which all objects including the manipulator are referenced. In this case we establish the base coordinate frame $o_0x_0y_0$ at the base of the robot, as shown in Figure 1-26, and the coordinates (x,y) of the tool are expressed in this coordinate frame as

$$x = a_1\cos\theta_1 + a_2\cos(\theta_1+\theta_2) \tag{1.5.1}$$

$$y = a_1\sin\theta_1 + a_2\sin(\theta_1+\theta_2) \tag{1.5.2}$$

Also the **orientation of the tool frame** relative to the base frame is given by the direction cosines of the x_2 and y_2 axes relative to the x_0 and y_0 axes, that is,

$$\begin{aligned} \mathbf{i}_2\cdot\mathbf{i}_0 &= \cos(\theta_1+\theta_2); \ \mathbf{i}_2\cdot\mathbf{j}_0 = -\sin(\theta_1+\theta_2) \\ \mathbf{j}_2\cdot\mathbf{i}_0 &= \sin(\theta_1+\theta_2); \ \mathbf{j}_2\cdot\mathbf{j}_0 = \cos(\theta_1+\theta_2) \end{aligned} \tag{1.5.3}$$

which we may combine into an **orientation matrix**

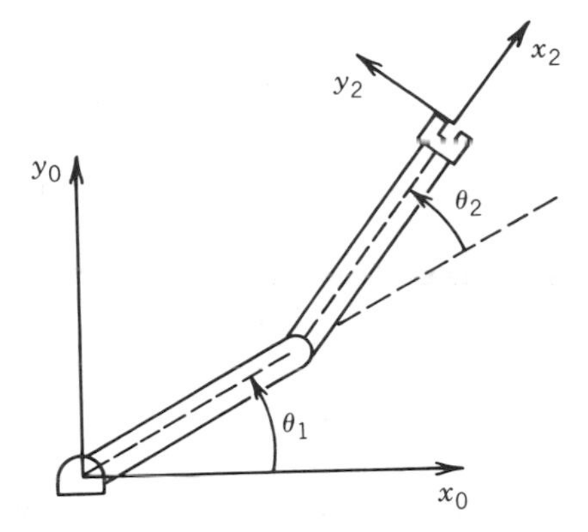

FIGURE 1-26
Coordinate frames for two-link planar robot.

$$\begin{bmatrix} \mathbf{i}_2 \cdot \mathbf{i}_0 & \mathbf{j}_2 \cdot \mathbf{i}_0 \\ \mathbf{i}_2 \cdot \mathbf{j}_0 & \mathbf{j}_2 \cdot \mathbf{j}_0 \end{bmatrix} = \begin{bmatrix} \cos(\theta_1+\theta_2) & -\sin(\theta_1+\theta_2) \\ \sin(\theta_1+\theta_2) & \cos(\theta_1+\theta_2) \end{bmatrix} \tag{1.5.4}$$

where $\mathbf{i}_0$, $\mathbf{j}_0$ are the standard orthonormal unit vectors in the base frame, and $\mathbf{i}_2$, $\mathbf{j}_2$ are the standard orthonormal unit vectors in the tool frame.

These equations (1.5.1–1.5.4) are called the **forward kinematic equations**. For a six degree-of-freedom robot these equations are quite complex and cannot be written down as easily as for the two-link manipulator. The general procedure that we discuss in Chapter Three establishes coordinate frames at each joint and allows one to transform systematically among these frames using matrix transformations. The procedure that we use is referred to as the **Denavit–Hartenberg** convention. We then use **homogeneous coordinates** and **homogeneous transformations** to simplify the transformation among coordinate frames.

1.5.2 PROBLEM 2: INVERSE KINEMATICS

Now, given the joint angles θ_1,θ_2 we can determine the end-effector coordinates x and y. In order to command the robot to move to location B we need the inverse; that is, we need the joint variables θ_1,θ_2 in terms of the x and y coordinates of B. This is the problem of **Inverse Kinematics**. In other words, given x and y in the forward kinematic equations 1.5.1–1.5.2, we wish to solve for the joint angles. Since the forward kinematic equations are nonlinear, a solution may not be easy to find nor is there a unique solution in general. We can see, for example, in the case of a two-link planar mechanism that there may be no solution, if the given (x,y) coordinates are out of reach of the manipulator. If the given (x,y) coordinates are within the manipulators reach there may be two solutions as shown in Figure 1-27, the so-called **elbow up** and **elbow down** configurations, or there may be exactly one solution if the manipulator must be fully extended to reach the point. There may even be an infinite number of solutions in some cases (Problem 1-25).

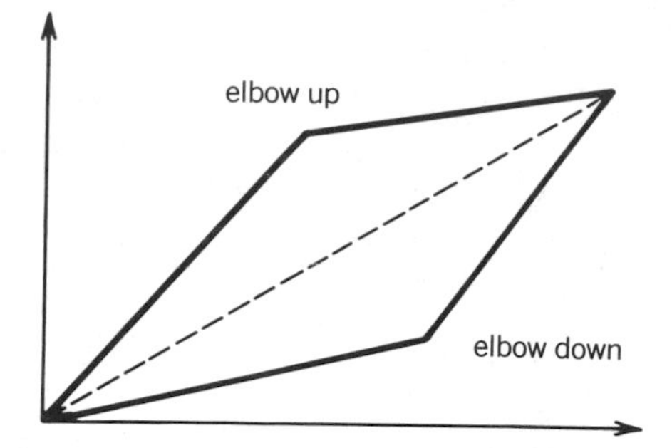

FIGURE 1-27
Multiple inverse kinematic solutions.

Consider the diagram of Figure 1-28. Using the **Law of Cosines** we see that the angle θ_2 is given by

$$\cos\theta_2 = \frac{x^2+y^2-a_1^2-a_2^2}{2a_1a_2} := D \tag{1.5.5}$$

We could now determine θ_2 as

$$\theta_2 = \cos^{-1}(D) \tag{1.5.6}$$

However, a better way to find θ_2 is to notice that if $\cos(\theta_2)$ is given by (1.5.5) then $\sin(\theta_2)$ is given as

$$\sin(\theta_2) = \pm\sqrt{1-D^2} \tag{1.5.7}$$

and, hence, θ_2 can be found by

$$\theta_2 = \tan^{-1}\frac{\pm\sqrt{1-D^2}}{D} \tag{1.5.8}$$

The advantage of this latter approach is that both the elbow-up and elbow-down solutions are recovered by choosing the positive and negative signs in (1.5.8), respectively.

It is left as an exercise (Problem 1-12) to show that θ_1 is now given as

$$\theta_1 = \tan^{-1}(y/x) - \tan^{-1}\left(\frac{a_2\sin\theta_2}{a_1+a_2\cos\theta_2}\right) \tag{1.5.9}$$

Notice that the angle θ_1 depends on θ_2. This makes sense physically since we would expect to require a different value for θ_1 depending on which solution is chosen for θ_2.

1.5.3 PROBLEM 3: VELOCITY KINEMATICS

In order to follow a contour at constant velocity, or at any prescribed velocity, we must know the relationship between the velocity of the tool and the joint velocities. In this case we can differentiate Equations 1.5.1 and 1.5.2 to obtain

$$\dot{x} = -a_1\sin\theta_1\cdot\dot{\theta}_1 - a_2\sin(\theta_1+\theta_2)(\dot{\theta}_1+\dot{\theta}_2) \tag{1.5.10}$$

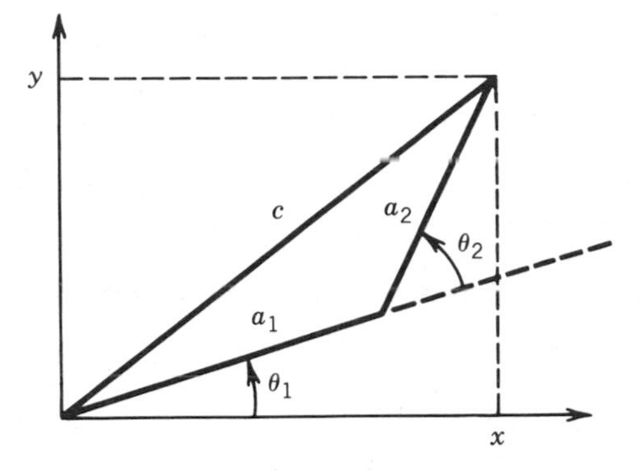

FIGURE 1-28
Solving for the joint angles of a two-link planar arm.

$$\dot{y} = a_1\cos\theta_1 \cdot \dot{\theta}_1 + a_2\cos(\theta_1+\theta_2)(\dot{\theta}_1 + \dot{\theta}_2)$$

Using the vector notation $\mathbf{x} = \begin{pmatrix} x \\ y \end{pmatrix}$ and $\boldsymbol{\theta} = \begin{bmatrix} \theta_1 \\ \theta_2 \end{bmatrix}$ we may write these cquations as

$$\dot{\mathbf{x}} = \begin{bmatrix} -a_1\sin\theta_1 - a_2\sin(\theta_1+\theta_2) & -a_2\sin(\theta_1+\theta_2) \\ a_1\cos\theta_1 + a_2\cos(\theta_1+\theta_2) & a_2\cos(\theta_1+\theta_2) \end{bmatrix} \dot{\boldsymbol{\theta}} \qquad (1.5.11)$$
$$= J\dot{\boldsymbol{\theta}}$$

The matrix J defined by (1.5.11) is called the **Jacobian** of the manipulator and is a fundamental object to determine for any manipulator. In Chapter Five we present a systematic procedure for deriving the Jacobian for any manipulator in the so-called **cross-product form**.

The determination of the joint velocities from the end-effector velocities is conceptually simple since the velocity relationship is linear. Thus the joint velocities are found from the end-effector velocities via the inverse Jacobian

$$\dot{\boldsymbol{\theta}} = J^{-1}\dot{\mathbf{x}} \qquad (1.5.12)$$

or

$$\begin{bmatrix} \dot{\theta}_1 \\ \dot{\theta}_2 \end{bmatrix} = \frac{1}{a_1 a_2 \sin\theta_2} \begin{bmatrix} a_2\cos(\theta_1+\theta_2) & a_2\sin(\theta_1+\theta_2) \\ -a_1\cos\theta_1 - a_2\cos(\theta_1+\theta_2) & -a_1\sin\theta_1 - a_2\sin(\theta_1+\theta_2) \end{bmatrix} \begin{bmatrix} x \\ y \end{bmatrix}$$

(1.5.13)

The determinant $\det J$ of the Jacobian in (1.5.11) is $a_1 a_2 \sin\theta_2$. The Jacobian does not have an inverse, therefore, when $\theta_2 = 0$ or π, in which case the manipulator is said to be in a **singular configuration**, such as shown in Figure 1-29 for $\theta_2 = 0$. The determination of such singular configurations is important for several reasons. At singular

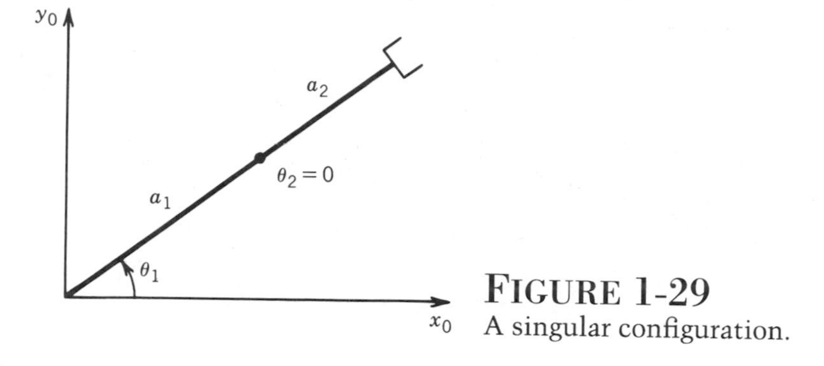

FIGURE 1-29
A singular configuration.

configurations there are infinitesimal motions that are unachievable; that is, the manipulator end-effector cannot move in certain directions. In the above cases the end effector cannot move in the direction parallel to a_1 from a singular configuration. Singular configurations are also related to the non-uniqueness of solutions of the inverse kinematics. For example, for a given end-effector position, there are in general two possible solutions to the inverse kinematics. Note that the singular configuration separates these two solutions in the sense that the manipulator cannot go from one configuration to the other without passing through the singularity. For many applications it is important to plan manipulator motions in such a way that singular configurations are avoided.

1.5.4 PROBLEM 4: DYNAMICS

A robot manipulator is basically a positioning device. To control the position we must know the dynamic properties of the manipulator in order to know how much force to exert on it to cause it to move. Too little force and the manipulator is slow to react. Too much force and the arm may crash into objects or oscillate about its desired position.

Deriving the dynamic equations of motion for robots is not a simple task due to the large number of degrees of freedom and nonlinearities present in the system. In Chapter Six we develop techniques based on Lagrangian dynamics for systematically deriving the equations of motion of such a system. In addition to the rigid links, the complete description of robot dynamics includes the dynamics of the actuators that produce the forces and torques to drive the robot, and the dynamics of the drive trains that transmit the power from the actuators to the links. Thus, in Chapter Seven we also discuss actuator and drive train dynamics and their effects on the control problem.

1.5.5 PROBLEM 5: POSITION CONTROL

Control theory is used in Chapters Seven and Eight to design control algorithms for the execution of programmed tasks. The motion control problem consists of the **Tracking and Disturbance Rejection Problem**, which is the problem of determining the control inputs necessary to follow, or **track**, a desired trajectory that has been planned for the manipulator, while simultaneously **rejecting** disturbances due to unmodeled dynamic effects such as friction and noise. We detail the standard approaches to robot control based on frequency domain techniques. We also introduce the notion of **feedforward control** and the techniques of **computed torque** and **inverse dynamics** as a means for compensating the complex nonlinear interaction forces among the links of the manipulator. **Robust control** is introduced in Chapter Eight using the **Second Method of Lyapunov**. Chapters Ten and Eleven provide some additional advanced techniques from nonlinear control theory that are useful for controlling high performance robots.

1.5.6 PROBLEM 6: FORCE CONTROL

Once the manipulator has reached location B, it must follow the contour S maintaining a constant force normal to the surface. Conceivably, knowing the location of the object and the shape of the contour, we could carry out this task using position control alone. This would be quite difficult to accomplish in practice, however. Since the manipulator itself possess high rigidity, any errors in position due to uncertainty in the exact location of the surface or tool would give rise to extremely large forces at the end-effector that could damage the tool, the surface, or the robot. A better approach is to measure the forces of interaction directly and use a **force control** scheme to accomplish the task. In Chapter Nine we discuss force control and compliance and discuss the two most common approaches to force control, **hybrid control** and **impedance control**.

REFERENCES AND SUGGESTED READING

[1] ARBIB, M., *Computers and the Cybernetic Society*, Academic Press, New York, 1977.

[2] ASADA, H., and SLOTINE, J-J. E., *Robot Analysis and Control*, Wiley, New York, 1986.

[3] BENI, G., and HACKWOOD, S., eds., *Recent Advances in Robotics*, Wiley, New York, 1985.

[4] BRADY, M., et. al., eds., *Robot Motion: Planning and Control,* MIT Press, Cambridge, MA, 1983.

[5] CRAIG, J., *Introduction to Robotics: Mechanics and Control,* Addison–Wesley, Reading, MA, 1986.

[6] CRITCHLOW, A.J., *Introduction to Robotics,* Macmillan, New York, 1985.

[7] DORF, R., *Robotics and Automated Manufacturing,* Reston, VA, 1983.

[8] ENGLEBERGER, J., *Robotics in Practice,* Kogan Page, London, 1980.

[9] FU, K.S., GONZALEZ, R.C., and LEE, C.S.G., *Robotics: Control Sensing, Vision, and Intelligence,* McGraw–Hill, St Louis, 1987.

[10] GROOVER, M., et. al., *Industrial Robotics: Technology, Programming, and Applications,* McGraw–Hill, St. Louis, 1986.

[11] KOREN, Y., *Robotics for Engineers,* McGraw-Hill, St. Louis, 1985.

[12] LEE, C.S.G., et. al., eds., *Tutorial on Robotics,* IEEE Computer Society Press, Silver Spring, MD, 1983.

[13] MCCLOY, D., and HARRIS, M., *Robotics: An Introduction,* Halstead Press, New York, 1986.

[14] MCCORDUCK, P., *Machines Who Think,* W.H. Freeman, San Francisco, 1979.

[15] MINSKY, M., ed., *Robotics,* Omni Publications International, Ltd., New York, 1985.

[16] *Oxford English Dictionary,* Oxford University Press, Oxford, 1971.

[17] PAUL, R., *Robot Manipulators: Mathematics, Programming and Control,* MIT Press, Cambridge, MA, 1982.

[18] REID, *Robotics: A Systems Approach,* Prentice–Hall, Englewood Cliffs, NJ, 1985.

[19] SHAHINPOOR, M., *A Robot Engineering Textbook,* Harper & Row, New York, 1987.

[20] SNYDER, W., *Industrial Robots: Computer Interfacing and Control,* Prentice–Hall, Englewood Cliffs, NJ, 1985.

[21] WOLOVICH, W., *Robotics: Basic Analysis and Design,* Holt, Rinehart, & Winston, New York, 1985.

PROBLEMS

1-1 What are the key features that distinguish robots from other forms of "automation," such as CNC milling machines?

1-2 Briefly define each of the following terms: forward kinematics, inverse kinematics, trajectory planning, workspace, accuracy, repeatability, resolution, joint variable, spherical wrist, end-effector.

1-3 What are the main ways to classify robots?

1-4 Make a list of robotics related magazines and journals carried by the university library.

1-5 From the list of references at the end of this chapter make a list of 20 robot applications. For each application discuss which type of manipulator would be best suited; which least suited. Justify your choices in each case.

1-6 List several applications for non-servo robots; for point-to-point robots, for continuous path robots.

1-7 List five applications that a continuous path robot could do that a point-to-point robot could not do.

1-8 List five applications where computer vision would be useful in robotics.

1-9 List five applications where either tactile sensing or force feedback control would be useful in robotics.

1-10 Find out how many industrial robots are currently in operation in the United States. How many are in operation in Japan? What country ranks third in the number of industrial robots in use?

1-11 Suppose we could close every factory in the United States today and reopen then tomorrow fully automated with robots. What would be some of the economic and social consequences of such a development?

1-12 Suppose a law were passed banning all future use of industrial robots in the United States. What would be some of the economic and social consequences of such an act?

1-13 Discuss possible applications where redundant manipulators would be useful.

1-14 Referring to Figure 1-30 suppose that the tip of a single link travels a distance d between two points. A linear axis would travel the distance d while a rotational link would travel through an arclength $\ell\theta$ as shown. Using the law of cosines show that the distance d is given by

$$d = \ell\sqrt{2(1-\cos(\theta))}$$

which is of course less than $\ell\theta$. With 10-bit accuracy and $\ell = 1$ m, $\theta = 90^o$ what is the resolution of the linear link? of the rotational link?

1-15 A single-link revolute arm is shown in Figure 1-30. If the length of the link is 50 cm and the arm travels 180° what is the control resolution obtained with an 8-bit encoder?

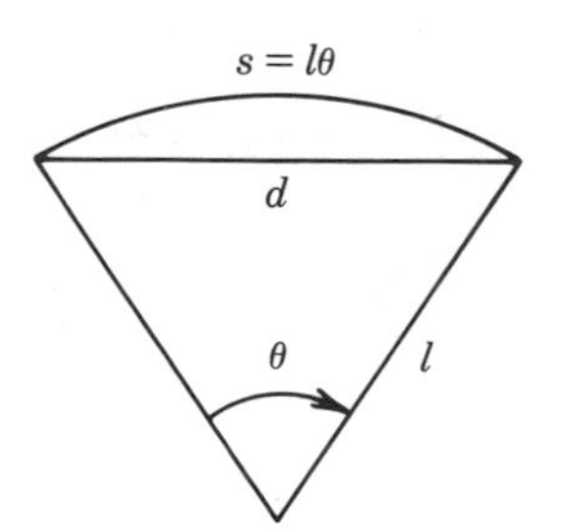

FIGURE 1-30
Diagram for Problem 1-15.

1-16 Repeat Problem 1-15 assuming that the 8-bit encoder is located on the motor shaft that is connected to the link through a 50:1 gear reduction. Assume perfect gears.

1-17 Why is accuracy generally less than repeatability?

1-18 How could manipulator accuracy be improved using direct end-point sensing? What other difficulties might direct end-point sensing introduce into the control problem?

1-19 Derive Equation 1.5.9.

1-20 For the two-link manipulator of Figure 1-25 suppose $a_1 = a_2 = 1$. Find the coordinates of the tool when $\theta_1 = \frac{\pi}{6}$ and $\theta_2 = \frac{\pi}{2}$.

1-21 Find the joint angles θ_1, θ_2 when the tool is located at coordinates $(\frac{1}{2}, \frac{1}{2})$.

1-22 If the joint velocities are constant at $\dot{\theta}_1 = 1$, $\dot{\theta}_2 = 2$, what is the velocity of the tool? What is the instantaneous tool velocity when $\theta_1 = \theta_2 = \frac{\pi}{4}$?

1-23 Write a computer program to plot the joint angles as a function of time given the tool locations and velocities as a function of time in cartesian coordinates.

1-24 Suppose we desire that the tool follow a straight line between the points (0,2) and (2,0) at constant velocity v. Plot the time history of joint angles.

1-25 For the two-link planar manipulator of Figure 1-25 is it possible for there to be an infinite number of solutions to the inverse kinematic equations? If so, explain how this can occur.

CHAPTER TWO

RIGID MOTIONS AND HOMOGENEOUS TRANSFORMATIONS

A large part of robot kinematics is concerned with the establishment of various coordinate systems to represent the positions and orientations of rigid objects and with transformations among these coordinate systems. Indeed, the geometry of three-dimensional space and of rigid motions plays a central role in all aspects of robotic manipulation. In this chapter we study the operations of rotation and translation and introduce the notion of homogeneous transformations.[1] Homogeneous transformations combine the operations of rotation and translation into a single matrix multiplication, and are used in Chapter Three to derive the so-called forward kinematic equations of rigid manipulators. We also investigate the transformation of velocities and accelerations among coordinate systems. These latter quantities are used in subsequent chapters to study the velocity kinematics in Chapter Five, including the derivation of the manipulator Jacobian, and also to derive the dynamic equations of motion of rigid manipulators in Chapter Six.

[1] Since we make extensive use of elementary matrix theory, the reader may wish to review Appendix A before beginning this chapter.

2.1 ROTATIONS

Figure 2-1 shows a rigid object S to which a coordinate frame $ox_1y_1z_1$ is attached. We wish to relate the coordinates of a point **p** on S in the $ox_1y_1z_1$ frame to the coordinates of **p** in a fixed (or nonrotated) reference frame $ox_0y_0z_0$. Let $\{ \mathbf{i}_0, \mathbf{j}_0, \mathbf{k}_0 \}$ denote the standard orthonormal basis in $ox_0y_0z_0$; thus $\mathbf{i}_0, \mathbf{j}_0, \mathbf{k}_0$ are unit vectors along the x_0, y_0, z_0 axes, respectively. Similarly, let $\{ \mathbf{i}_1, \mathbf{j}_1, \mathbf{k}_1 \}$ be the standard orthonormal basis in $ox_1y_1z_1$. Then the vector from the common origin to the point **p** on the object can be represented either with respect to $ox_0y_0z_0$ as

$$\mathbf{p}_0 = p_{0x}\mathbf{i}_0 + p_{0y}\mathbf{j}_0 + p_{0z}\mathbf{k}_0 \tag{2.1.1}$$

or with respect to $ox_1y_1z_1$ as

$$\mathbf{p}_1 = p_{1x}\mathbf{i}_1 + p_{1y}\mathbf{j}_1 + p_{1z}\mathbf{k}_1 \tag{2.1.2}$$

Since $\mathbf{p}_0$ and $\mathbf{p}_1$ are representations of the same vector **p**, the relationship between the components of **p** in the two coordinate frames can be obtained as follows.

$$\begin{aligned} p_{0x} &= \mathbf{p}_0 \cdot \mathbf{i}_0 = \mathbf{p}_1 \cdot \mathbf{i}_0 \\ &= p_{1x}\mathbf{i}_1 \cdot \mathbf{i}_0 + p_{1y}\mathbf{j}_1 \cdot \mathbf{i}_0 + p_{1z}\mathbf{k}_1 \cdot \mathbf{i}_0 \end{aligned} \tag{2.1.3}$$

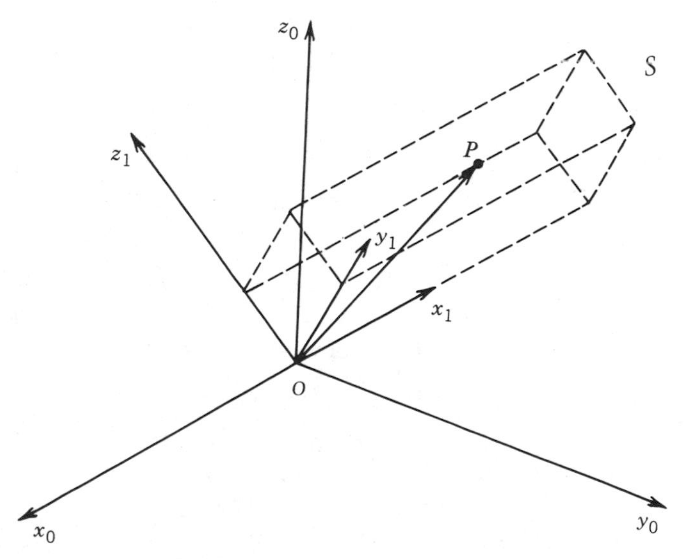

FIGURE 2-1
Coordinates frame attached to a rigid body

We have similar formulas for p_{0y} and p_{0z}, namely

$$p_{0y} = p_{1x}\mathbf{i}_1 \cdot \mathbf{j}_0 + p_{1y}\mathbf{j}_1 \cdot \mathbf{j}_0 + p_{1z}\mathbf{k}_1 \cdot \mathbf{j}_0 \tag{2.1.4}$$

$$p_{0z} = p_{1x}\mathbf{i}_1 \cdot \mathbf{k}_0 + p_{1z}\mathbf{j}_1 \cdot \mathbf{k}_0 + p_{1z}\mathbf{k}_1 \cdot \mathbf{k}_0 \tag{2.1.5}$$

We may write the above three equations together as

$$\mathbf{p}_0 = R_0^1 \mathbf{p}_1 \tag{2.1.6}$$

where

$$R_0^1 = \begin{bmatrix} \mathbf{i}_1 \cdot \mathbf{i}_0 & \mathbf{j}_1 \cdot \mathbf{i}_0 & \mathbf{k}_1 \cdot \mathbf{i}_0 \\ \mathbf{i}_1 \cdot \mathbf{j}_0 & \mathbf{j}_1 \cdot \mathbf{j}_0 & \mathbf{k}_1 \cdot \mathbf{j}_0 \\ \mathbf{i}_1 \cdot \mathbf{k}_0 & \mathbf{j}_1 \cdot \mathbf{k}_0 & \mathbf{k}_1 \cdot \mathbf{k}_0 \end{bmatrix} \tag{2.1.7}$$

The 3×3 matrix represents the transformation matrix from the coordinates of $\mathbf{p}$ with respect to the frame $ox_1y_1z_1$ to the coordinates with respect to the frame $ox_0y_0z_0$. Thus, if a given point is expressed in $ox_1y_1z_1$-coordinates as $\mathbf{p}_1$ then $R_0^1\mathbf{p}_1$ represents the **same vector** expressed relative to the $ox_0y_0z_0$-coordinate frame.

Similarly we can write

$$\begin{aligned} p_{1x} &= \mathbf{p}_1 \cdot \mathbf{i}_1 = \mathbf{p}_0 \cdot \mathbf{i}_1 \\ &= p_{0x}\mathbf{i}_0 \cdot \mathbf{i}_1 + p_{0y}\mathbf{j}_0 \cdot \mathbf{i}_1 + p_{0z}\mathbf{k}_0 \cdot \mathbf{i}_1 \end{aligned} \tag{2.1.8}$$

etc., or in matrix form

$$\mathbf{p}_1 = R_1^0 \mathbf{p}_0 \tag{2.1.9}$$

where

$$R_1^0 = \begin{bmatrix} \mathbf{i}_0 \cdot \mathbf{i}_1 & \mathbf{j}_0 \cdot \mathbf{i}_1 & \mathbf{k}_0 \cdot \mathbf{i}_1 \\ \mathbf{i}_0 \cdot \mathbf{j}_1 & \mathbf{j}_0 \cdot \mathbf{j}_1 & \mathbf{k}_0 \cdot \mathbf{j}_1 \\ \mathbf{i}_0 \cdot \mathbf{k}_1 & \mathbf{j}_0 \cdot \mathbf{k}_1 & \mathbf{k}_0 \cdot \mathbf{k}_1 \end{bmatrix} \tag{2.1.10}$$

Thus the matrix R_1^0 represents the inverse of the transformation R_0^1. Since the inner product is commutative, i.e., $\mathbf{i}_0 \cdot \mathbf{j}_0 = \mathbf{j}_0 \cdot \mathbf{i}_0$, etc., we see that

$$R_1^0 = (R_0^1)^{-1} = (R_0^1)^T \tag{2.1.11}$$

Such a matrix R_0^1 whose inverse is its transpose is said to be **orthogonal**. The column vectors of R_0^1 are of unit length and mutually orthogonal (Problem 2-1). It can also be shown (Problem 2-2) that $\det R_0^1 = \pm 1$. If we restrict ourselves to right-handed coordinate systems, as defined in Appendix A, then $\det R_0^1 = +1$ (Problem 2-3). For simplicity we refer to orthogonal matrices with determinant +1 as **rotation matrices**. It is customary to refer to the set of **all** 3×3 rotation matrices by the symbol $SO(3)$.[2]

[2]The notation $SO(3)$ stands for Special Orthogonal group of order 3.

(i) *Example 2.1.1*

Suppose the frame $ox_1y_1z_1$ is rotated through an angle θ about the z_0 axis, and it is desired to find the resulting transformation matrix R_0^1. Note that by convention the positive sense for the angle θ is given by the right hand rule; that is, a positive rotation of θ degrees about the z-axis would advance a right-hand threaded screw along the positive z-axis. From Figure 2-2 we see that

$$\begin{aligned} \mathbf{i}_0 \cdot \mathbf{i}_1 &= \cos\theta \qquad & \mathbf{j}_1 \cdot \mathbf{i}_0 &= -\sin\theta \\ \mathbf{j}_0 \cdot \mathbf{j}_1 &= \cos\theta \qquad & \mathbf{i}_1 \cdot \mathbf{j}_0 &= \sin\theta \\ & & \mathbf{k}_0 \cdot \mathbf{k}_1 = 1 & \end{aligned} \tag{2.1.12}$$

and all other dot products are zero. Thus the transformation R_0^1 has a particularly simple form in this case, namely

$$R_0^1 = \begin{bmatrix} \cos\theta & -\sin\theta & 0 \\ \sin\theta & \cos\theta & 0 \\ 0 & 0 & 1 \end{bmatrix} \tag{2.1.13}$$

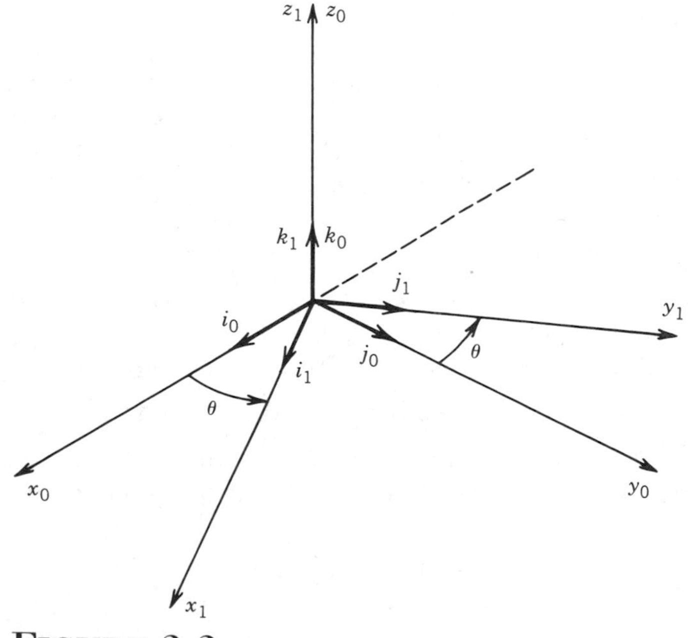

FIGURE 2-2
Rotation about the z_0 axis.

The transformation (2.1.13) is called a **basic rotation matrix** (about the z-axis). In this case we find it useful to use the more descriptive notation $R_{z,\theta}$ instead of R_0^1 to denote the matrix (2.1.13). It is easy to verify that the basic rotation matrix $R_{z,\theta}$ has the properties

$$R_{z,0} = I \tag{2.1.14}$$

$$R_{z,\theta}R_{z,\phi} = R_{z,\theta+\phi} \tag{2.1.15}$$

which together imply

$$R_{z,\theta}^{-1} = R_{z,-\theta} \tag{2.1.16}$$

Similarly the basic rotation matrices representing rotations about the x and y axes are given as (Problem 2-5)

$$R_{x,\theta} = \begin{bmatrix} 1 & 0 & 0 \\ 0 & \cos\theta & -\sin\theta \\ 0 & \sin\theta & \cos\theta \end{bmatrix} \tag{2.1.17}$$

$$R_{y,\theta} = \begin{bmatrix} \cos\theta & 0 & \sin\theta \\ 0 & 1 & 0 \\ -\sin\theta & 0 & \cos\theta \end{bmatrix} \tag{2.1.18}$$

which also satisfy properties analogous to (2.1.14)–(2.1.16).

We may also interpret a given rotation matrix as specifying the *orientation* of the coordinate frame $ox_1y_1z_1$ relative to the frame $ox_0y_0z_0$. In fact, the columns of R_0^1 are the direction cosines of the coordinate axes in $ox_1y_1z_1$ relative to the coordinate axes of $ox_0y_0z_0$. For example, the first column $(\mathbf{i}_1\cdot\mathbf{i}_0, \mathbf{i}_1\cdot\mathbf{j}_0, \mathbf{i}_1\cdot\mathbf{k}_0)^T$ of R_0^1 specifies the direction of the x_1-axis relative to the $ox_0y_0z_0$ frame.

(ii) *Example 2.1.2*

Consider the frames $ox_0y_0z_0$ and $ox_1y_1z_1$ shown in Figure 2-3. Projecting the unit vectors $\mathbf{i}_1, \mathbf{j}_1, \mathbf{k}_1$ onto $\mathbf{i}_0, \mathbf{j}_0, \mathbf{k}_0$ gives the coordinates of $\mathbf{i}_1, \mathbf{j}_1, \mathbf{k}_1$ in the $ox_0y_0z_0$ frame. We see that the coordinates of $\mathbf{i}_1$ are $(\frac{1}{\sqrt{2}}, 0, \frac{1}{\sqrt{2}})^T$, the coordinates of $\mathbf{j}_1$ are $(\frac{1}{\sqrt{2}}, 0, \frac{-1}{\sqrt{2}})^T$ and the coordinates of $\mathbf{k}_1$ are $(0, 1, 0)^T$. The rotation matrix R_0^1 specifying the orientation of $ox_1y_1z_1$ relative to $ox_0y_0z_0$ has these as its column vectors, that is,

$$R_0^1 = \begin{bmatrix} \frac{1}{\sqrt{2}} & \frac{1}{\sqrt{2}} & 0 \\ 0 & 0 & 1 \\ \frac{1}{\sqrt{2}} & \frac{-1}{\sqrt{2}} & 0 \end{bmatrix} \tag{2.1.19}$$

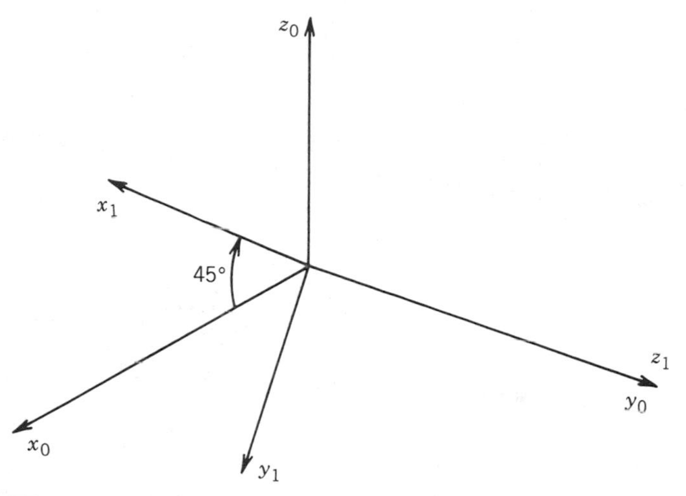

FIGURE 2-3
Defining the relative orientation of two frames.

A third interpretation of a rotation matrix $R \in SO(3)$ is as an operator acting on vectors in a fixed frame $ox_0y_0z_0$. In other words, instead of relating the coordinates of a fixed vector with respect to two different coordinate frames, the expression (2.1.10) can represent the coordinates in $ox_0y_0z_0$ of a point $\mathbf{p}_1$ which is obtained from a point $\mathbf{p}_0$ by a given rotation.

(*iii*) *Example 2.1.3*

The vector $\mathbf{p}_0 = (1, 1, 0)^T$ is rotated about the y_0-axis by $\frac{\pi}{2}$ as shown in Figure 2-4. The resulting vector $\mathbf{p}_1$ is given by

$$\mathbf{p}_1 = R_{y,\frac{\pi}{2}}\mathbf{p}_0 \tag{2.1.20}$$

$$= \begin{bmatrix} 0 & 0 & 1 \\ 0 & 1 & 0 \\ -1 & 0 & 0 \end{bmatrix} \begin{bmatrix} 1 \\ 1 \\ 0 \end{bmatrix} = \begin{bmatrix} 0 \\ 1 \\ -1 \end{bmatrix}$$

2.1.1 SUMMARY

We have seen that a rotation matrix $R \in SO(3)$ can be interpreted in three distinct ways:

1. It represents a coordinate transformation relating the coordinates of a point $\mathbf{p}$ in two different frames.

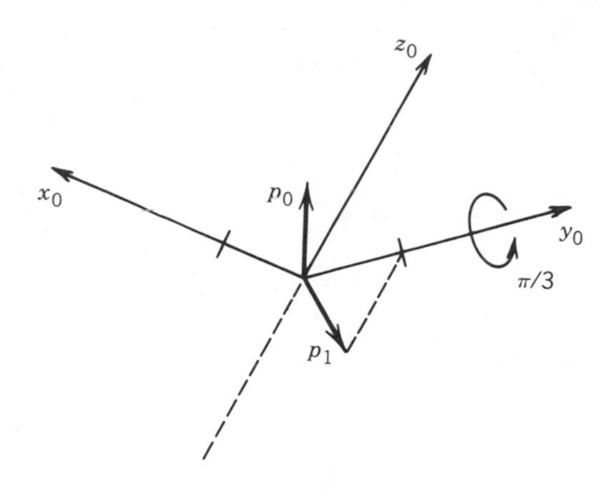

FIGURE 2-4
Rotating a vector about an axis.

2. It gives the orientation of a transformed coordinate frame with respect to a fixed coordinate frame.
3. It is an operator taking a vector **p** and rotating it to a new vector $R\mathbf{p}$ in the same coordinate system.

The particular interpretation of a given rotation matrix R that is being used must then be made clear by the context.

2.2 COMPOSITION OF ROTATIONS

In this section we discuss the composition of rotations. It is important for subsequent chapters that the reader understand the material in this section thoroughly before moving on. Recall that the matrix R_0^1 in equation (2.1.6) represents a rotational transformation between the frames $ox_0y_0z_0$ and $ox_1y_1z_1$. Suppose we now add a third coordinate frame $ox_2y_2z_2$ related to the frames $ox_0y_0z_0$ and $ox_1y_1z_1$ by rotational transformations. A given point **p** can then be represented in three ways: $\mathbf{p}_0$, $\mathbf{p}_1$, and $\mathbf{p}_2$ in the three frames. The relationship between these representations of **p** is

$$\mathbf{p}_0 = R_0^1\mathbf{p}_1 \tag{2.2.1}$$

$$\mathbf{p}_0 = R_0^2\mathbf{p}_2 \tag{2.2.2}$$

$$\mathbf{p}_1 = R_1^2\mathbf{p}_2 \tag{2.2.3}$$

where each R_i^j is a rotation matrix. Note that R_0^1 and R_0^2 represent rotations relative to the $ox_0y_0z_0$ axes, while R_1^2 represents a rotation relative to the $ox_1y_1z_1$ frame. Substituting (2.2.3) into (2.2.1) yields

$$\mathbf{p}_0 = R_0^1 R_1^2\mathbf{p}_2 \tag{2.2.4}$$

Comparing (2.2.2) and (2.2.4) we have the identity

$$R_0^2 = R_0^1 R_1^2 \tag{2.2.5}$$

Equation 2.2.5 is the composition law for rotational transformations. It states that, in order to transform the coordinates of a point **p** from its representation $\mathbf{p}_2$ in the $ox_2y_2z_2$-frame to its representation $\mathbf{p}_0$ in the $ox_0y_0z_0$-frame, we may first transform to its coordinates $\mathbf{p}_1$ in the $ox_1y_1z_1$-frame using R_1^2 and then transform $\mathbf{p}_1$ to $\mathbf{p}_0$ using R_0^1.

We may interpret Equation 2.2.4 as follows. Suppose initially that all three of the coordinate frames coincide. We first rotate the frame $ox_1y_1z_1$ relative to $ox_0y_0z_0$ according to the transformation R_0^1. Then, with the frames $ox_1y_1z_1$ and $ox_2y_2z_2$ coincident, we rotate $ox_2y_2z_2$ relative to $ox_1y_1z_1$ according to the transformation R_1^2. In each case we call the frame relative to which the rotation occurs the **current frame**.

(i) *Example 2.2.1*

Henceforth, whenever convenient we use the shorthand notation $c_\theta = \cos\theta$, $s_\theta = \sin\theta$ for trigonometric functions. Suppose a rotation matrix R represents a rotation of ϕ degrees about the current y-axis followed by a rotation of θ degrees about the current z axis. Then the matrix R is given by

$$R = R_{y,\phi} R_{z,\theta} \tag{2.2.6}$$

$$= \begin{bmatrix} c_\phi & 0 & s_\phi \\ 0 & 1 & 0 \\ -s_\phi & 0 & c_\phi \end{bmatrix} \begin{bmatrix} c_\theta & -s_\theta & 0 \\ s_\theta & c_\theta & 0 \\ 0 & 0 & 1 \end{bmatrix}$$

$$= \begin{bmatrix} c_\phi c_\theta & -c_\phi s_\theta & s_\phi \\ s_\theta & c_\theta & 0 \\ -s_\phi c_\theta & s_\phi s_\theta & c_\phi \end{bmatrix}$$

It is important to remember that the order in which a sequence of rotations are carried out, and consequently the order in which the rotation matrices are multiplied together, is crucial. The reason is that rotation, unlike position, is not a vector quantity and is therefore **not** subject to the laws of vector addition, and so rotational transformations do not commute in general.

(ii) *Example 2.2.2*

Suppose that the above rotations are performed in the reverse order, that is, first a rotation about the current z-axis followed by a rotation about the current y-axis.

Then the resulting rotation matrix is given by

$$R' = R_{z,\theta}R_{y,\phi} \tag{2.2.7}$$

$$= \begin{bmatrix} c_\theta & -s_\theta & 0 \\ s_\theta & c_\theta & 0 \\ 0 & 0 & 1 \end{bmatrix} \begin{bmatrix} c_\phi & 0 & s_\phi \\ 0 & 1 & 0 \\ -s_\phi & 0 & c_\phi \end{bmatrix}$$

$$= \begin{bmatrix} c_\phi c_\theta & -s_\theta & s_\phi c_\theta \\ s_\theta c_\phi & c_\theta & s_\theta s_\phi \\ -s_\phi & 0 & c_\phi \end{bmatrix}$$

Comparing (2.2.6) and (2.2.7) we see that $R \neq R'$.

Many times it is desired to perform a sequence of rotations, each about a given fixed coordinate frame, rather than about successive current frames. For example we may wish to perform a rotation about the x_0 axis followed by a rotation about the y_0 (and not y_1!) axis. We will refer to $ox_0y_0z_0$ as the **fixed frame**. In this case the above composition law is not valid. It turns out that the correct composition law in this case is simply to multiply the successive rotation matrices **in the reverse order** from that given by (2.2.5). Note that the rotations themselves are not performed in reverse order. Rather they are performed about the fixed frame rather than about the current frame.

(iii) Example 2.2.3

Suppose that a rotation matrix R represents a rotation of ϕ degrees about the y_0–axis followed by a rotation of θ about the fixed z_0-axis. Refer to Figure 2-5. Let $\mathbf{p}_0$, $\mathbf{p}_1$, and $\mathbf{p}_2$ be representations of a vector $\mathbf{p}$ as shown.

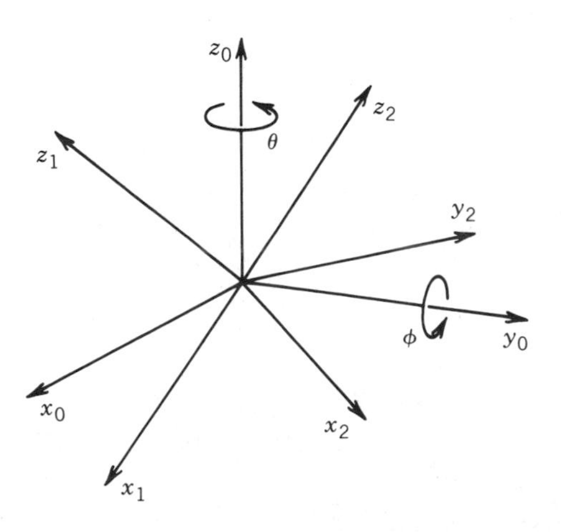

FIGURE 2-5
Composition of rotations.

Initially the fixed and current axes are the same, namely $ox_0y_0z_0$, and therefore we can write as before

$$\mathbf{p}_0 = R_{y,\phi}\mathbf{p}_1 \tag{2.2.8}$$

where $R_{y,\phi}$ is the basic rotation matrix about the y-axis. Now, since the second rotation is about the fixed frame $ox_0y_0z_0$ and not the current frame $ox_1y_1z_1$, we **cannot conclude** that

$$\mathbf{p}_1 = R_{z,\theta}\mathbf{p}_2 \tag{2.2.9}$$

since this would require that we interpret $R_{z,\theta}$ as being a rotation about z_1. In order to use our previous composition law we need somehow to have the fixed and current frames, in this case z_0 and z_1, coincident. Therefore we need first to **undo** the previous rotation, then rotate about z_0 and finally reinstate the original transformation, that is,

$$\mathbf{p}_1 = R_{y,-\phi}R_{z,\theta}R_{y,\phi}\mathbf{p}_2 \tag{2.2.10}$$

This is the correct expression, and not (2.2.9). Now, substituting (2.2.10) into (2.2.8) we obtain

$$\begin{aligned}\mathbf{p}_0 &= R_{y,\phi}\mathbf{p}_1 &(2.2.11)\\ &= R_{y,\phi}R_{y,-\phi}R_{z,\theta}R_{y,\phi}\mathbf{p}_2\\ &= R_{z,\theta}R_{y,\phi}\mathbf{p}_2\end{aligned}$$

It is not necessary to remember the above derivation, only to note by comparing (2.2.11) with (2.2.6) that we obtain the same basic rotation matrices in the reverse order.

We can be summarize the rule of composition of rotational transformations by the following recipe.

Given a fixed frame $ox_0y_0z_0$, a current frame $ox_1y_1z_1$, together with rotation matrix R_0^1 relating them, if a third frame $ox_2y_2z_2$ is obtained by a rotation R_1^2 performed relative to the **current frame** then **postmultiply** R_0^1 by R_1^2 to obtain

$$R_0^2 = R_0^1 R_1^2 \tag{2.2.12}$$

If the second rotation is to be performed relative to the **fixed frame** then **premultiply** R_0^1 by R_1^2 to obtain

$$R_0^2 = R_1^2 R_0^1 \tag{2.2.13}$$

In each case R_0^2 represents the transformation between the frames $ox_0y_0z_0$ and $ox_2y_2z_2$. The frame $ox_2y_2z_2$ that results in (2.2.12) will be different from that resulting from (2.2.13).

2.2.1 ROTATION ABOUT AN ARBITRARY AXIS

Rotations are not always performed about the principal coordinate axes. We are often interested in a rotation about an arbitrary axis in space. Therefore let $\mathbf{k} = (k_x, k_y, k_z)^T$, expressed in the frame $ox_0y_0z_0$, be a unit vector defining an axis. We wish to derive the rotation matrix $R_{\mathbf{k},\theta}$ representing a rotation of θ degrees about this axis.

There are several ways in which the matrix $R_{\mathbf{k},\theta}$ can be derived. Perhaps the simplest way is to rotate the vector **k** into one of the coordinate axes, say z_0, then rotate about z_0 by θ and finally rotate **k** back to its original position. Referring to Figure 2-6 we see that we can rotate **k** into z_0 by first rotating about z_0 by $-\alpha$, then rotating about y_0 by $-\beta$. Since all rotations are performed relative to the fixed frame $ox_0y_0z_0$ the matrix $R_{\mathbf{k},\theta}$ is obtained as

$$R_{\mathbf{k},\theta} = R_{z,\alpha}R_{y,\beta}R_{z,\theta}R_{y,-\beta}R_{z,-\alpha} \tag{2.2.14}$$

From Figure 2-6, since **k** is a unit vector, we see that

$$\sin\alpha = \frac{k_y}{\sqrt{k_x^2+k_y^2}} \tag{2.2.15}$$

$$\cos\alpha = \frac{k_x}{\sqrt{k_x^2+k_y^2}}$$

$$\sin\beta = \sqrt{k_x^2+k_y^2}$$

$$\cos\beta = k_z$$

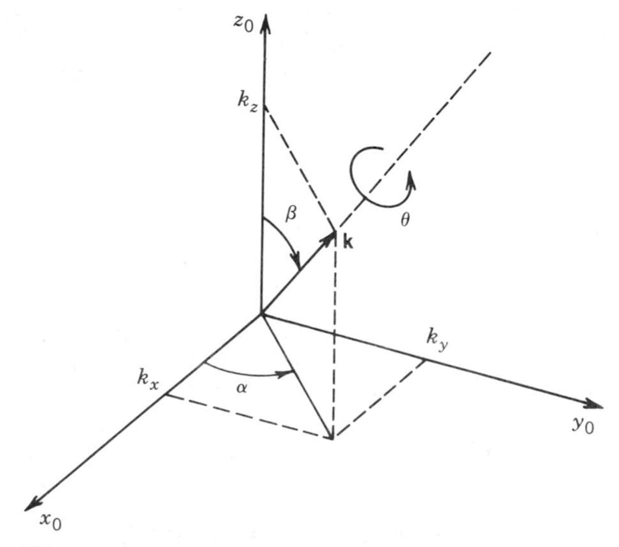

FIGURE 2-6
Rotation about an arbitrary axis.

Substituting (2.2.15) into (2.2.14) we obtain after some lengthy calculation (Problem 2-9)

$$R_{\mathbf{k},\theta} = \begin{bmatrix} k_x^2 v_\theta + c_\theta & k_x k_y v_\theta \; k_z s_\theta & k_x k_z v_\theta + k_y s_\theta \\ k_x k_y v_\theta + k_z s_\theta & k_y^2 v_\theta + c_\theta & k_y k_z v_\theta - k_x s_\theta \\ k_x k_z v_\theta - k_y s_\theta & k_y k_z v_\theta + k_x s_\theta & k_z^2 v_\theta + c_\theta \end{bmatrix} \tag{2.2.16}$$

where $v_\theta = \text{vers}\,\theta = 1 - c_\theta$.

2.3 FURTHER PROPERTIES OF ROTATIONS

The nine elements r_{ij} in a general rotational transformation R as in (2.1.7) are not independent quantities. Indeed a rigid body possess at most three rotational degrees-of-freedom and thus at most three quantities are required to specify its orientation. In this section we derive three ways in which an arbitrary rotation can be represented using only three independent quantities. The first is the **axis/angle** representation. The second is the **Euler Angle** representation and the third is the **roll-pitch-yaw** representation.

2.3.1 AXIS/ANGLE REPRESENTATION

A rotation matrix $R \in SO(3)$ can always be represented by a single rotation about a suitable axis in space by a suitable angle as

$$R = R_{\mathbf{k},\theta} \tag{2.3.1}$$

where **k** is a unit vector defining the axis of rotation, and θ is the angle of rotation about **k**. Equation (2.3.1) is called the **axis-angle representation** of R. Given an arbitrary rotation matrix R with components (r_{ij}), the equivalent angle θ and equivalent axis **k** are given by the expressions [2]

$$\theta = \cos^{-1}\left(\frac{Tr(R) - 1}{2}\right) \tag{2.3.2}$$

$$= \cos^{-1}\left(\frac{r_{11}+r_{22}+r_{33} - 1}{2}\right)$$

where Tr denotes the trace of R, and

$$\mathbf{k} = \frac{1}{2\sin\theta}\begin{bmatrix} r_{32} - r_{23} \\ r_{13} - r_{31} \\ r_{21} - r_{12} \end{bmatrix} \tag{2.3.3}$$

The axis/angle representation is not unique since a rotation of $-\theta$ about $-\mathbf{k}$ is the same as a rotation of θ about **k**, that is,

$$R_{\mathbf{k},\theta} = R_{-\mathbf{k},-\theta} \tag{2.3.4}$$

If $\theta = 0$ then R is the identity matrix and the axis of rotation is undefined.

(i) *Example 2.3.1*

Suppose R is generated by a rotation of 90° about z_0 followed by a rotation of 30° about y_0 followed by a rotation of 60° about x_0. Then

$$R = R_{x,60}R_{y,30}R_{z,90} \tag{2.3.5}$$

$$= \begin{bmatrix} 0 & -\frac{\sqrt{3}}{2} & \frac{1}{2} \\ \frac{1}{2} & -\frac{\sqrt{3}}{4} & -\frac{3}{4} \\ \frac{\sqrt{3}}{2} & \frac{1}{4} & \frac{\sqrt{3}}{4} \end{bmatrix}$$

We see that $Tr(R) = 0$ and hence the equivalent angle is given by (2.3.2) as

$$\theta = \cos^{-1}(-\frac{1}{2}) = 120^o \tag{2.3.6}$$

The equivalent axis is given from (2.3.3) as

$$\mathbf{k} = (\frac{1}{\sqrt{3}}, \frac{1}{2\sqrt{3}} - \frac{1}{2}, \frac{1}{2\sqrt{3}} + \frac{1}{2})^T \tag{2.3.7}$$

The above axis/angle representation characterizes a given rotation by four quantities, namely the three components of the equivalent axis **k** and the equivalent angle θ. However, since the equivalent axis **k** is given as a unit vector only two of its components are independent. The third is constrained by the condition that **k** is of unit length, Therefore, only three independent quantities are required in this representation of a rotation R. We can represent the equivalent angle/axis by a single vector **r** as

$$\mathbf{r} = (r_x , r_y , r_z)^T = (\theta k_x , \theta k_y , \theta k_z)^T \tag{2.3.8}$$

Note, since **k** is a unit vector, that the length of the vector **r** is the equivalent angle θ and the direction of **r** is the equivalent axis **k**.

2.3.2 EULER ANGLES

A more common method of specifying a rotation matrix in terms of three independent quantities is to use the so-called **Euler Angles**. Consider again the fixed coordinate frame $ox_0y_0z_0$ and the rotated frame $ox_1y_1z_1$ shown in Figure 2-7.

We can specify the orientation of the frame $ox_1y_1z_1$ relative to the frame $ox_0y_0z_0$ by three angles (θ, ϕ, ψ), known as Euler Angles, and obtained by three successive rotations as follows: First rotate about the z axis by the angle ϕ. Next rotate about the current y axis by the angle θ.

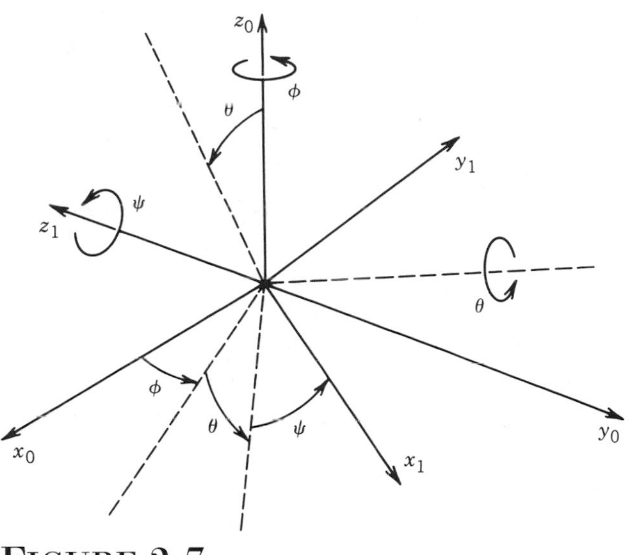

FIGURE 2-7
Euler angle representation.

Finally rotate about the current z by the angle ψ. In terms of the basic rotation matrices (2.1.14)–(2.1.16) the resulting rotational transformation R_0^1 can be generated as the product

$$R_0^1 = R_{z,\phi} R_{y,\theta} R_{z,\psi} \tag{2.3.9}$$

$$= \begin{bmatrix} c_\phi & -s_\phi & 0 \\ s_\phi & c_\phi & 0 \\ 0 & 0 & 1 \end{bmatrix} \begin{bmatrix} c_\theta & 0 & s_\theta \\ 0 & 1 & 0 \\ -s_\theta & 0 & c_\theta \end{bmatrix} \begin{bmatrix} c_\psi & -s_\psi & 0 \\ s_\psi & c_\psi & 0 \\ 0 & 0 & 1 \end{bmatrix}$$

$$= \begin{bmatrix} c_\phi c_\theta c_\psi - s_\phi s_\psi & -c_\phi c_\theta s_\psi - s_\phi c_\psi & c_\phi s_\theta \\ s_\phi c_\theta c_\psi + c_\phi s_\psi & -s_\phi c_\theta s_\psi + c_\phi c_\psi & s_\phi s_\theta \\ -s_\theta s_\psi & s_\theta c_\psi & c_\theta \end{bmatrix}$$

In Chapter Four we study the inverse problem of finding the Euler Angles (θ,ϕ,ψ) given an arbitrary rotation matrix R.

2.3.3 ROLL, PITCH, YAW ANGLES

A rotation matrix R can also be described as a product of successive rotations about the principal coordinate axes x_0, y_0, and z_0 taken in a specific order. These rotations define the **roll**, **pitch**, and **yaw** angles, which we shall also denote ϕ,θ,ψ, and which are shown in Figure 2-8.

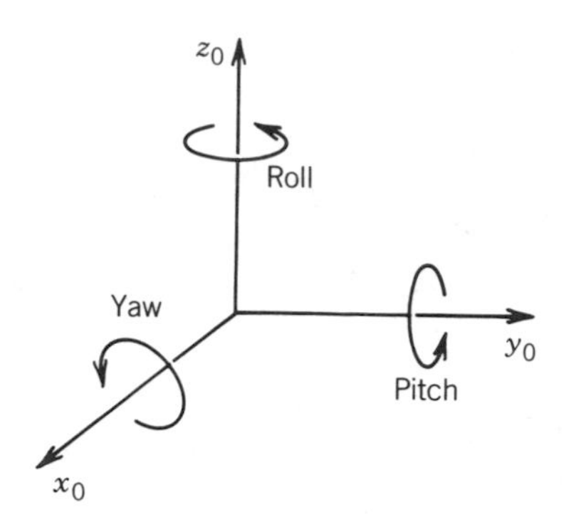

FIGURE 2-8
Roll, pitch and yaw angles.

We specify the order of rotation as $x-y-z$, in other words, first a yaw about the x_0-axis through an angle ψ, then pitch about the y_0-axis an angle θ, and finally roll about the z_0-axis an angle ϕ. Since the successive rotations are relative to the fixed frame, the resulting transformation matrix is given by

$$R_0^1 = R_{z,\phi}R_{y,\theta}R_{x,\psi} \tag{2.3.10}$$

$$= \begin{bmatrix} c_\phi & -s_\phi & 0 \\ s_\phi & c_\phi & 0 \\ 0 & 0 & 1 \end{bmatrix} \begin{bmatrix} c_\theta & 0 & s_\theta \\ 0 & 1 & 0 \\ -s_\theta & 0 & c_\theta \end{bmatrix} \begin{bmatrix} 1 & 0 & 0 \\ 0 & c_\psi & -s_\psi \\ 0 & s_\psi & c_\psi \end{bmatrix}$$

$$= \begin{bmatrix} c_\phi c_\theta & -s_\phi c_\psi + c_\phi s_\theta s_\psi & s_\phi s_\psi + c_\phi s_\theta c_\psi \\ s_\phi c_\theta & c_\phi c_\psi + s_\phi s_\theta s_\psi & -c_\phi s_\psi + s_\phi s_\theta c_\psi \\ -s_\theta & c_\theta s_\psi & c_\theta c_\psi \end{bmatrix}$$

Of course, instead of yaw-pitch-roll relative to the fixed frames we could also interpret the above transformation as roll-pitch-yaw, in that order, each taken with respect to the current frame. The end result is the same matrix (2.3.10).

2.4 HOMOGENEOUS TRANSFORMATIONS

Consider now a coordinate system $o_1x_1y_1z_1$ obtained from $o_0x_0y_0z_0$ by a parallel translation of distance $|\mathbf{d}|$ as shown in Figure 2-9. Thus $\mathbf{i}_0, \mathbf{j}_0, \mathbf{k}_0$ are parallel to $\mathbf{i}_1, \mathbf{j}_1, \mathbf{k}_1$, respectively. The vector $\mathbf{d}_0^1$ is the vector from the origin o_0 to the origin o_1 expressed in the coordinate system $o_0x_0y_0z_0$.

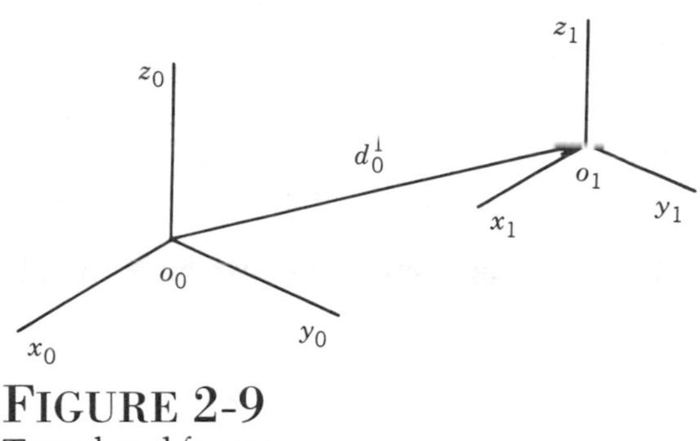

FIGURE 2-9
Translated frame.

Then any point $\mathbf{p}$ has representation $\mathbf{p}_0$ and $\mathbf{p}_1$ as before. Since the respective coordinate axes in the two frames are parallel the vectors $\mathbf{p}_0$ and $\mathbf{p}_1$ are related by

$$\mathbf{p}_0 = \mathbf{p}_1 + \mathbf{d}_0^1 \tag{2.4.1}$$

or

$$\begin{aligned} p_{0x} &= p_{1x} + d_{0x}^1 \\ p_{0y} &= p_{1y} + d_{0y}^1 \\ p_{0z} &= p_{1z} + d_{0z}^1 \end{aligned} \tag{2.4.2}$$

The most general relationship between the coordinate systems $o_0x_0y_0z_0$ and $o_1x_1y_1z_1$ that we consider can be expressed as the combination of a pure rotation and a pure translation, which is referred to as a **rigid motion**.

(i) *Definition 2.4.1*

A transformation

$$\mathbf{p}_0 = R\mathbf{p}_1 + \mathbf{d} \tag{2.4.3}$$

is said to define a **rigid motion** if R is orthogonal. Note that the definition of a rigid motion includes **reflections** when $\det R = -1$. In our case we will never have need for the most general rigid motion, so we assume always that $R \in SO(3)$.

If we have the two rigid motions

$$\mathbf{p}_0 = R_0^1\mathbf{p}_1 + \mathbf{d}_0^1 \tag{2.4.4}$$

and

$$\mathbf{p}_1 = R_1^2\mathbf{p}_2 + \mathbf{d}_1^2 \tag{2.4.5}$$

then their composition defines a third rigid motion, which we can describe by substituting the expression for $\mathbf{p}_1$ from (2.4.5) into (2.4.4).

$$\mathbf{p}_0 = R_0^1 R_1^2 \mathbf{p}_2 + R_0^1 \mathbf{d}_1^2 + \mathbf{d}_0^1 \tag{2.4.6}$$

Since the relationship between $\mathbf{p}_0$ and $\mathbf{p}_2$ is also a rigid motion, we can equally describe it as

$$\mathbf{p}_0 = R_0^2 \mathbf{p}_2 + \mathbf{d}_0^2 \tag{2.4.7}$$

Comparing equations (2.4.6) and (2.4.7) we have the identities

$$R_0^2 = R_0^1 R_1^2 \tag{2.4.8}$$

$$\mathbf{d}_0^2 = \mathbf{d}_0^1 + R_0^1 \mathbf{d}_1^2 \tag{2.4.9}$$

Equation 2.4.8 shows that the orientation transformations can simply be multiplied together and Equation 2.4.9 shows that the vector from the origin o_0 to the origin o_2 is the vector sum of the vectors $\mathbf{d}_0^1$ from o_0 to o_1 and the vector $R_0^1 \mathbf{d}_1^2$ from o_1 to o_2 expressed in the orientation of the coordinate system $o_0x_0y_0z_0$.

A comparison of this with the matrix identity

$$\begin{bmatrix} R_0^1 & \mathbf{d}_0^1 \\ \mathbf{0} & 1 \end{bmatrix} \begin{bmatrix} R_1^2 & \mathbf{d}_1^2 \\ \mathbf{0} & 1 \end{bmatrix} = \begin{bmatrix} R_0^1 R_1^2 & R_0^1 \mathbf{d}_1^2 + \mathbf{d}_0^1 \\ \mathbf{0} & 1 \end{bmatrix} \tag{2.4.10}$$

where $\mathbf{0}$ denotes $(0\,0\,0)$, shows that the rigid motions can be represented by the set of matrices of the form

$$H = \begin{bmatrix} R & \mathbf{d} \\ \mathbf{0} & 1 \end{bmatrix}; \; R \in SO(3) \tag{2.4.11}$$

Using the fact that R is orthogonal it is an easy exercise to show that the inverse transformation H^{-1} is given by

$$H^{-1} = \begin{bmatrix} R^T & -R^T\mathbf{d} \\ \mathbf{0} & 1 \end{bmatrix} \tag{2.4.12}$$

Transformation matrices of the form (2.4.11) are called **homogeneous transformations**. In order to represent the transformation (2.4.3) by a matrix multiplication, one needs to augment the vectors $\mathbf{p}_0$ and $\mathbf{p}_1$ by the addition of a fourth component of 1 as follows. Set

$$\mathbf{P}_0 = \begin{pmatrix} \mathbf{p}_0 \\ 1 \end{pmatrix} \tag{2.4.13}$$

$$\mathbf{P}_1 = \begin{pmatrix} \mathbf{p}_1 \\ 1 \end{pmatrix} \tag{2.4.14}$$

The vectors $\mathbf{P}_0$ and $\mathbf{P}_1$ are known as **homogeneous representations** of the vectors $\mathbf{p}_0$ and $\mathbf{p}_1$, respectively. It can now be seen directly that the transformation (2.4.3) is equivalent to the (homogeneous) matrix equation

$$\mathbf{P}_0 = H\mathbf{P}_1 = H_0^1\mathbf{P}_1 \tag{2.4.15}$$

The set of all 4×4 matrices H of the form (2.4.11) is denoted by $E(3)$.[3] A set of **basic homogeneous transformations** generating $E(3)$ is given by

$$Trans_{x,a} = \begin{bmatrix} 1 & 0 & 0 & a \\ 0 & 1 & 0 & 0 \\ 0 & 0 & 1 & 0 \\ 0 & 0 & 0 & 1 \end{bmatrix}; \; Trans_{y,b} = \begin{bmatrix} 1 & 0 & 0 & 0 \\ 0 & 1 & 0 & b \\ 0 & 0 & 1 & 0 \\ 0 & 0 & 0 & 1 \end{bmatrix}; \; Trans_{z,c} = \begin{bmatrix} 1 & 0 & 0 & 0 \\ 0 & 1 & 0 & 0 \\ 0 & 0 & 1 & c \\ 0 & 0 & 0 & 1 \end{bmatrix} \tag{2.4.16}$$

for translation, and

$$Rot_{x,\alpha} = \begin{bmatrix} 1 & 0 & 0 & 0 \\ 0 & c_\alpha & -s_\alpha & 0 \\ 0 & s_\alpha & c_\alpha & 0 \\ 0 & 0 & 0 & 1 \end{bmatrix}; \; Rot_{y,\phi} = \begin{bmatrix} c_\phi & 0 & s_\phi & 0 \\ 0 & 1 & 0 & 0 \\ -s_\phi & 0 & c_\phi & 0 \\ 0 & 0 & 0 & 1 \end{bmatrix}; \; Rot_{z,\theta} = \begin{bmatrix} c_\theta & -s_\theta & 0 & 0 \\ s_\theta & c_\theta & 0 & 0 \\ 0 & 0 & 1 & 0 \\ 0 & 0 & 0 & 1 \end{bmatrix} \tag{2.4.17}$$

for rotation about the x,y,z axes respectively.

The most general homogeneous transformation that we will consider may be written now as

$$T = \begin{bmatrix} n_x & s_x & a_x & d_x \\ n_y & s_y & a_y & d_y \\ n_z & s_z & a_z & d_z \\ 0 & 0 & 0 & 1 \end{bmatrix} = \begin{bmatrix} \mathbf{n} & \mathbf{s} & \mathbf{a} & \mathbf{d} \\ 0 & 0 & 0 & 1 \end{bmatrix} \tag{2.4.18}$$

In the above equation $\mathbf{n} = (n_x, n_y, n_z)^T$ is a vector representing the direction of the o_1x_1 axis in the $o_0x_0y_0z_0$ system, $\mathbf{s} = (s_x, s_y, s_z)^T$ represents the direction of the o_1y_1 axis, and $\mathbf{a} = (a_x, a_y, a_z)^T$ represents the direction of the o_1z_1 axis. The vector $\mathbf{d} = (d_x, d_y, d_z)^T$ represents the vector from the origin o_0 to the origin o_1 expressed in the $o_0x_0y_0z_0$ frame. The rationale behind the choice of letters $\mathbf{n}$, $\mathbf{s}$ and $\mathbf{a}$ is explained in Chapter Three. *NOTE: The same interpretation regarding composition and ordering of transformations holds for* 4×4 *homogeneous transformations as for* 3×3 *rotations.*

(ii) *Example 2.4.2*

The homogeneous transformation matrix H that represents a rotation of α degrees about the current x-axis followed by a translation of b units along the current x-axis, followed by a translation of d units

[3]The notation $E(3)$ stands for Euclidean group of order 3.

along the current z-axis, followed by a rotation of θ degrees about the current z-axis, is given by

$$H = Rot_{x,\alpha} Trans_{x,b} Trans_{z,d} Rot_{z,\theta} \tag{2.4.19}$$

$$= \begin{bmatrix} 1 & 0 & 0 & 0 \\ 0 & c_\alpha & -s_\alpha & 0 \\ 0 & s_\alpha & c_\alpha & 0 \\ 0 & 0 & 0 & 1 \end{bmatrix} \begin{bmatrix} 1 & 0 & 0 & b \\ 0 & 1 & 0 & 0 \\ 0 & 0 & 1 & 0 \\ 0 & 0 & 0 & 1 \end{bmatrix} \begin{bmatrix} 1 & 0 & 0 & 0 \\ 0 & 1 & 0 & 0 \\ 0 & 0 & 1 & d \\ 0 & 0 & 0 & 1 \end{bmatrix} \begin{bmatrix} c_\theta & -s_\theta & 0 & 0 \\ s_\theta & c_\theta & 0 & 0 \\ 0 & 0 & 1 & 0 \\ 0 & 0 & 0 & 1 \end{bmatrix}$$

$$= \begin{bmatrix} c_\theta & -s_\theta & 0 & b \\ c_\alpha s_\theta & c_\alpha c_\theta & -s_\alpha & -s_\alpha d \\ s_\alpha s_\theta & s_\alpha c_\theta & c_\alpha & c_\alpha d \\ 0 & 0 & 0 & 1 \end{bmatrix}$$

The homogeneous representation (2.4.11) is a special case of homogeneous coordinates, which have been extensively used in the field of computer graphics. There one is, in addition, interested in scaling and/or perspective transformations. The most general homogeneous transformation takes the form

$$H = \begin{bmatrix} R_{3x3} & | & \mathbf{d}_{3x1} \\ --- & | & --- \\ \mathbf{f}_{1x3} & | & s_{1x1} \end{bmatrix} = \begin{bmatrix} \textit{Rotation} & | & \textit{Translation} \\ --- & | & --- \\ \textit{perspective} & | & \textit{scale factor} \end{bmatrix} \tag{2.4.20}$$

For our purposes we always take the last row vector of H to be $(0,0,0,1)$, although the more general form given by (2.4.20) could be useful for interfacing a vision system into the overall robotic system or for graphic simulation.

2.5 SKEW SYMMETRIC MATRICES

In this section we derive some further properties of rotation matrices that are useful for computing relative velocity and acceleration transformations between coordinate frames. Such transformations involve computing derivatives of rotation matrices. By introducing the notion of a skew symmetric matrix it is possible to simplify many of the computations involved.

(iii) Definition 2.5.1

A matrix S is said to be **skew symmetric** if and only if

$$S^T + S = 0 \tag{2.5.1}$$

We denote the set of all 3×3 skew symmetric matrices by $SS(3)$. If $S \in SS(3)$ has components s_{ij}, $i,j = 1,2,3$ then (2.5.1) is equivalent to the nine equations

$$s_{ij} + s_{ji} = 0 \quad i,j = 1,2,3 \tag{2.5.2}$$

From (2.5.2) we see that $s_{ii} = 0$; that is, the diagonal terms of S are zero and the off diagonal terms s_{ij} $i \neq j$ satisfy $s_{ij} = -s_{ji}$. Thus S contains only three independent entries and every 3×3 skew symmetric matrix has the form

$$S = \begin{bmatrix} 0 & -s_1 & s_2 \\ s_1 & 0 & -s_3 \\ -s_2 & s_3 & 0 \end{bmatrix} \tag{2.5.3}$$

If $\mathbf{a} = (a_x, a_y, a_z)^T$ is a 3-vector, we define the skew symmetric matrix $S(\mathbf{a})$ as

$$S(\mathbf{a}) = \begin{bmatrix} 0 & -a_z & a_y \\ a_z & 0 & -a_x \\ -a_y & a_x & 0 \end{bmatrix} \tag{2.5.4}$$

(iv) *Example 2.5.2*

The skew symmetric matrices $S(\mathbf{i})$, $S(\mathbf{j})$, and $S(\mathbf{k})$ that result when $\mathbf{a}$ equals the unit normal vectors $\mathbf{i}$, $\mathbf{j}$, and $\mathbf{k}$, respectively, are given by

$$S(\mathbf{i}) = \begin{bmatrix} 0 & 0 & 0 \\ 0 & 0 & -1 \\ 0 & 1 & 0 \end{bmatrix}; \; S(\mathbf{j}) = \begin{bmatrix} 0 & 0 & 1 \\ 0 & 0 & 0 \\ -1 & 0 & 0 \end{bmatrix}; \; S(\mathbf{k}) = \begin{bmatrix} 0 & -1 & 0 \\ 1 & 0 & 0 \\ 0 & 0 & 0 \end{bmatrix} \tag{2.5.5}$$

An important property possessed by the matrix $S(\mathbf{a})$ is linearity. Thus for any vectors $\mathbf{a}$ and $\mathbf{b}$ belonging to $\mathbb{R}^3$ and scalars α and β we have

$$S(\alpha\mathbf{a} + \beta\mathbf{b}) = \alpha S(\mathbf{a}) + \beta S(\mathbf{b}) \tag{2.5.6}$$

Another important property of $S(\mathbf{a})$ is that for any vector $\mathbf{p} = (p_x, p_y, p_z)^T$

$$S(\mathbf{a})\mathbf{p} = \mathbf{a} \times \mathbf{p} \tag{2.5.7}$$

where $\mathbf{a} \times \mathbf{p}$ denotes the vector cross product defined in Appendix A. Equation 2.5.7 can be verified by direct calculation.

If $R \in SO(3)$ and $\mathbf{a},\mathbf{b}$ are vectors in $\mathbb{R}^3$ it can also be shown by direct calculation that

$$R(\mathbf{a} \times \mathbf{b}) = R\mathbf{a} \times R\mathbf{b} \tag{2.5.8}$$

Equation 2.5.8 is **not** true in general unless R is orthogonal. Equation 2.5.8 says that if we first rotate the vectors $\mathbf{a}$ and $\mathbf{b}$ using the rotation transformation R and then form the cross product of the rotated vectors $R\mathbf{a}$ and $R\mathbf{b}$, the result is the same as that obtained by first forming the cross product $\mathbf{a} \times \mathbf{b}$ and then rotating to $R(\mathbf{a} \times \mathbf{b})$.

For any $R \in SO(3)$ and any $\mathbf{b} \in \mathbb{R}^3$, it follows from (2.5.7) and (2.5.8) that

$$\begin{aligned} RS(\mathbf{a})R^T\mathbf{b} &= R(\mathbf{a}\times R^T\mathbf{b}) \\ &= (R\mathbf{a})\times(RR^T\mathbf{b}) \\ &= (R\mathbf{a})\times\mathbf{b} \quad \text{since } R \text{ is orthogonal} \\ &= S(R\mathbf{a})\mathbf{b} \end{aligned} \tag{2.5.9}$$

Thus we have shown the useful fact that

$$RS(\mathbf{a})R^T = S(R\mathbf{a}) \tag{2.5.10}$$

for $R \in SO(3)$, $\mathbf{a} \in \mathbb{R}^3$. As we will see, (2.5.10) is one of the most useful expressions that we will derive. The left hand side of Equation 2.5.10 represents a similarity transformation of the matrix $S(\mathbf{a})$. The equation says therefore that the matrix representation of $S(\mathbf{a})$ in a coordinate frame rotated by R is the same as the skew symmetric matrix $S(R\mathbf{a})$ corresponding to the vector $\mathbf{a}$ rotated by R.

Suppose now that a rotation matrix R is a function of the single variable θ. Hence $R = R(\theta) \in SO(3)$ for every θ. Since R is orthogonal for all θ it follows that

$$R(\theta)R(\theta)^T = I \tag{2.5.11}$$

Differentiating both sides of (2.5.11) with respect to θ using the product rule gives

$$\frac{dR}{d\theta}R(\theta)^T + R(\theta)\frac{dR^T}{d\theta} = 0 \tag{2.5.12}$$

Let us define the matrix

$$S := \frac{dR}{d\theta}R(\theta)^T \tag{2.5.13}$$

Then the transpose of S is

$$S^T = \left(\frac{dR}{d\theta}R(\theta)^T\right)^T = R(\theta)\frac{dR^T}{d\theta} \tag{2.5.14}$$

Equation 2.5.12 says therefore that

$$S + S^T = 0 \tag{2.5.15}$$

In other words, the matrix S defined by (2.5.13) is skew symmetric. Multiplying both sides of (2.5.13) on the right by R and using the fact that $R^TR = I$ yields

$$\frac{dR}{d\theta} = S\,R(\theta) \tag{2.5.16}$$

Equation 2.5.16 is very important. It says that computing the derivative of the rotation matrix R is equivalent to a matrix multiplication by a skew symmetric matrix S. The most commonly encountered situation is the case where R is a basic rotation matrix or a product of basic rotation matrices.

(v) *Example 2.5.3*

If $R = R_{x,\theta}$, the basic rotation matrix given by (2.1.16), then direct computation shows that

$$S = \frac{dR}{d\theta}R^T = \begin{bmatrix} 0 & 0 & 0 \\ 0 & -\sin\theta & -\cos\theta \\ 0 & \cos\theta & -\sin\theta \end{bmatrix} \begin{bmatrix} 1 & 0 & 0 \\ 0 & \cos\theta & \sin\theta \\ 0 & -\sin\theta & \cos\theta \end{bmatrix} \tag{2.5.17}$$

$$= \begin{bmatrix} 0 & 0 & 0 \\ 0 & 0 & -1 \\ 0 & 1 & 0 \end{bmatrix} = S(\mathbf{i})$$

Thus we have shown that

$$\frac{dR_{x,\theta}}{d\theta} = S(\mathbf{i})R_{x,\theta} \tag{2.5.18}$$

Similar computations show that

$$\frac{dR_{y,\theta}}{d\theta} = S(\mathbf{j})R_{y,\theta} \ ; \quad \frac{dR_{z,\theta}}{d\theta} = S(\mathbf{k})R_{z,\theta} \tag{2.5.19}$$

(vi) *Example 2.5.4*

Let $R_{\mathbf{k},\theta}$ be a rotation about the axis defined by **k** as in (2.3.3). It is easy to check that $S(\mathbf{k})^3 = -S(\mathbf{k})$. Using this fact together with Problem 2-25 it follows that

$$\frac{dR_{\mathbf{k},\theta}}{d\theta} = S(\mathbf{k})R_{\mathbf{k},\theta} \tag{2.5.20}$$

Other examples are given in the next section and also in Chapter Five.

2.6 ANGULAR VELOCITY AND ACCELERATION

In the previous sections we derived expressions relating position and orientation of various coordinate frames via the introduction of homogeneous transformations. In this section we discuss relative velocities and accelerations in the same context.

Suppose that a rotation matrix R is time varying, so that $R = R(t) \in SO(3)$ for every $t \in R$. An argument identical to the one in the previous section shows that the time derivative $\dot{R}(t)$ of $R(t)$ is given by

$$\dot{R}(t) = S(t)R(t) \tag{2.6.1}$$

where the matrix $S(t)$ is skew symmetric. Now, since $S(t)$ is skew symmetric, it can be represented as $S(\boldsymbol{\omega}(t))$ for a unique vector $\boldsymbol{\omega}(t)$. This vector $\boldsymbol{\omega}(t)$ is the **angular velocity** of the rotating frame with respect to the fixed frame at time t.

(vii) Example 2.6.1

Suppose that $R(t) = R_{x,\theta(t)}$. Then $\dot{R}(t) = \frac{dR}{dt}$ is computed using the chain rule as

$$\dot{R} = \frac{dR}{d\theta}\frac{d\theta}{dt} = \dot{\theta}S(\mathbf{i})R(t) = S(\boldsymbol{\omega}(t))R(t) \tag{2.6.2}$$

where $\boldsymbol{\omega} = \mathbf{i}\dot{\theta}$ is the **angular velocity** about the x-axis.

Suppose $\mathbf{p}_1$ represents a vector fixed in a coordinate frame $o_1x_1y_1z_1$, and the frame $o_1x_1y_1z_1$ is rotating relative to the frame $o_0x_0y_0z_0$. Then the coordinates of $\mathbf{p}_1$ in $o_0x_0y_0z_0$ are given by

$$\mathbf{p}_0 = R(t)\mathbf{p}_1 \tag{2.6.3}$$

The velocity $\dot{\mathbf{p}}_0$ is then given as

$$\begin{aligned}\dot{\mathbf{p}}_0 &= S(\boldsymbol{\omega})R(t)\mathbf{p}_1 \\ &= S(\boldsymbol{\omega})\mathbf{p}_0 = \boldsymbol{\omega}\times\mathbf{p}_0\end{aligned} \tag{2.6.4}$$

which is the familiar expression for the velocity in terms of the vector cross product. Now suppose that the motion of the frame $o_1x_1y_1z_1$ relative to $o_0x_0y_0z_0$ is more general. Suppose that the homogeneous transformation relating the two frames is time-dependent, so that

$$H_0^1(t) = \begin{bmatrix} R_0^1(t) & \mathbf{d}_0^1(t) \\ \mathbf{0} & 1 \end{bmatrix} \tag{2.6.5}$$

For simplicity we omit the argument t and the subscripts and superscripts on R_0^1 and $\mathbf{d}_0^1$, and write

$$\mathbf{p}_0 = R\,\mathbf{p}_1 + \mathbf{d} \tag{2.6.6}$$

Differentiating the above expression using the product rule gives

$$\begin{aligned}\dot{\mathbf{p}}_0 &= \dot{R}\,\mathbf{p}_1 + \dot{\mathbf{d}} \quad \text{since } \mathbf{p}_1 \text{ is constant} \\ &= S(\boldsymbol{\omega})R\,\mathbf{p}_1 + \dot{\mathbf{d}} \\ &= \boldsymbol{\omega}\times\mathbf{r} + \mathbf{v}\end{aligned} \tag{2.6.7}$$

where $\mathbf{r} = R\,\mathbf{p}_1$ is the vector from o_1 to $\mathbf{p}$ expressed in the orientation of the frame $o_0x_0y_0z_0$, and $\mathbf{v}$ is the rate at which the origin o_1 is moving.

If the vector $\mathbf{p}_1$ is also changing relative to the frame $o_1x_1y_1z_1$ then we must add to the term $\mathbf{v}$ the term $R(t)\dot{\mathbf{p}}_1$, which is the rate of change of $\mathbf{p}_1$ expressed in the frame $o_0x_0y_0z_0$.

We may also derive the expression for the relative acceleration in the two coordinate frames as follows. First, recall that the cross product satisfies the product rule for differentiation (Appendix A)

$$\frac{d}{dt}(\mathbf{a}\times\mathbf{b}) = \frac{d\mathbf{a}}{dt}\times\mathbf{b} + \mathbf{a}\times\frac{d\mathbf{b}}{dt} \tag{2.6.8}$$

If we now rewrite Equation 2.6.7 as

$$\dot{\mathbf{p}}_0 - \dot{\mathbf{d}} = \dot{R}\,\mathbf{p}_1 \qquad (2.6.9)$$
$$= \boldsymbol{\omega} \times R\,\mathbf{p}_1$$

and differentiate both sides with respect to t we obtain

$$\ddot{\mathbf{p}}_0 - \ddot{\mathbf{d}} = \dot{\boldsymbol{\omega}} \times R\,\mathbf{p}_1 + \boldsymbol{\omega} \times (\dot{R}\,\mathbf{p}_1) \qquad (2.6.10)$$
$$= \dot{\boldsymbol{\omega}} \times \mathbf{r} + \boldsymbol{\omega} \times (\boldsymbol{\omega} \times \mathbf{r})$$

Thus (2.6.10) may be written as

$$\ddot{\mathbf{p}}_0 = \dot{\boldsymbol{\omega}} \times \mathbf{r} + \boldsymbol{\omega} \times (\boldsymbol{\omega} \times \mathbf{r}) + \mathbf{a} \qquad (2.6.11)$$

where $\mathbf{a}$ is the linear acceleration. The term $\boldsymbol{\omega} \times (\boldsymbol{\omega} \times \mathbf{r})$ is called the **centripetal acceleration** of the particle. It is always directed toward the axis of rotation and is perpendicular to that axis. The term $\dot{\boldsymbol{\omega}} \times \mathbf{r}$ is called the **transverse acceleration**.

Again, if the vector $\mathbf{p}_1$ is changing with respect to $o_1x_1y_1z_1$, the above expression must be modified. In this case it is left as an exercise to show that Equation 2.6.11 is replaced by

$$\ddot{\mathbf{p}}_0 = \dot{\boldsymbol{\omega}} \times \mathbf{r} + \boldsymbol{\omega} \times (\boldsymbol{\omega} \times \mathbf{r}) + 2\boldsymbol{\omega} \times R\,\dot{\mathbf{p}}_1 + \mathbf{a} \qquad (2.6.12)$$

where $\mathbf{a} = R\,\ddot{\mathbf{p}}_1 + \ddot{\mathbf{d}}$. The term $2\boldsymbol{\omega} \times R\,\dot{\mathbf{p}}_1$ is known as the **Coriolis acceleration**.

2.7 ADDITION OF ANGULAR VELOCITIES

We are often interested in finding the resultant angular velocity due to the relative rotation of several coordinate frames. We now derive the expressions for the composition of angular velocities of two moving frames $o_1x_1y_1z_1$ and $o_2x_2y_2z_2$ relative to a fixed frame $o_0x_0y_0z_0$.

Given a point $\mathbf{p}$ with representations $\mathbf{p}_0$, $\mathbf{p}_1$, $\mathbf{p}_2$ in the respective frames we have the relationships

$$\mathbf{p}_0 = R_0^1\mathbf{p}_1 + \mathbf{d}_0^1 \qquad (2.7.1)$$
$$\mathbf{p}_1 = R_1^2\mathbf{p}_2 + \mathbf{d}_1^2 \qquad (2.7.2)$$
$$\mathbf{p}_0 = R_0^2\mathbf{p}_2 + \mathbf{d}_0^2 \qquad (2.7.3)$$

where as before

$$R_0^2 = R_0^1 R_1^2 \qquad (2.7.4)$$

and

$$\mathbf{d}_0^2 = \mathbf{d}_0^1 + R_0^1\,\mathbf{d}_1^2 \qquad (2.7.5)$$

As before, all of the above quantities are functions of time. Taking derivatives of both sides of (2.7.4) yields

$$\dot{R}_0^2 = \dot{R}_0^1 R_1^2 + R_0^1 \dot{R}_1^2 \tag{2.7.6}$$

The term $\dot{R}_0^2$ on the left-hand side of (2.7.6) can be written

$$\dot{R}_0^2 = S(\boldsymbol{\omega}_0^2) R_0^2 \tag{2.7.7}$$

The first term on the right-hand side of (2.7.6) is simply

$$\dot{R}_0^1 R_1^2 = S(\boldsymbol{\omega}_0^1) R_0^1 R_1^2 = S(\boldsymbol{\omega}_0^1) R_0^2 \tag{2.7.8}$$

Let us examine the second term on the right hand side of (2.7.6). Using the expression (2.5.10) we have

$$\begin{aligned} R_0^1 \dot{R}_1^2 &= R_0^1 S(\boldsymbol{\omega}_1^2) R_1^2 \\ &= R_0^1 S(\boldsymbol{\omega}_1^2) R_0^{1T} R_0^1 R_1^2 = S(R_0^1 \boldsymbol{\omega}_1^2) R_0^1 R_1^2 \\ &= S(R_0^1 \boldsymbol{\omega}_1^2) R_0^2 \end{aligned} \tag{2.7.9}$$

Combining the above expressions we have shown that

$$S(\boldsymbol{\omega}_0^2) R_0^2 = \{S(\boldsymbol{\omega}_0^1) + S(R_0^1 \boldsymbol{\omega}_1^2)\} R_0^1 R_1^2 \tag{2.7.10}$$

Since $S(\mathbf{a}) + S(\mathbf{b}) = S(\mathbf{a}+\mathbf{b})$, we see that

$$\boldsymbol{\omega}_0^2 = \boldsymbol{\omega}_0^1 + R_0^1 \boldsymbol{\omega}_1^2 \tag{2.7.11}$$

In other words, the angular velocities can be added once they are expressed relative to the same coordinate frame, in this case $o_0x_0y_0z_0$.

The above expression can be extended to any number of coordinate systems. For example, if

$$R_0^n = R_0^1 R_1^2 \cdots R_{n-1}^n \tag{2.7.12}$$

then

$$\dot{R}_0^n = S(\boldsymbol{\omega}_0^n) R_0^n \tag{2.7.13}$$

where

$$\boldsymbol{\omega}_0^n = \boldsymbol{\omega}_0^1 + R_0^1 \boldsymbol{\omega}_1^2 + R_0^2 \boldsymbol{\omega}_1^2 + R_0^3 \boldsymbol{\omega}_2^3 + \cdots + R_0^{n-1} \boldsymbol{\omega}_{n-1}^n \tag{2.7.14}$$

REFERENCES AND SUGGESTED READING

[1] BARNETT, S., *Matrix Methods for Engineers and Scientists*, McGraw-Hill, London, 1979.

[2] CHEN, Y.C., and VIDYASAGAR, M., "On the Axis–Angle Parametrization of the Rotation Group and its Applications to Robotics," preprint 1987.

[3] CURTISS, M.L., *Matrix Groups*, Second Edition, Springer–Verlag, New York, 1984.

[4] FRIEDBERG, S.H., INSEL, A.J., and SPENCE, L.E., *Linear Algebra*, Prentice-Hall, Englewood Cliffs, NJ, 1979.

[5] REDDY, J.N., and RASMUSSEN, M.L., *Advanced Engineering Analysis*, Wiley, New York, 1982.

[6] SOKOLNIKOFF, I.S., and REDHEFFER, R.M., *Mathematical Methods of Physics and Modern Engineering*, McGraw-Hill, New York, 1958.

[7] WHITTAKER, E.T., *Dynamics of Particles and Rigid Bodies*, Cambridge University Press, London, 1904.

PROBLEMS

2-1 If R is an orthogonal matrix show that the column vectors of R are of unit length and mutually perpendicular.

2-2 If R is an orthogonal matrix show that $\det R = \pm 1$.

2-3 Show that $\det R = +1$ if we restrict ourselves to right-handed coordinate systems.

2-4 Verify Equations 2.1.14–2.1.16.

2-5 Derive Equations 2.1.17 and 2.1.18.

2-6 Suppose A is a 2×2 rotation matrix. In other words $A^T A = I$ and $\det A = 1$. Show that there exists a unique θ such that A is of the form

$$A = \begin{bmatrix} \cos\theta & -\sin\theta \\ \sin\theta & \cos\theta \end{bmatrix}$$

2-7 Find the rotation matrix representing a roll of $\frac{\pi}{4}$ followed by a yaw of $\frac{\pi}{2}$ followed by a pitch of $\frac{\pi}{2}$.

2-8 If the coordinate frame $o_1x_1y_1z_1$ is obtained from the coordinate frame $o_0x_0y_0z_0$ by a rotation of $\frac{\pi}{2}$ about the x-axis followed by a rotation of $\frac{\pi}{2}$ about the fixed y-axis, find the rotation matrix R representing the composite transformation. Sketch the initial and final frames.

2-9 Suppose that three coordinate frames $o_1x_1y_1z_1$, $o_2x_2y_2z_2$, and $o_3x_3y_3z_3$ are given, and suppose

$$R_1^2 = \begin{bmatrix} 1 & 0 & 0 \\ 0 & \frac{1}{2} & -\frac{\sqrt{3}}{2} \\ 0 & \frac{\sqrt{3}}{2} & \frac{1}{2} \end{bmatrix}; \quad R_1^3 = \begin{bmatrix} 0 & 0 & -1 \\ 0 & 1 & 0 \\ 1 & 0 & 0 \end{bmatrix}$$

Find the matrix R_2^3.

2-10 Verify Equation 2.2.16.

2-11 If R is a rotation matrix show that +1 is an eigenvalue of R. Let $\mathbf{k}$ be a unit eigenvector corresponding to the eigenvalue +1. Give a physical interpretation of $\mathbf{k}$.

2-12 Let $\mathbf{k} = \frac{1}{\sqrt{3}}(1,1,1)^T$, $\theta = 90^o$. Find $R_{\mathbf{k},\theta}$.

2-13 Show by direct calculation that $R_{\mathbf{k},\theta}$ given by (2.2.16) is equal to R given by (2.3.5) if θ and $\mathbf{k}$ are given by (2.3.6) and (2.3.7), respectively.

2-14 Suppose R represents a rotation of 90° about y_o followed by a rotation of 45° about z_1. Find the equivalent axis/angle to represent R. Sketch the initial and final frames and the equivalent axis vector $\mathbf{k}$.

2-15 Find the rotation matrix corresponding to the set of Euler angles $\{ \frac{\pi}{2}, 0, \frac{\pi}{4} \}$. What is the direction of the x_1 axis relative to the base frame?

2-16 Compute the homogeneous transformation representing a translation of 3 units along the x-axis followed by a rotation of $\frac{\pi}{2}$ about the current z-axis followed by a translation of 1 unit along the fixed y-axis. Sketch the frame. What are the coordinates of the origin o_1 with respect to the original frame in each case?

2-17 Consider the diagram of Figure 2-10. Find the homogeneous transformations H_0^1, H_0^2, H_1^2 representing the transformations among the three frames shown. Show that $H_0^2 = H_0^1 H_1^2$.

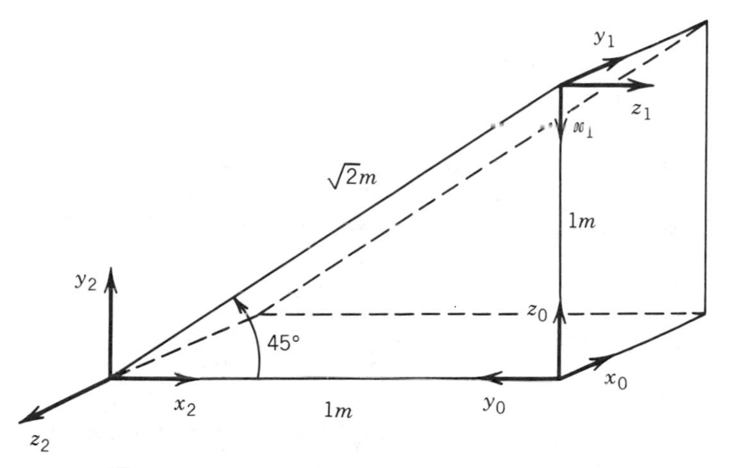

FIGURE 2-10
Diagram for Problem 2-17.

2-18 Consider the diagram of Figure 2-11. A robot is set up 1 meter from a table, two of whose legs are on the y_0 axis as shown. The table top is 1 meter high and 1 meter square. A frame $o_1x_1y_1z_1$ is fixed to the edge of the table as shown. A cube measuring 20 cm on a side is placed in the center of the table with frame $o_2x_2y_2z_2$ established at the center of the cube as shown. A camera is situated directly above the center of the block 2 m above the table top with frame $o_3x_3y_3z_3$ attached as shown. Find the homogeneous transformations relating each of these frames to the base frame $o_0x_0y_0z_0$. Find the homogeneous transformation relating the frame $o_2x_2y_2z_2$ to the camera frame $o_3x_3y_3z_3$.

2-19 In Problem 2-18, suppose that, after the camera is calibrated, it is rotated 90° about the axis z_3. Recompute the above coordinate transformations.

2-20 If the block on the table is rotated 90° degrees about the axis z_2 and moved so that its center has coordinates $(0,.8,.1)^T$ relative to the frame $o_1x_1y_1z_1$, compute the homogeneous transformation relating the block frame to the camera frame; the block frame to the base frame.

2-21 Verify Equation 2.5.7 by direct calculation.

2-22 Prove assertion (2.5.8) that $R(\mathbf{a}\times\mathbf{b}) = R\mathbf{a}\times R\mathbf{b}$, for $R \in SO(3)$.

2-23 Suppose that $\mathbf{a} = (1 , -1 , 2)^T$ and that $R = R_{x,90}$. Show by direct calculation that

$$RS(\mathbf{a})R^T = S(R\mathbf{a})$$

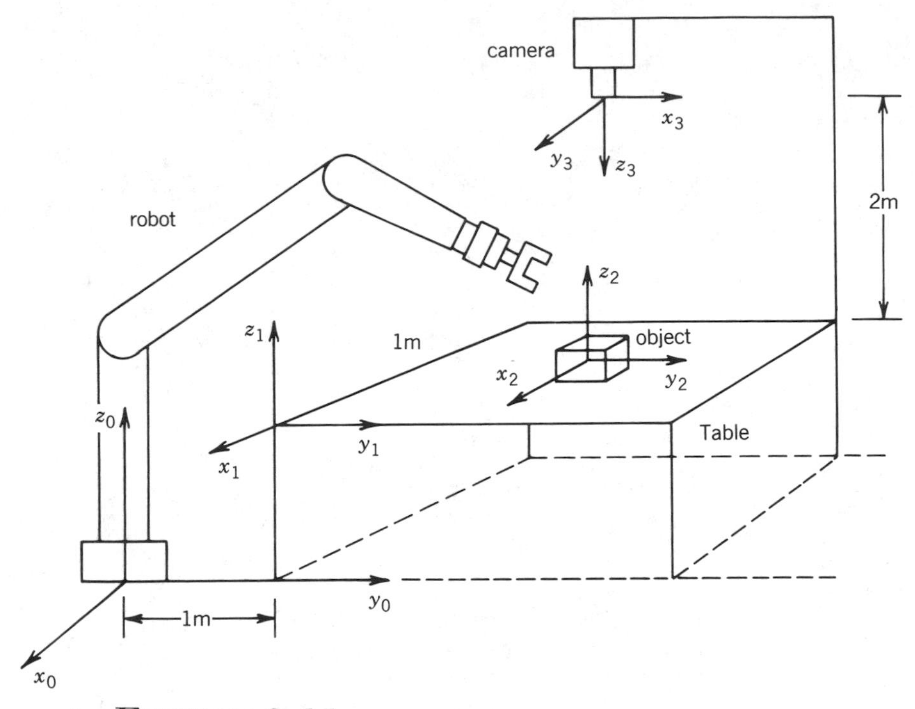

FIGURE 2-11
Diagram for Problem 2-18.

2-24 Given $R_0^1 = R_{x,\theta} R_{y,\phi}$, compute $\dfrac{\partial R_0^1}{\partial \phi}$. Evaluate $\dfrac{\partial R_0^1}{\partial \phi}$ at $\theta = \dfrac{\pi}{2}$, $\phi = \dfrac{\pi}{2}$.

2-25 Use Equation 2.2.16 to show that

$$R_{\mathbf{k},\theta} = I + S(\mathbf{k})\sin(\theta) + S^2(\mathbf{k})\text{vers}(\theta)$$

2-26 Verify (2.5.19) by direct calculation.

2-27 Show that $S(\mathbf{k})^3 = -S(\mathbf{k})$. Use this and Problem 2-25 to verify Equation 2.5.20.

2-28 Given any square matrix A, the exponential of A is a matrix defined as

$$e^A = I + A + \frac{1}{2}A^2 + \frac{1}{3!}A^3 + \cdots$$

Given $S \in SS(3)$ show that $e^S \in SO(3)$.
[Hint: Verify the facts that $e^A e^B = e^{A+B}$ provided that A and B commute, that is, $AB = BA$, and also that $\det(e^A) = e^{Tr(A)}$.]

2-29 Show that $R_{\mathbf{k},\theta} = e^{S(\mathbf{k})\theta}$.
[Hint: Use the series expansion for the matrix exponential together with Problems 2-23 and 2-24. Alternatively use the fact $R_{\mathbf{k},\theta}$ satisfies the differential equation

$$\frac{dR}{d\theta} = S(\mathbf{k})R$$

2-30 Use Problem 2-29 to show the converse of 2-28, that is, if $R \in SO(3)$ then there exists $S \in SS(3)$ such that $R = e^S$.

2-31 Given the Euler angle transformation

$$R = R_{z,\psi}R_{y,\theta}R_{z,\phi}$$

show that $\frac{d}{dt}R = S(\boldsymbol{\omega})R$ where

$$\boldsymbol{\omega} = \{c_\psi s_\theta \dot{\phi} - s_\psi \dot{\theta}\}\mathbf{i} + \{s_\psi s_\theta \dot{\phi} + c_\psi \dot{\theta}\}\mathbf{j} + \{\dot{\psi} + c_\theta \dot{\phi}\}\mathbf{k}.$$

The components of **i**,**j**,**k**, respectively, are called the **nutation, spin, and precession.**

2-32 Repeat Problem 2-31 for the Roll-Pitch-Yaw transformation. In other words, find an explicit expression for $\boldsymbol{\omega}$ such that $\frac{d}{dt}R = S(\boldsymbol{\omega})R$, where R is given by (2.3.10).

2-33 Two frames $o_0x_0y_0z_0$ and $o_1x_1y_1z_1$ are related by the homogeneous transformation

$$H = \begin{bmatrix} 0 & -1 & 0 & 1 \\ 1 & 0 & 0 & -1 \\ 0 & 0 & 1 & 0 \\ 0 & 0 & 0 & 1 \end{bmatrix}$$

A particle has velocity $\mathbf{v}_1(t) = (3,1,0)^T$ relative to frame $o_1x_1y_1z_1$. What is the velocity of the particle in frame $o_0x_0y_0z_0$?

2-34 Three frames $o_0x_0y_0z_0$, $o_1x_1y_1z_1$, and $o_2x_2y_2z_2$ are given below. If the angular velocities $\boldsymbol{\omega}_0^1$ and $\boldsymbol{\omega}_1^2$ are given as

$$\boldsymbol{\omega}_0^1 = \begin{bmatrix} 1 \\ 1 \\ 0 \end{bmatrix} \quad ; \quad \boldsymbol{\omega}_1^2 = \begin{bmatrix} 2 \\ 0 \\ 1 \end{bmatrix}$$

what is the angular velocity $\boldsymbol{\omega}_0^2$ at the instant when

$$R_0^1 = \begin{bmatrix} 1 & 0 & 0 \\ 0 & 0 & -1 \\ 0 & 1 & 0 \end{bmatrix}$$

CHAPTER THREE

FORWARD KINEMATICS: THE DENAVIT–HARTENBERG REPRESENTATION

In this chapter we develop the **forward** or **configuration kinematic equations** for rigid robots. The forward kinematics problem can be stated as follows: Given the joint variables of the robot, determine the position and orientation of the end-effector. The joint variables are the angles between the links in the case of revolute or rotational joints, and the link extension in the case of prismatic or sliding joints. The forward kinematics problem is to be contrasted with the **inverse** kinematics problem, which is studied in the next chapter, and which can be stated as follows: Given a desired position and orientation for the end-effector of the robot, determine a set of joint variables that achieve the desired position and orientation.

3.1 KINEMATIC CHAINS

For the purposes of kinematic analysis, one can think of a robot as a set of rigid links connected together at various joints. The joints can either be very simple, such as a revolute joint or a prismatic joint, or else they can be more complex, such as a ball and socket joint. (Recall that a revolute joint is like a hinge and allows a relative rotation about a single axis, and a prismatic joint permits a linear motion along a single axis, namely an extension or retraction.) The difference between the two situations is that, in the first instance, the joint has only a single degree-of-freedom of motion: the angle of rotation in the case of a revolute joint, and the amount of linear displacement in the case of a prismatic joint. In contrast, a ball and socket joint has two degrees-of-freedom. In this book it is assumed throughout that all joints have only a single degree-of-freedom. Note that the assumption does not involve any real loss of generality, since joints such as a ball and socket joint (two degrees-of-freedom) or a spherical wrist (three degrees-of-freedom) can always be thought of as a succession of single degree-of-freedom joints with links of length zero in between.

With the assumption that each joint has a single degree-of-freedom, the action of each joint can be described by a single real number; the angle of rotation in the case of a revolute joint or the displacement in the case of a prismatic joint. The objective of forward kinematic analysis is to determine the *cumulative* effect of the entire set of joint variables. To do this in a systematic manner, one should really introduce some conventions. It is of course possible to carry out forward kinematics analysis even without respecting these conventions as we did for the two-link planar example in Chapter One. However, the kinematic analysis of an n-link manipulator is extremely complex and the conventions introduced below simplify the equations considerably. Moreover, they give rise to a universal language with which robot engineers can communicate.

Suppose a robot has $n+1$ links numbered from 0 to n starting from the base of the robot, which is taken as link 0. The joints are numbered from 1 to n, and the i-th joint is the point in space where links $i-1$ and i are connected. The i-th joint variable is denoted by q_i. In the case of a revolute joint, q_i is the angle of rotation, and in the case of a prismatic joint, q_i is the joint displacement. Next, a coordinate frame is attached rigidly to each link. To be specific, we attach an inertial frame to the base and call it frame 0. Then we choose frames 1 through n such that frame i is rigidly attached to link i. This means that, whatever motion the robot executes, the coordinates of each point on link i are constant when expressed in the i-th coordinate frame. Figure 3-1 illustrates the idea of attaching frames rigidly to links in the case of an elbow manipulator.

Now suppose A_i is the homogeneous matrix that transforms the coordinates of a point from frame i to frame $i-1$. The matrix A_i is not constant, but varies as the configuration of the robot is changed. However, the assumption that all joints are either revolute or prismatic means that A_i is a function of only a single joint variable, namely q_i. In other words,

$$A_i = A_i(q_i) \tag{3.1.1}$$

Now the homogeneous matrix that transforms the coordinates of a point from frame j to frame i is called, by convention, a **transformation matrix**, and is usually denoted by T_i^j. From Chapter 2 we see that

$$T_i^j = A_{i+1}A_{i+2}\ldots A_{j-1}A_j \quad \text{if } i < j$$

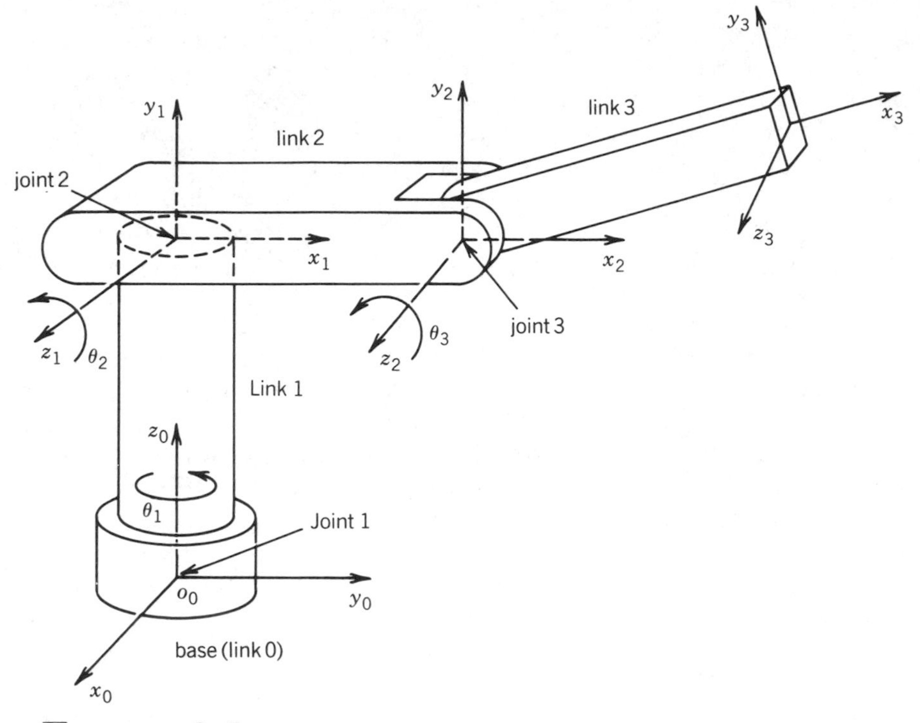

FIGURE 3-1
Coordinate frames attached to elbow manipulator.

$$\begin{aligned} T_i^j &= I \quad \text{if } i = j \\ T_i^j &= (T_j^i)^{-1} \quad \text{if } j > i \end{aligned} \tag{3.1.2}$$

By the manner in which we have rigidly attached the various frames to the corresponding links, it follows that the position of any point on the end-effector, when expressed in frame n, is a constant independent of the configuration of the robot. Denote the position and orientation of the end-effector with respect to the inertial or base frame by a three-vector $\mathbf{d}_0^n$ and a 3×3 rotation matrix R_0^n, respectively, and define the homogeneous matrix

$$H = \begin{bmatrix} R_0^n & \mathbf{d}_0^n \\ 0 & 1 \end{bmatrix} \tag{3.1.3}$$

Then the position and orientation of the end-effector in the inertial frame are given by

$$H = T_0^n = A_1(q_1) \cdots A_n(q_n) \tag{3.1.4}$$

Each homogeneous transformation A_i is of the form

$$A_i = \begin{bmatrix} R_{i-1}^{i} & \mathbf{d}_{i-1}^{i} \\ 0 & 1 \end{bmatrix} \tag{3.1.5}$$

Hence

$$T_i^j = A_{i+1} \cdots A_j = \begin{bmatrix} R_i^j & \mathbf{d}_i^j \\ 0 & 1 \end{bmatrix} \tag{3.1.6}$$

The matrix R_i^j expresses the orientation of frame j relative to frame i and is given by the rotational parts of the A-matrices as

$$R_i^j = R_i^{i+1} \cdots R_{j-1}^j \tag{3.1.7}$$

The vectors $\mathbf{d}_i^j$ are given recursively by the formula

$$\mathbf{d}_i^j = \mathbf{d}_i^{j-1} + R_i^{j-1}\mathbf{d}_{j-1}^j \tag{3.1.8}$$

These expressions will be useful in Chapter Five when we study Jacobian matrices.

In principle, that is all there is to forward kinematics! Determine the functions $A_i(q_i)$, and multiply them together as needed. However, it is possible to achieve a considerable amount of streamlining and simplification by introducing further conventions, such as the Denavit–Hartenberg representation of a joint, and this is the objective of the remainder of the chapter.

3.2 DENAVIT-HARTENBERG REPRESENTATION

While it is possible to carry out all of the analysis in this chapter using an arbitrary frame attached to each link, it is helpful to be systematic in the choice of these frames. A commonly used convention for selecting frames of reference in robotic applications is the Denavit–Hartenberg, or D–H convention. In this convention, each homogeneous transformation A_i is represented as a product of four "basic" transformations

$$A_i = Rot_{z,\theta_i}\, Trans_{z,d_i}\, Trans_{x,a_i}\, Rot_{x,\alpha_i} \tag{3.2.1}$$

$$= \begin{bmatrix} c_{\theta_i} & -s_{\theta_i} & 0 & 0 \\ s_{\theta_i} & c_{\theta_i} & 0 & 0 \\ 0 & 0 & 1 & 0 \\ 0 & 0 & 0 & 1 \end{bmatrix} \begin{bmatrix} 1 & 0 & 0 & 0 \\ 0 & 1 & 0 & 0 \\ 0 & 0 & 1 & d_i \\ 0 & 0 & 0 & 1 \end{bmatrix} \begin{bmatrix} 1 & 0 & 0 & a_i \\ 0 & 1 & 0 & 0 \\ 0 & 0 & 1 & 0 \\ 0 & 0 & 0 & 1 \end{bmatrix} \begin{bmatrix} 1 & 0 & 0 & 0 \\ 0 & c_{\alpha_i} & -s_{\alpha_i} & 0 \\ 0 & s_{\alpha_i} & c_{\alpha_i} & 0 \\ 0 & 0 & 0 & 1 \end{bmatrix}$$

$$= \begin{bmatrix} c_{\theta_i} & -s_{\theta_i} c_{\alpha_i} & s_{\theta_i} s_{\alpha_i} & a_i c_{\theta_i} \\ s_{\theta_i} & c_{\theta_i} c_{\alpha_i} & -c_{\theta_i} s_{\alpha_i} & a_i s_{\theta_i} \\ 0 & s_{\alpha_i} & c_{\alpha_i} & d_i \\ 0 & 0 & 0 & 1 \end{bmatrix}$$

where the four quantities θ_i, a_i, d_i, α_i are parameters of link i and joint i. The various parameters in (3.2.1) are generally given the following names: a_i is called the **length,** α_i is called the **twist,** d_i is called the **offset,** and θ_i is called the **angle**. Since the matrix A_i is a function of a single variable it turns out that three of the above four quantities are constant for a given link, while the fourth parameter, θ_i for a revolute joint and d_i for a prismatic joint, is the joint variable.

From Chapter Two one can see that an arbitrary homogeneous matrix can be characterized by six numbers, such as, for example, the three components of the displacement vector **d** and three Euler angles corresponding to the rotation matrix R. In the D–H representation, in contrast, there are only *four* parameters. How is this possible? The answer is that, while frame i is required to be rigidly attached to link i, we have considerable freedom in choosing the origin and the coordinate axes of the frame. For example, it is not necessary that the origin o_i of frame i should correspond to joint i or to joint $i + 1$, that is, to either end of link i. Thus, by a clever choice of the origin and the coordinate axes, it is possible to cut down the number of parameters needed from six to four (or even fewer in some cases). Let us see how this can be done.

We begin by determining just which homogeneous transformations can be expressed in the form (3.2.1). Suppose we are given two frames, denoted by frames 0 and 1, respectively. Then there exists a unique homogeneous transformation matrix A that takes the coordinates from frame 1 into those of frame 0. Now suppose the two frames have two additional features, namely:

(DH1) The axis x_1 is perpendicular to the axis z_0

(DH2) The axis x_1 intersects the axis z_0

as shown in Figure 3-2. Under these conditions, we claim that there exist *unique* numbers a, d, θ, α such that

$$A = Rot_{z,\theta} Trans_{z,d_i} Trans_{x,a_i} Rot_{x,\alpha} \tag{3.2.2}$$

Of course, since θ and α are angles, we really mean that they are unique to within a multiple of 2π.

To show that the matrix A can be written in this form, write A as

$$A = \begin{bmatrix} R & \mathbf{d} \\ 0 & 1 \end{bmatrix} \tag{3.2.3}$$

and let $\mathbf{r}_i$ denote the i-th column of the rotation matrix R. Referring to Figure 3-2 we see that assumption (DH1) above means that the vector $\mathbf{r}_1$ (which is the representation of the unit vector $\mathbf{i}_1$ in frame 0) is orthogonal to $\mathbf{k}_0 = [0\ 0\ 1]^T$, that is, $r_{31} = 0$. From this we claim that there exist *unique* angles θ and α such that

$$R = R_{z,\theta}\, R_{x,\alpha} = \begin{bmatrix} c_\theta & -s_\theta c_\alpha & s_\theta s_\alpha \\ s_\theta & c_\theta c_\alpha & -c_\theta s_\alpha \\ 0 & s_\alpha & c_\alpha \end{bmatrix} \tag{3.2.4}$$

The only information we have is that $r_{31} = 0$. But this is enough. First, since each row and column of R must have unit length, $r_{31} = 0$ implies that

$$r_{11}^2 + r_{21}^2 = 1, \ r_{32}^2 + r_{33}^2 = 1 \tag{3.2.5}$$

Hence there exist unique θ, α such that

$$(r_{11}, r_{21}) = (c_\theta, s_\theta), \ (r_{33}, r_{32}) = (c_\alpha, s_\alpha) \tag{3.2.6}$$

Once θ and α are found, it is routine to show that the remaining elements of R must have the form shown in (3.2.4), using the fact that R is a rotation matrix.

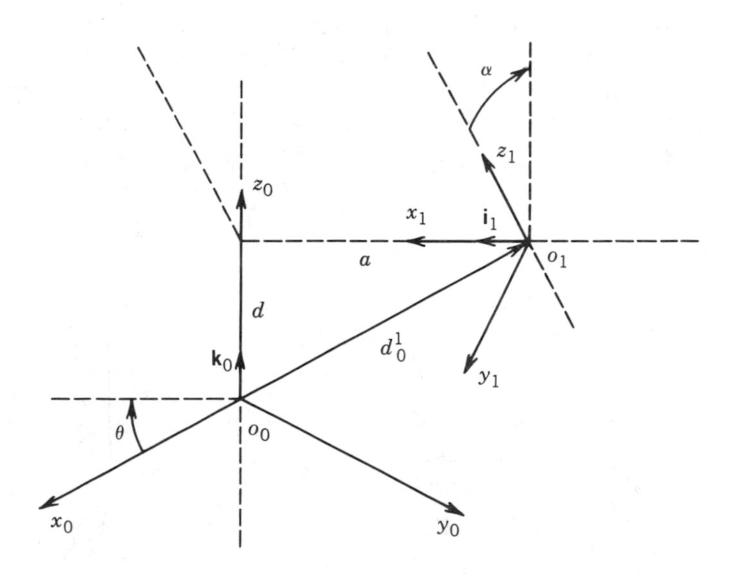

FIGURE 3-2
Coordinate frames satisfying assumptions DH-1 and DH-2.

Next, assumption (DH2) means that the vector $\mathbf{d} = \mathbf{d}_0^1$ (which is the coordinate vector of the origin of frame 1 in terms of frame 0) is a linear combination of $\mathbf{k}_0$ and $R\mathbf{i}_1$. Therefore, since $r_{31} = 0$, we can express $\mathbf{d}_0^1$ *uniquely* as

$$\mathbf{d}_0^1 = d\mathbf{k}_0 + aR\mathbf{i}_1 \tag{3.2.7}$$

$$= d\mathbf{k}_0 + a\mathbf{r}_1 = \begin{bmatrix} ac_\theta \\ as_\theta \\ d \end{bmatrix}$$

Substituting R from (3.2.5) and $\mathbf{d}$ from (3.2.8) into (3.2.4) we obtain (3.2.1) as claimed.

Now that we have established that each homogeneous matrix satisfying conditions (DH1) and (DH2) above can be represented in the form (3.2.1), we can in fact give a physical interpretation to each of the four quantities in (3.2.1). The parameter a is the distance between the axes z_0 and z_1, and is measured along the axis x_1. The angle α is the angle between the axes z_0 and z_1, measured in a plane normal to x_1. The positive sense for α is determined from z_0 to z_1 by the right-hand rule as shown in Figure 3-3. The parameter d is the distance between the origin o_0 and the intersection of the x_1 axis with z_0 measured along the z_0 axis. Finally, θ is the angle between the x_0 and x_1 axes measured in a plane normal to the z_0 axis.

It only remains now to show that, for a robot manipulator, one can always choose the frames $0, \ldots, n$ in such a way that the above two conditions are satisfied, provided one is willing to accept the possibility that the origin o_i of frame i need not lie at joint i. Recall the two conditions: (DH1) x_i is perpendicular to z_{i-1} and (DH2) x_i intersects z_{i-1}. In reading the material below, it is important to keep in mind that the

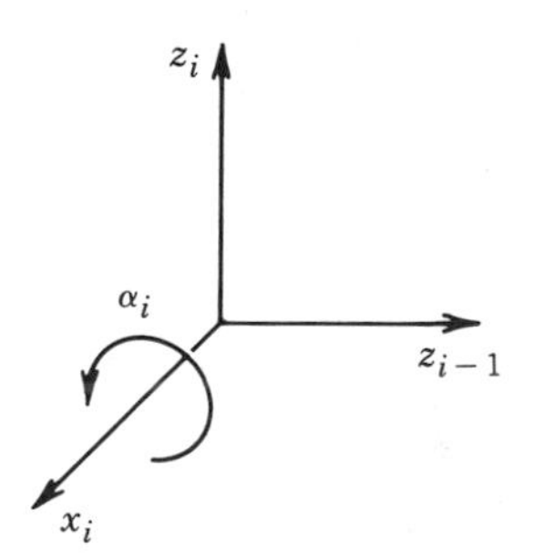

z_{i-1} θ_i x_{i-1} x_i

FIGURE 3-3
Positive sense for α_i and θ_i.

choices of the various coordinate frames are not unique, even when constrained by the requirements above. Once we have illustrated the general procedure, we will discuss various common special cases where it is possible to simplify the homogeneous matrix further.

To start, it is helpful to identify all of the joint axes and label them $z_0, \ldots, z_{n-1}$. z_i is the axis of revolution of joint $i+1$ if joint $i+1$ is revolute, and is the axis of translation of joint $i+1$ if joint $i+1$ is prismatic. Next choose the origin o_0 of the base frame. This point can be chosen anywhere along the z_0 axis. Finally, choose x_0, y_0 in any convenient manner so long as the resulting frame is right-handed. This sets up frame 0.

Now suppose frames $0, \ldots, i-1$ have been set up. To understand the following it will be helpful to consider Figure 3-4. In order to set up frame i it is necessary to consider two cases: (a) the axes z_{i-1}, z_i are not coplanar, and (b) they are coplanar. If the axes z_{i-1}, z_i are not coplanar, then there exists a *unique* line segment perpendicular to both such that it connects both lines and it has minimum length. The line containing this common normal to z_{i-1} and z_i is then defined to be x_i, and the point where it intersects z_i is the origin o_i. Then by construction, both conditions (DH1) and (DH2) are satisfied and the vector from o_{i-1} to o_i is a linear combination of z_{i-1} and x_i. The specification of frame i is completed by choosing the axis y_i to form a right-hand frame. Since assumptions (DH1) and (DH2) are satisfied the homogeneous matrix A_i is of the form (3.2.1).

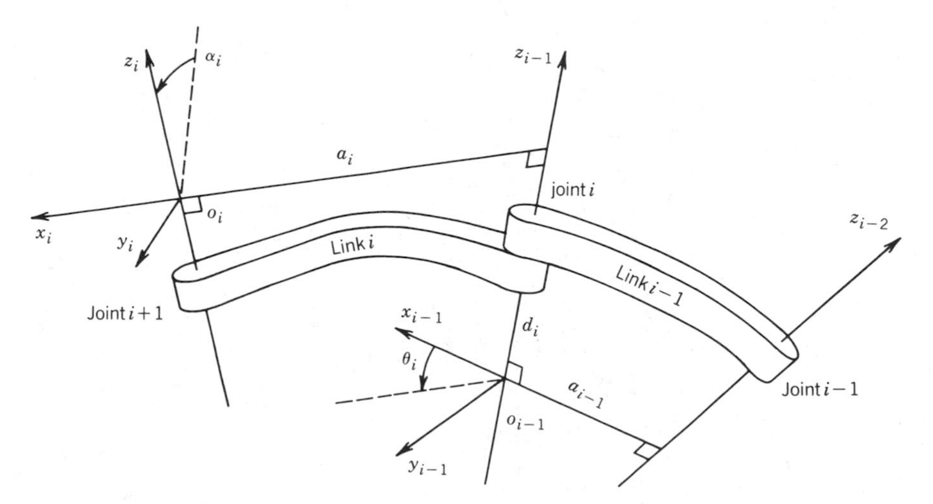

FIGURE 3-4
Denavit–Hartenberg frame assignment.

Now consider case (b), that is, the axes z_{i-1} and z_i are coplanar. This means either that they are parallel, or that they intersect. This situation is in fact quite common, and deserves some detailed analysis. If the axes z_{i-1} and z_i are parallel, there are infinitely many common normals between them and condition (DH1) does not specify x_i completely. In this case we choose the origin o_i to be at joint i so that x_i is that normal from z_{i-1} which passes through o_i. Note that the choice of o_i is arbitrary in this case. We could just as well choose the normal that passes through o_{i-1} as the x_i axis in which case d_i would be equal to zero. In fact, the latter choice is common in much of the robotics literature. Since the axes z_{i-1} and z_i are parallel, α_i will be zero in this case. Once x_i is fixed, y_i is determined, as usual by the right-hand rule.

Finally, consider the case where z_i intersects the axis z_{i-1}. In this case x_i is chosen normal to the plane formed by z_i and z_{i-1}. The positive direction of x_i is arbitrary. The most natural choice for the origin o_i in this case is at the point of intersection of z_i and z_{i-1}. However, any convenient point along the axis z_i suffices. Note that in this case the parameter a_i equals 0.

This constructive procedure works for frames $0, \cdots, n-1$ in an n-link robot. To complete the construction, it is necessary to specify frame n. The final coordinate system $o_n x_n y_n z_n$ is commonly referred to as the **end-effector** or **tool** frame (see Figure 3-5). The origin o_n is most often placed symmetrically between the fingers of the gripper. The unit vectors along the x_n, y_n, and z_n axes are labeled as **n**, **s**, and **a**, respectively. The terminology arises from fact that the direction **a** is the **approach** direction, in the sense that the gripper typically approaches an object along the **a** direction. Similarly the **s** direction is the

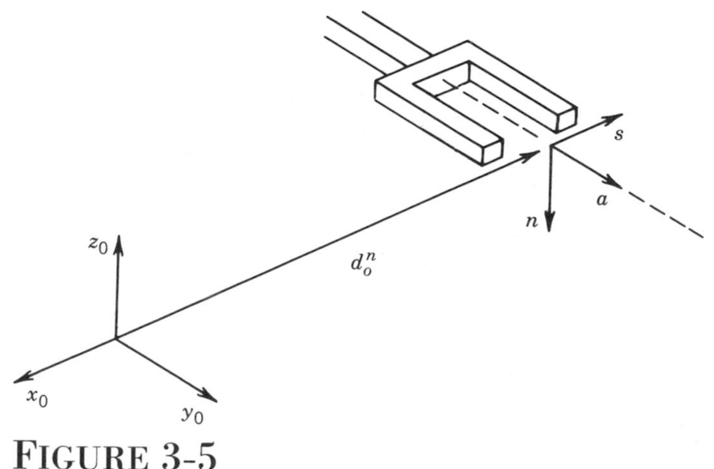

FIGURE 3-5
Tool frame assignemnt.

sliding direction, the direction along which the fingers of the gripper slide to open and close, and **n** is the direction **normal** to the plane formed by **a** and **s**.

In all contemporary robots the final joint motion is a rotation of the end-effector θ_n and the final two joint axes, z_{n-1} and z_n, coincide. Therefore the transformation between the final two coordinate frames is always a translation along z_{n-1} a distance d_6 followed (or preceded) by a rotation of θ_6 radians about z_{n-1}. This is an important observation and will simplify the computation of the inverse kinematics in the next chapter.

Finally, note the following important fact. In all cases, whether the joint in question is revolute or prismatic, the quantities a_i and α_i are always constant for all i and are characteristic of the manipulator. If joint i is prismatic, then θ_i is also a constant, while d_i is the i-th joint variable. Similarly, if joint i is revolute, then d_i is constant and θ_i is the i-th joint variable.

3.2.1 SUMMARY

We may summarize the above procedure based on the D–H convention in the following algorithm for deriving the forward kinematics for any manipulator.

Step 1: Locate and label the joint axes $z_0, \cdots z_{n-1}$.

Step 2: Establish the base frame. Set the origin anywhere on the z_0-axis. The x_0 and y_0 axes are chosen conveniently to form a right-hand frame.

For $i=1, \cdots n-1$, perform Steps 3 to 5.

Step 3: Locate the origin o_i where the common normal to z_i and z_{i-1} intersects z_i. If z_i intersects z_{i-1} locate o_i at this intersection. If z_i and z_{i-1} are parallel, locate o_i at joint i.

Step 4: Establish x_i along the common normal between z_{i-1} and z_i through o_i, or in the direction normal to the z_{i-1}-z_i plane if z_{i-1} and z_i intersect.

Step 5: Establish y_i to complete a right-hand frame.

Step 6: Establish the end-effector frame $o_n x_n y_n z_n$. Assuming the n-th joint is revolute, set $\mathbf{k}_n = \mathbf{a}$ along the direction z_{n-1}. Establish the origin o_n conveniently along z_n, preferably at the center of the gripper or at the tip of any tool that the manipulator may be carrying. Set $\mathbf{j}_n = \mathbf{s}$ in the direction of the gripper closure and set $\mathbf{i}_n = \mathbf{n}$ as $\mathbf{s}\times\mathbf{a}$. If the tool is not a simple gripper set x_n and y_n conveniently to form a right-hand frame.

Step 7: Create a table of link parameters a_i, d_i, α_i, θ_i.

a_i = distance along x_i from o_i to the intersection of the x_i and z_{i-1} axes.

d_i = distance along z_{i-1} from o_{i-1} to the intersection of the x_i and z_{i-1} axes. d_i is variable if joint i is prismatic.

α_i = the angle between z_{i-1} and z_i measured about x_i (See Figure 3-3).

θ_i = the angle between x_{i-1} and x_i measured about z_{i-1} (See Figure 3-3). θ_i is variable is joint i is revolute.

Step 8: Form the homogeneous transformation matrices A_i by substituting the above parameters into (3.2.1).

Step 9: Form $T_0^n = A_1 \cdots A_n$. This then gives the position and orientation of the tool frame expressed in base coordinates.

3.3 EXAMPLES

In the D-H convention the only variable angle is θ, so we simplify notation by writing c_i for $\cos\theta_i$, etc. We also denote $\theta_1 + \theta_2$ by θ_{12}, and $\cos(\theta_1 + \theta_2)$ by c_{12} and so on. In the following examples it is important to remember that the D–H convention, while systematic, still allows considerable freedom in the choice of some of the manipulator parameters. This is particularly true in the case of parallel joint axes or when prismatic joints are involved.

(i) *Example 3.3.1 Planar Elbow Manipulator*

Consider the two-link planar arm of Figure 3-6. The joint axes z_0 and z_1 are normal to the page. We establish the base frame $o_0x_0y_0z_0$ as shown. The origin is chosen at the point of intersection of the z_0 axis with the page and the direction of the x_0 axis is completely arbitrary. Once the base frame is established, the $o_1x_1y_1z_1$ frame is fixed as shown by the D–H convention, where the origin o_1 has been located at the intersection of z_1 and the page. The final frame $o_2x_2y_2z_2$ is fixed by choosing the origin o_2 at the end of link 2 as shown. The link parameters are shown in Table 3-1. The A-matrices are determined from (3.2.1) as

$$A_1 = \begin{bmatrix} c_1 & -s_1 & 0 & a_1c_1 \\ s_1 & c_1 & 0 & a_1s_1 \\ 0 & 0 & 1 & 0 \\ 0 & 0 & 0 & 1 \end{bmatrix} \tag{3.3.1}$$

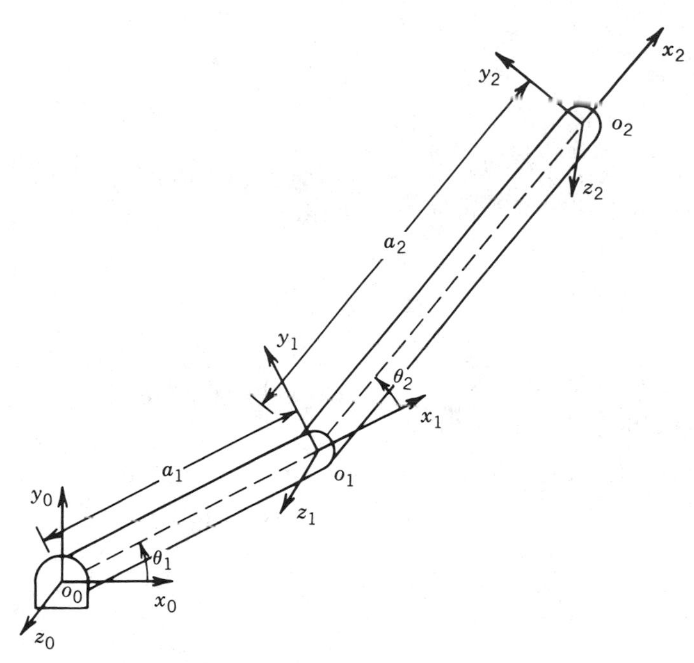

FIGURE 3-6
Two-link planar manipulator.

$$A_2 = \begin{bmatrix} c_2 & -s_2 & 0 & a_2c_2 \\ s_2 & c_2 & 0 & a_2s_2 \\ 0 & 0 & 1 & 0 \\ 0 & 0 & 0 & 1 \end{bmatrix} \tag{3.3.2}$$

The T-matrices are thus given by

$$T_0^1 = A_1 \tag{3.3.3}$$

TABLE 3-1
Link Parameters for 2-link Planar Manipulator

Link	a_i	α_i	d_i	θ_i
1	a_1	0	0	θ_1^*
2	a_2	0	0	θ_2^*

*variable

$$T_0^2 = A_1 A_2 = \begin{bmatrix} c_{12} & -s_{12} & 0 & a_1c_1 + a_2c_{12} \\ s_{12} & c_{12} & 0 & a_1s_1 + a_2s_{12} \\ 0 & 0 & 1 & 0 \\ 0 & 0 & 0 & 1 \end{bmatrix} \tag{3.3.4}$$

Notice that the first two entries of the last column of T_2 are the x and y components of the origin o_2 in the base frame; that is,

$$\begin{aligned} x &= a_1c_1 + a_2c_{12} \\ y &= a_1s_1 + a_2s_{12} \end{aligned} \tag{3.3.5}$$

are the coordinates of the end-effector in the base frame. The rotational part of T_2 gives the orientation of the frame $o_2x_2y_2z_2$ relative to the base frame.

(ii) *Example 3.3.2 Three-Link Cylindrical Robot*

Consider now the three-link cylindrical robot represented symbolically by Figure 3-7. We establish o_0 as shown at joint 1. Note that the placement of the origin o_0 along z_0 as well as the direction of the x_0 axis are arbitrary. Our choice of o_0 is the most natural but o_0 could just as well be placed at joint 2. The x_0-axis is chosen normal to the page. Next, since z_0 and z_1 coincide, the origin o_1 is chosen at joint 1 as shown. The x_1 axis is normal to the page when $\theta_1 = 0$ but, of course its direction will change since θ_1 is variable. Since z_2 and z_1 intersect, the origin o_2 is placed at this intersection. The direction of x_2 is chosen parallel to x_1 so that θ_2 is zero. Finally, the third frame is chosen at the end of link 3 as shown.

The link parameters are now shown in Table 3-2. The corresponding A and T matrices are

$$A_1 = \begin{bmatrix} c_1 & -s_1 & 0 & 0 \\ s_1 & c_1 & 0 & 0 \\ 0 & 0 & 1 & d_1 \\ 0 & 0 & 0 & 1 \end{bmatrix} \tag{3.3.6}$$

$$A_2 = \begin{bmatrix} 1 & 0 & 0 & 0 \\ 0 & 0 & 1 & 0 \\ 0 & -1 & 0 & d_2 \\ 0 & 0 & 0 & 1 \end{bmatrix} \qquad A_3 = \begin{bmatrix} 1 & 0 & 0 & 0 \\ 0 & 1 & 0 & 0 \\ 0 & 0 & 1 & d_3 \\ 0 & 0 & 0 & 0 \end{bmatrix}$$

$$T_0^3 = A_1A_2A_3 = \begin{bmatrix} c_1 & 0 & -s_1 & -s_1d_3 \\ s_1 & 0 & c_1 & c_1d_3 \\ 0 & -1 & 0 & d_1 + d_2 \\ 0 & 0 & 0 & 1 \end{bmatrix} \tag{3.3.7}$$

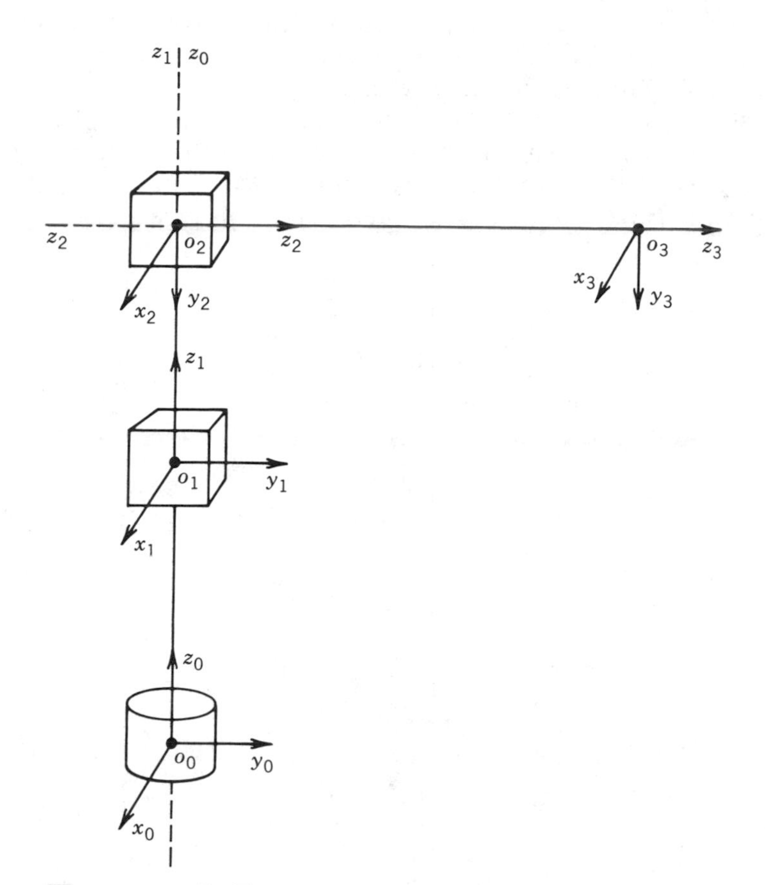

FIGURE 3-7
Three-link cylindrical manipulator.

(iii) *Example 3.3.3 Spherical Wrist*

The spherical wrist configuration is shown in Figure 3-8, in which the joint axes z_3, z_4, z_5 intersect at o. The Denavit–Hartenberg parameters

TABLE 3-2
Link Parameters for
3-link Cylindrical Manipulator

Link	a_i	α_i	d_i	θ_i
1	0	0	d_1	θ_1^*
2	0	-90	d_2^*	0
3	0	0	d_3^*	0

* variable

are shown in Table 3-3. The Stanford manipulator is an example of a manipulator that possesses a wrist of this type. In fact, the following analysis applies to virtually all spherical wrists.

We show now that the final three joint variables, θ_4, θ_5, θ_6 are the Euler angles ϕ, θ, ψ, respectively, with respect to the coordinate frame $o_3x_3y_3z_3$. To see this we need only compute the matrices A_4, A_5, and A_6 using Table 3-3 and the expression (3.2.1). This gives

$$A_4 = \begin{bmatrix} c_4 & 0 & -s_4 & 0 \\ s_4 & 0 & c_4 & 0 \\ 0 & -1 & 0 & 0 \\ 0 & 0 & 0 & 1 \end{bmatrix} \tag{3.3.8}$$

$$A_5 = \begin{bmatrix} c_5 & 0 & s_5 & 0 \\ s_5 & 0 & -c_5 & 0 \\ 0 & 1 & 0 & 0 \\ 0 & 0 & 0 & 1 \end{bmatrix} \tag{3.3.9}$$

$$A_6 = \begin{bmatrix} c_6 & -s_6 & 0 & 0 \\ s_6 & c_6 & 0 & 0 \\ 0 & 0 & 1 & d_6 \\ 0 & 0 & 0 & 1 \end{bmatrix} \tag{3.3.10}$$

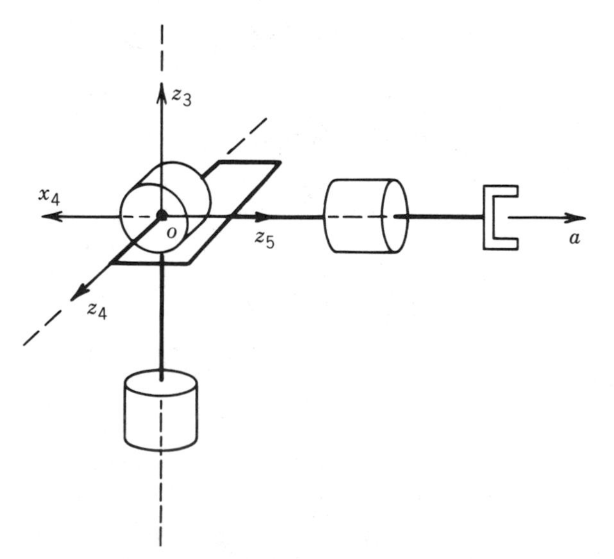

FIGURE 3-8
The spherical wrist frame assignment.

TABLE 3-3
DH-parameters for spherical wrist

Link	a_i	α_i	d_i	θ_i
4	0	-90	0	θ_4^*
5	0	90	0	θ_5^*
6	0	0	d_6^*	θ_6^*

* variable

Multiplying these together yields

$$T_3^6 = A_4 A_5 A_6 = \begin{bmatrix} R_3^6 & \mathbf{d}_3^6 \\ 0 & 1 \end{bmatrix} \tag{3.3.11}$$

$$= \begin{bmatrix} c_4c_5c_6 - s_4s_6 & -c_4c_5s_6 - s_4c_6 & c_4s_5 & c_4s_5d_6 \\ s_4c_5c_6 + c_4s_6 & -s_4c_5s_6 + c_4c_6 & s_4s_5 & s_4s_5d_6 \\ -s_5c_6 & s_5s_6 & c_5 & c_5d_6 \\ 0 & 0 & 0 & 1 \end{bmatrix}$$

Comparing the rotational part R_3^6 of T_3^6 with the Euler angle transformation (2.3.9) shows that θ_4, θ_5, θ_6 can indeed be identified as the Euler angles ϕ, θ, and ψ with respect to the coordinate frame $o_3x_3y_3z_3$.

(iv) Example 3.3.4 Cylindrical Manipulator with Spherical Wrist

Suppose that we now attach a spherical wrist to the cylindrical manipulator of Example 3.3.2 as shown in Figure 3-9. Note that the axis of rotation of joint 4 is parallel to z_2 and thus coincides with the axis z_3 of Example 3.3.2. The implication of this is that we can immediately combine the two previous expression (3.3.7) and (3.3.11) to derive the forward kinematics as

$$T_0^6 = T_0^3 T_3^6 \tag{3.3.12}$$

with T_0^3 given by (3.3.7) and T_3^6 given by (3.3.11). Therefore the forward kinematics of this manipulator is described by

$$T_0^6 = \begin{bmatrix} c_1 & 0 & -s_1 & -s_1d_3 \\ s_1 & 0 & c_1 & c_1d_3 \\ 0 & -1 & 0 & d_1 + d_2 \\ 0 & 0 & 0 & 1 \end{bmatrix} \begin{bmatrix} c_4c_5c_6 - s_4s_6 & -c_4c_5s_6 - s_4c_6 & c_4s_5 & c_4s_5d_6 \\ s_4c_5c_6 + c_4s_6 & -s_4c_5s_6 + c_4c_6 & s_4s_5 & s_4s_5d_6 \\ -s_5c_6 & s_5s_6 & c_5 & c_5d_6 \\ 0 & 0 & 0 & 1 \end{bmatrix} \tag{3.3.13}$$

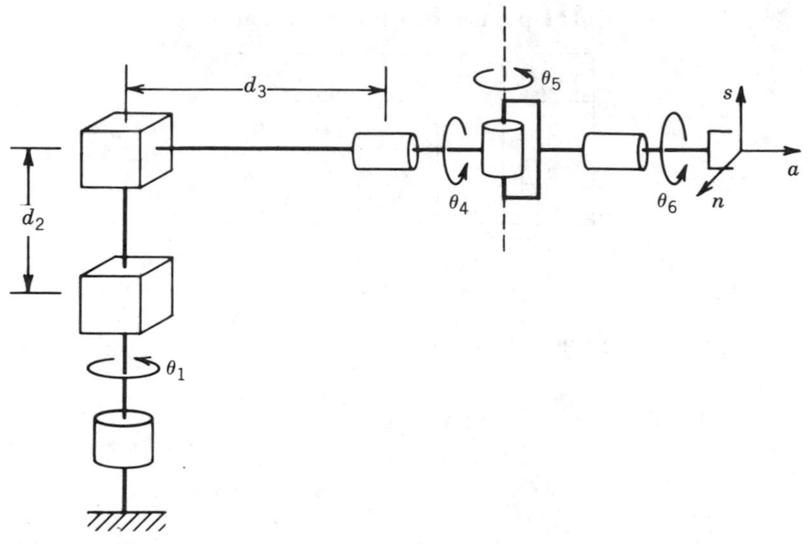

FIGURE 3-9
Cylindrical robot with spherical wrist.

$$= \begin{bmatrix} r_{11} & r_{12} & r_{13} & d_x \\ r_{21} & r_{22} & r_{23} & d_y \\ r_{31} & r_{32} & r_{33} & d_z \\ 0 & 0 & 0 & 1 \end{bmatrix}$$

where

$$\begin{aligned}
r_{11} &= c_1c_4c_5c_6 - c_1s_4s_6 + s_1s_5c_6 \\
r_{21} &= s_1c_4c_5c_6 - s_1s_4s_6 - c_1s_5c_6 \\
r_{31} &= -s_4c_5c_6 - c_4s_6 \\
r_{12} &= -c_1c_4c_5s_6 - c_1s_4c_6 - s_1s_5s_6 \\
r_{22} &= -s_1c_4c_5s_6 - s_1s_4c_6 + c_1s_5s_6 \\
r_{32} &= s_4c_5s_6 - c_4c_6 \\
r_{13} &= c_1c_4s_5 - s_1c_5 \\
r_{23} &= s_1c_4s_5 + c_1c_5 \\
r_{33} &= -s_4s_5 \\
d_x &= c_1c_4s_5d_6 - s_1c_5d_6 - s_1d_3
\end{aligned}$$

$$d_y = s_1c_4s_5d_6 + c_1c_5d_6 + c_1d_3$$

$$d_z = -s_4s_5d_6+d_1 + d_2$$

Notice how most of the complexity of the forward kinematics for this manipulator results from the orientation of the end-effector while the expression for the arm position from (3.3.7) is fairly simple. The spherical wrist assumption not only simplifies the derivation of the forward kinematics here, but will also greatly simplify the inverse kinematics problem in the next chapter.

(v) *Example 3.3.5 Stanford Manipulator*

Consider now the Stanford Manipulator shown in Figure 3-10. This manipulator is an example of a spherical (RRP) manipulator with a spherical wrist. This manipulator has an offset in the shoulder joint that slightly complicates both the forward and inverse kinematics problems.

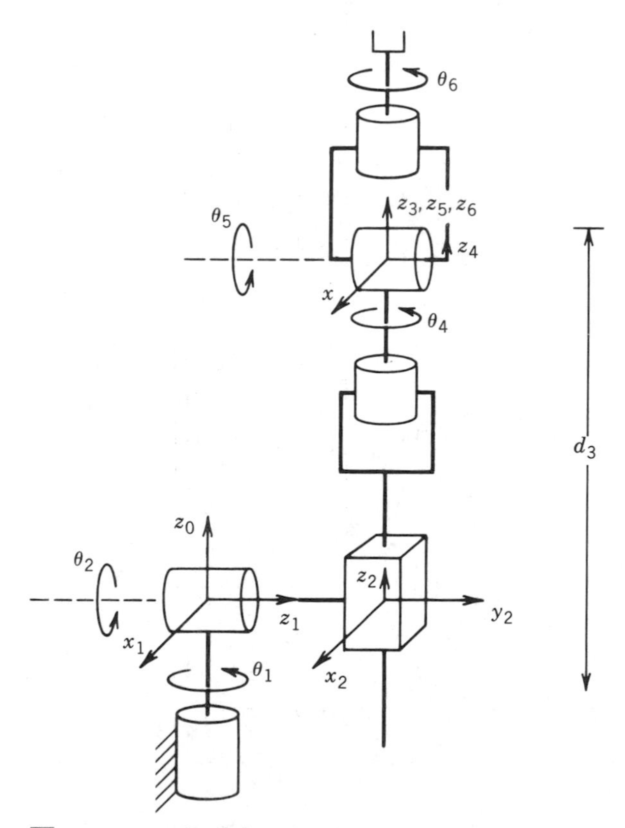

FIGURE 3-10
DH coordinate frame assignment for the Stanford manipulator.

We first establish the joint coordinate frames using the D–H convention as shown. The link parameters are shown in the Table 3-4. It is straightforward to compute the matrices A_i as

$$A_1 = \begin{bmatrix} c_1 & 0 & -s_1 & 0 \\ s_1 & 0 & c_1 & 0 \\ 0 & -1 & 0 & 0 \\ 0 & 0 & 0 & 1 \end{bmatrix} \tag{3.3.14}$$

$$A_2 = \begin{bmatrix} c_2 & 0 & s_2 & 0 \\ s_2 & 0 & -c_2 & 0 \\ 0 & 1 & 0 & d_2 \\ 0 & 0 & 0 & 1 \end{bmatrix} \tag{3.3.15}$$

$$A_3 = \begin{bmatrix} 1 & 0 & 0 & 0 \\ 0 & 1 & 0 & 0 \\ 0 & 0 & 1 & d_3 \\ 0 & 0 & 0 & 1 \end{bmatrix} \tag{3.3.16}$$

$$A_4 = \begin{bmatrix} c_4 & 0 & -s_4 & 0 \\ s_4 & 0 & c_4 & 0 \\ 0 & -1 & 0 & 0 \\ 0 & 0 & 0 & 1 \end{bmatrix} \tag{3.3.17}$$

$$A_5 = \begin{bmatrix} c_5 & 0 & s_5 & 0 \\ s_5 & 0 & -c_5 & 0 \\ 0 & 1 & 0 & 0 \\ 0 & 0 & 0 & 1 \end{bmatrix} \tag{3.3.18}$$

$$A_6 = \begin{bmatrix} c_6 & -s_6 & 0 & 0 \\ s_6 & c_6 & 0 & 0 \\ 0 & 0 & 1 & d_6 \\ 0 & 0 & 0 & 1 \end{bmatrix} \tag{3.3.19}$$

TABLE 3-4
DH-parameters for Stanford Manipulator

Link	d_i	a_i	α_i	θ_i
1	0	0	-90	*
2	d_2	0	+90	*
3	*	0	0	0
4	0	0	-90	*
5	0	0	+90	*
6	0	0	0	*

*joint variable

T_0^6 is then given as

$$T_0^6 = A_1 \cdots A_6 \tag{3.3.20}$$
$$= \begin{bmatrix} r_{11} & r_{12} & r_{13} & d_x \\ r_{21} & r_{22} & r_{32} & d_y \\ r_{31} & r_{32} & r_{33} & d_z \\ 0 & 0 & 0 & 1 \end{bmatrix}$$

where

$$r_{11} = c_1[c_2(c_4c_5c_6-s_4s_6)-s_2s_5c_6]-s_1(s_4c_5c_6 + c_4s_6)$$
$$r_{21} = s_1[c_2(c_4c_5c_6-s_4s_6)-s_2s_5c_6] + c_1(s_4c_5c_6 + c_4s_6)$$
$$r_{31} = -s_2(c_4c_5c_6-s_4s_6)-c_2s_5c_6$$
$$r_{12} = c_1[-c_2(c_4c_5s_6 + s_4c_6) + s_2s_5s_6]-s_1(-s_4c_5s_6 + c_4c_6)$$
$$r_{22} = s_1[-c_2(c_4c_5s_6 + s_4c_6) + s_2s_5s_6] + c_1(-s_4c_5s_6 + c_4c_6)$$
$$r_{32} = s_2(c_4c_5s_6 + s_4c_6) + c_2s_5s_6 \tag{3.3.21}$$
$$r_{13} = c_1(c_2c_4s_5 + s_2c_5)-s_1s_4s_5$$
$$r_{23} = s_1(c_2c_4s_5 + s_2c_5) + c_1s_4s_5$$
$$r_{33} = -s_2c_4s_5 + c_2c_5$$
$$d_x = c_1s_2d_3-s_1d_2 + d_6(c_1c_2c_4s_5 + c_1c_5s_2-s_1s_4s_5)$$
$$d_y = s_1s_2d_3 + c_1d_2 + d_6(c_1s_4s_5 + c_2c_4s_1s_5 + c_5s_1s_2)$$
$$d_z = c_2d_3 + d_6(c_2c_5-c_4s_2s_5)$$

(vi) *Example 3.3.6 SCARA Manipulator*

As another example of the general procedure, consider the SCARA manipulator of Figure 3-11. This manipulator, which is an abstraction of the AdeptOne robot of Figure 1-13, consists of an RRP arm and a one degree-of-freedom wrist, whose motion is a roll about the vertical

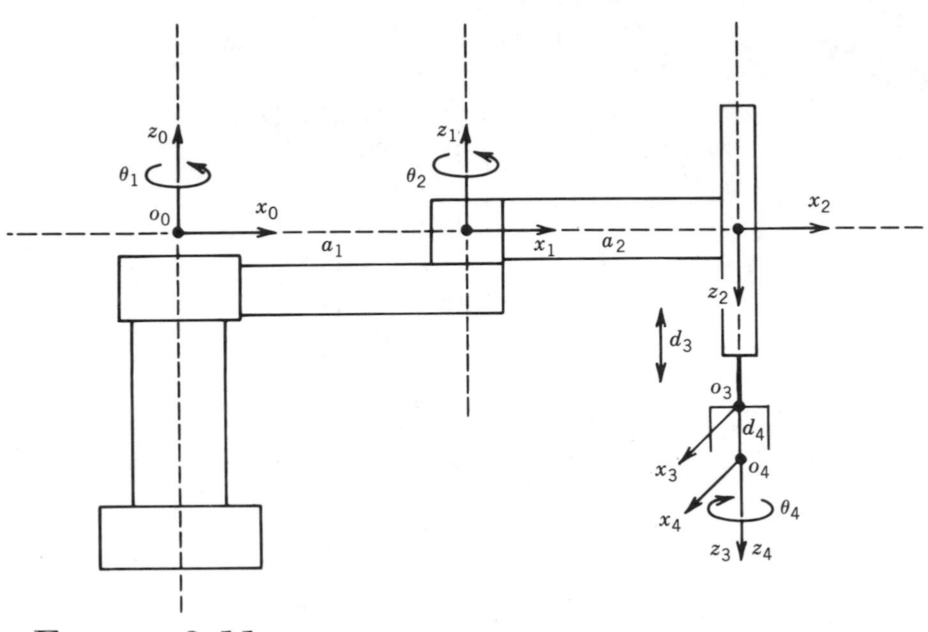

FIGURE 3-11
DH coordinate frame assignment for the SCARA manipulator.

axis. The first step is to locate and label the joint axes as shown. Since all joint axes are parallel we have some freedom in the placement of the origins. For convenience all origins are placed at the joints. We establish the x_0 axis in the plane of the page as shown. This is completely arbitrary and only affects the zero configuration of the manipulator, that is, the position of the manipulator when $\theta_1 = 0$.

The joint parameters are given in Table 3-5, and the A-matrices are as follows.

$$A_1 = \begin{bmatrix} c_1 & -s_1 & 0 & a_1 c_1 \\ s_1 & c_1 & 0 & a_1 s_1 \\ 0 & 0 & 1 & 0 \\ 0 & 0 & 0 & 1 \end{bmatrix} \tag{3.3.22}$$

$$A_2 = \begin{bmatrix} c_2 & -s_2 & 0 & a_2 c_2 \\ s_2 & c_2 & 0 & a_2 s_2 \\ 0 & 0 & 1 & 0 \\ 0 & 0 & 0 & 1 \end{bmatrix} \tag{3.3.23}$$

TABLE 3-5
Joint parameters
for SCARA

Joint	a_i	α_i	d_i	θ_i
1	a_1	0	0	*
2	a_2	0	0	*
3	0	180	*	0
4	0	0	d_4	*

*joint variable

$$A_3 = \begin{bmatrix} 1 & 0 & 0 & 0 \\ 0 & -1 & 0 & 0 \\ 0 & 0 & -1 & d_3 \\ 0 & 0 & 0 & 1 \end{bmatrix} \tag{3.3.24}$$

$$A_4 = \begin{bmatrix} c_4 & -s_4 & 0 & 0 \\ s_4 & c_4 & 0 & 0 \\ 0 & 0 & 1 & d_4 \\ 0 & 0 & 0 & 1 \end{bmatrix} \tag{3.3.25}$$

The forward kinematic equations are therefore given by

$$T_0^4 = A_1 \cdots A_4 = \begin{bmatrix} c_{12}c_4 + s_{12}s_4 & -c_{12}s_4 + s_{12}c_4 & 0 & a_1c_1 + a_2c_{12} \\ s_{12}c_4 - c_{12}s_4 & -s_{12}s_4 - c_{12}c_4 & 0 & a_1s_1 + a_2s_{12} \\ 0 & 0 & -1 & d_3 - d_4 \\ 0 & 0 & 0 & 1 \end{bmatrix} \tag{3.3.26}$$

REFERENCES AND SUGGESTED READING

[1] BOTEMA, O., and ROTH, B., *Theoretical Kinematics*, North Holland, Amsterdam, 1979.

[2] COLSON, J.C., and PERREIRA, N.D., "Kinematic Arrangements Used in Industrial Robots," *Proc. 13th International Symposium on Industrial Robots*, 1983.

[3] DENAVIT, J., and HARTENBERG, R.S., "A Kinematic Notation for Lower Pair Mechanisms," *J. Applied Mechanics*, Vol. 22, pp.215–221, 1955.

[4] DUFFY, J., *Analysis of Mechanisms and Robot Manipulators*, Wiley, New York, 1980.

[5] LEE, C.S.G., "Robot Arm Kinematics, Dynamics, and Control," *Computer*, Vol. 15, No. 12, pp. 62–80, 1982.

[6] LEE, C.S.G., GONZALES, R.C., and FU, K.S., *Tutorial on Robotics*, IEEE Computer Press, Silver Spring, MD, 1986.

[7] PAUL, R.P., SHIMANO, B.E., and MAYER, G., "Kinematic Control Equations for Simple Manipulators," *IEEE Trans. Systems, Man., and Cybernetics*, Vol. SMC-11, No. 6, pp. 339–455, 1981.

[8] SUH, C.H., and RADCLIFFE, C.W., *Kinematics and Mechanisms Design*, Wiley, New York, 1978.

[9] UICKER, J.J., Jr., DENAVIT, J., and HARTENBERG, R.S., "An Iterative Method for the Displacement Analysis of Spatial Mechanisms," *Trans. ASME, J. Applied Mechanics*, Vol. 31, Series E, pp. 309–314, 1964.

[10] WHITNEY, D.E., "The Mathematics of Coordinated Control of Prosthetic Arms and Manipulators," *J. Dyn. Sys., Meas. Cont.*, December 1972.

PROBLEMS

3-1 Verify the statement after equation (3.2.6) that the rotation matrix R has the form (3.2.4) provided assumptions DH1 and DH2 are satisfied.

3-2 Consider the three-link planar manipulator shown in Figure 3-12. Derive the forward kinematic equations using the DH-convention.

3-3 Consider the two-link cartesian manipulator of Figure 3-13. Derive the forward kinematic equations using the DH-convention.

3-4 Consider the two-link manipulator if Figure 3-14 which has joint 1 revolute and joint 2 prismatic. Derive the forward kinematic equations using the DH-convention.

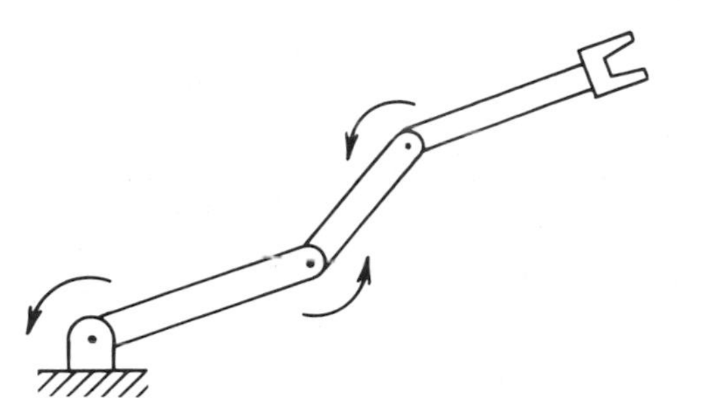

FIGURE 3-12
Three-link planar arm of Problem 3-2.

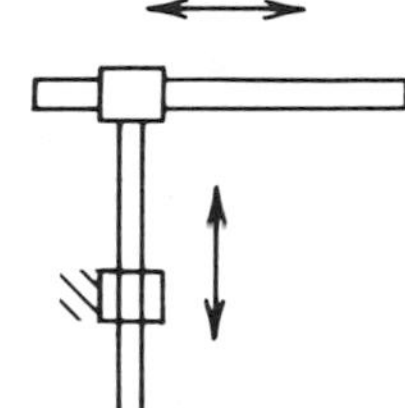

FIGURE 3-13
Two-link cartesian robot of Problem 3-3.

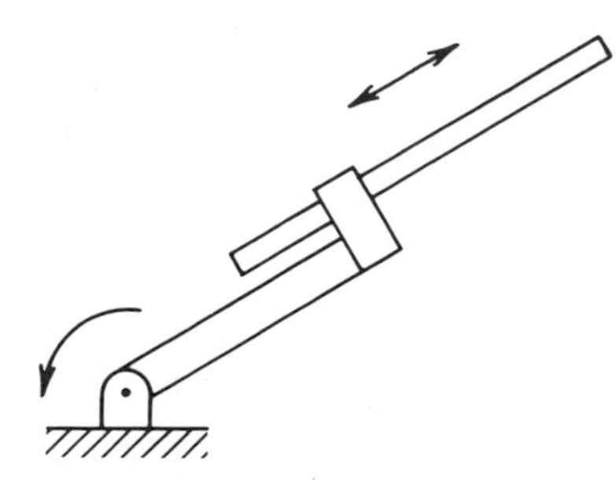

FIGURE 3-14
Two-link planar arm of Problem 3-4.

3-5 Consider the three-link planar manipulator of Figure 3-15. Derive the forward kinematic equations using the DH-convention.

3-6 Consider the three-link articulated robot of Figure 3-16. Derive the forward kinematic equations using the DH-convention.

3-7 Consider the three-link cartesian manipulator of Figure 3-17. Derive the forward kinematic equations using the DH-convention.

3-8 Attach a spherical wrist to the three-link articulated manipulator of Problem 3-6 as shown in Figure 3-18. Derive the forward kinematic equations for this manipulator.

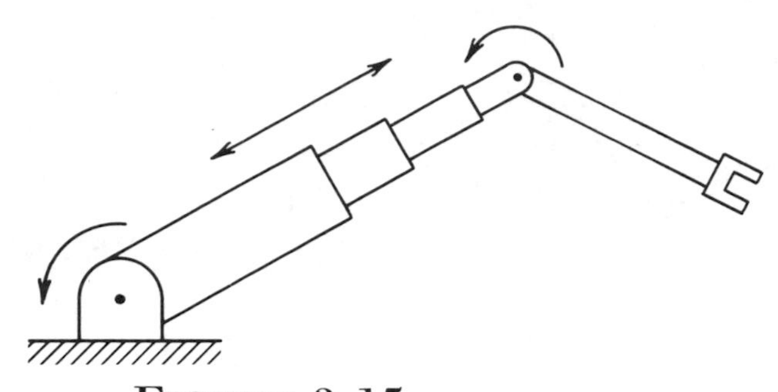

FIGURE 3-15
Three-link planar arm with prismatic joint of Problem 3-5.

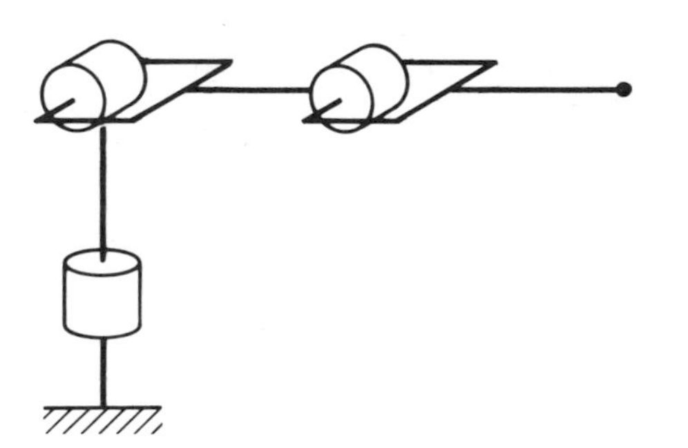

FIGURE 3-16
Three-link articulated robot.

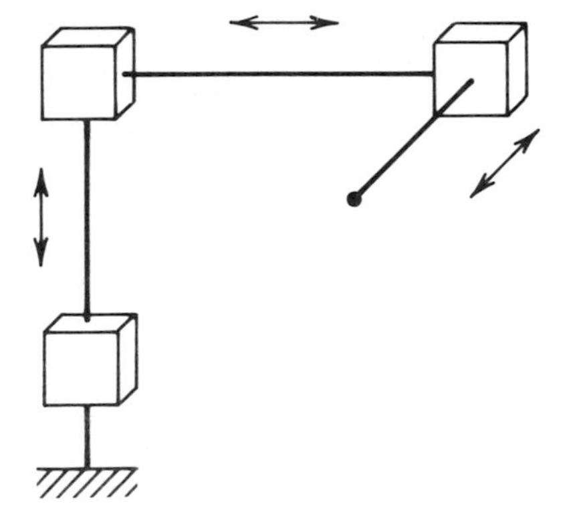

FIGURE 3-17
Three-link cartesian robot.

3-9 Attach a spherical wrist to the three-link cartesian manipulator of Problem 3-7 as shown in Figure 3-19. Derive the forward kinematic equations for this manipulator.

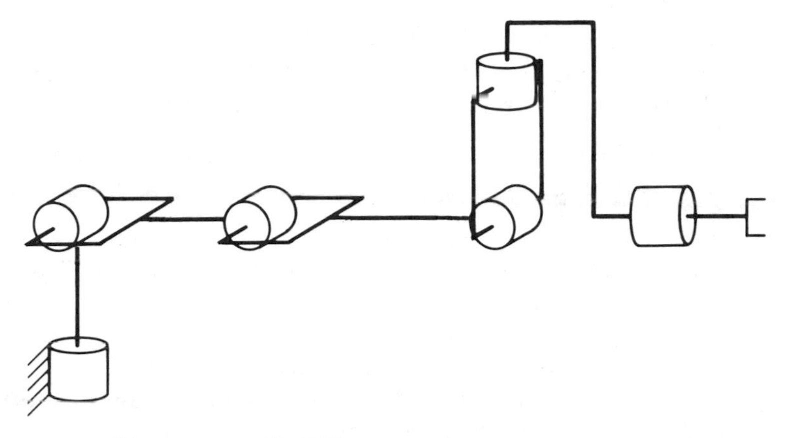

FIGURE 3-18
Elbow manipulator with spherical wrist.

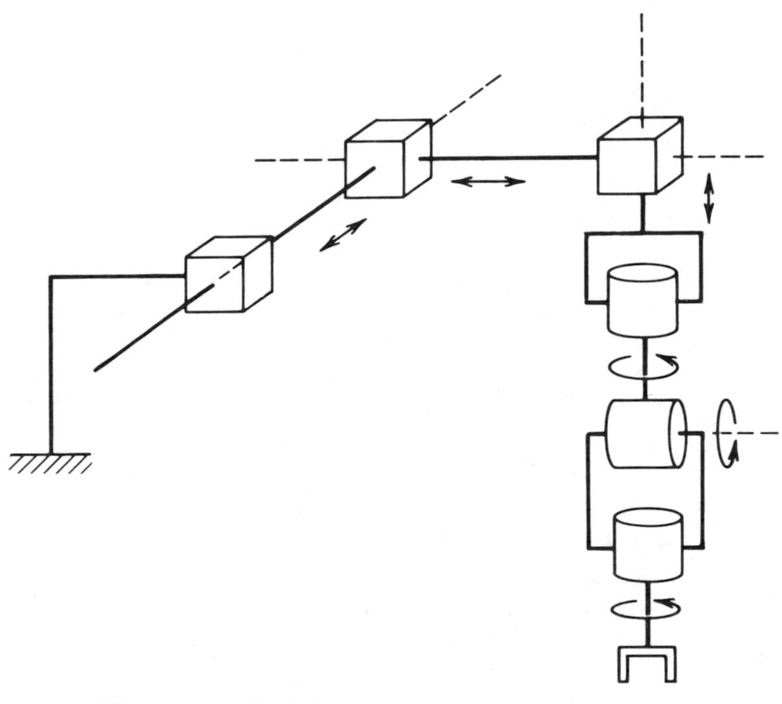

FIGURE 3-19
Cartesian manipulator with spherical wrist.

3-10 Consider the PUMA 260 manipulator shown in Figure 3-20. Derive the complete set of forward kinematic equations, by establishing appropriate D-H coordinate frames, constructing a table of link parameters, forming the A-matrices, etc.

3-11 Repeat Problem 3-9 for the five degree-of-freedom Rhino XR-3 robot shown in Figure 3-21.

3-12 Suppose that a Rhino XR-3 is bolted to a table upon which a coordinate frame $o_s x_s y_s z_s$ is established as shown in Figure 3-22. (The frame $o_s x_s y_s z_s$ is often referred to as the **station frame**.) Given the base frame that you established in Problem 3-7, find the homogeneous transformation T_s^0 relating the base frame to the station frame. Find the homogeneous transformation T_s^6 relating the end-effector frame to the station frame. What is the position and orientation of the end-effector in the station frame when $\theta_1 = \theta_2 = \cdots = \theta_5 = 0$?

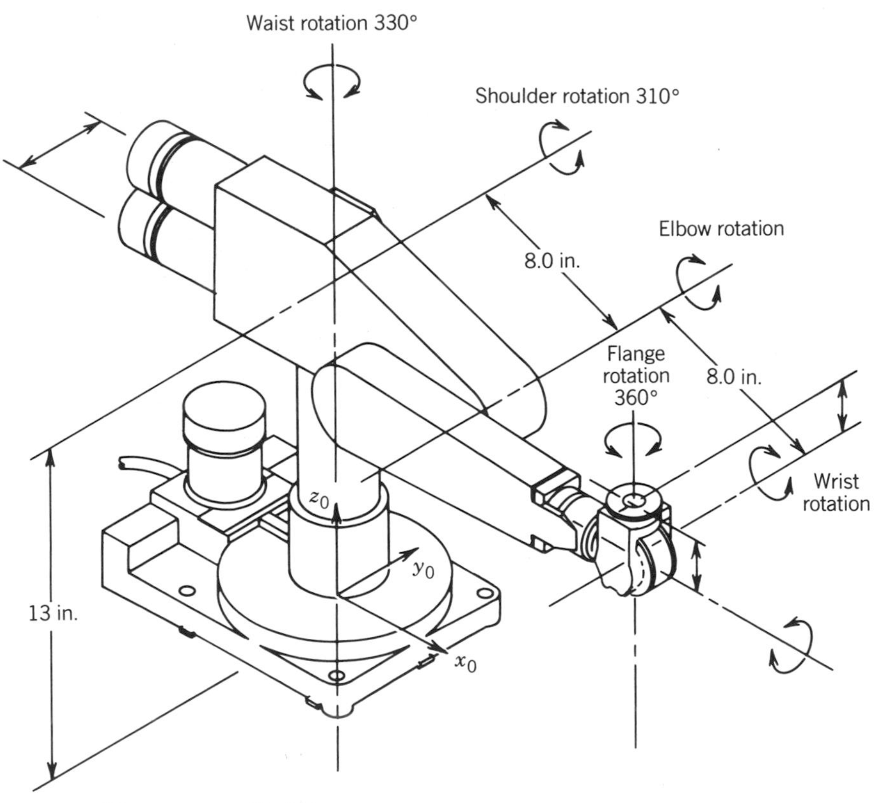

FIGURE 3-20
PUMA 260 manipulator.

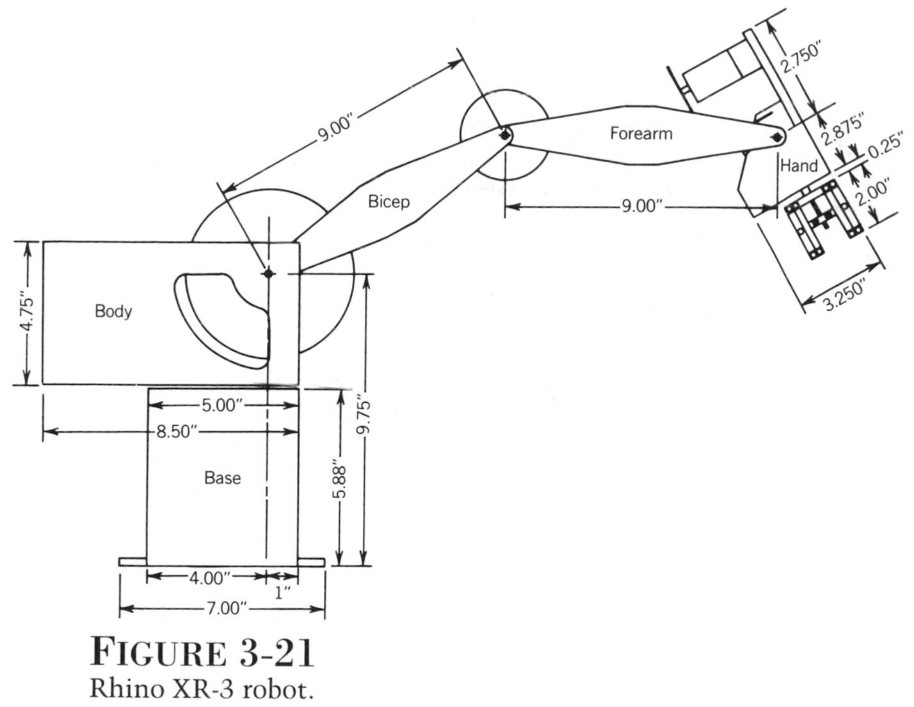

FIGURE 3-21
Rhino XR-3 robot.

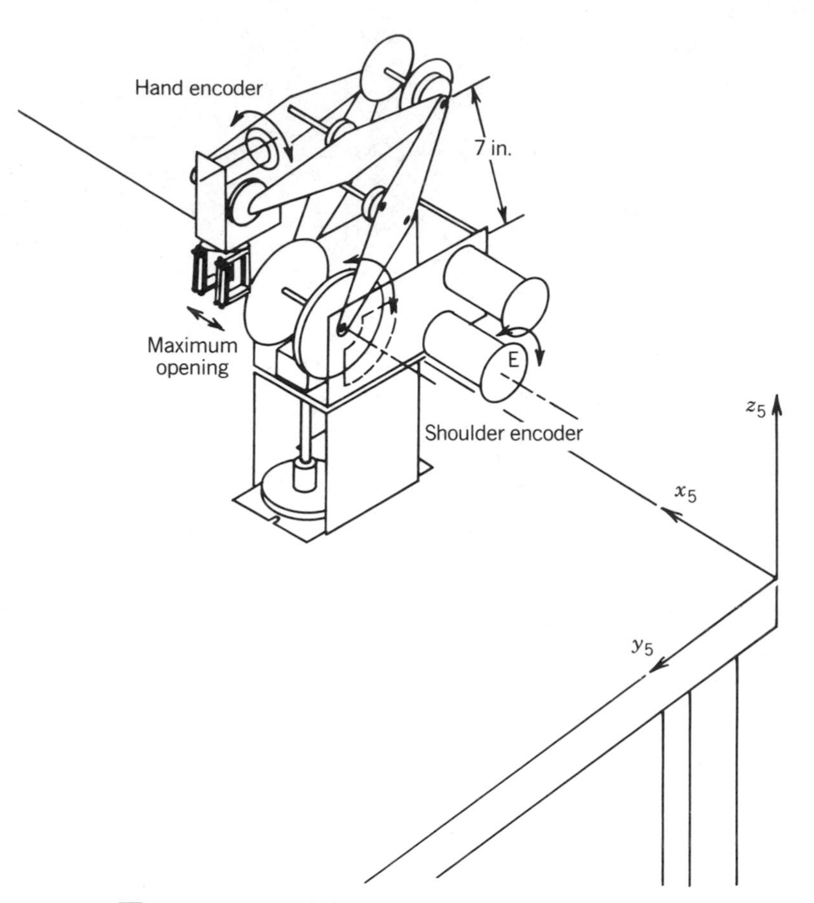

FIGURE 3-22
Rhino robot attached to a table. From: *A Robot Engineering Textbook*, by Mohsen Shahinpoor. Copyright 1987, Harper & Row Publishers, Inc.

3-13 Consider the GMF S-400 robot shown in Figure 3-23. Draw the symbolic representation for this manipulator. Establish DH-coordinate frames and write the forward kinematic equations.

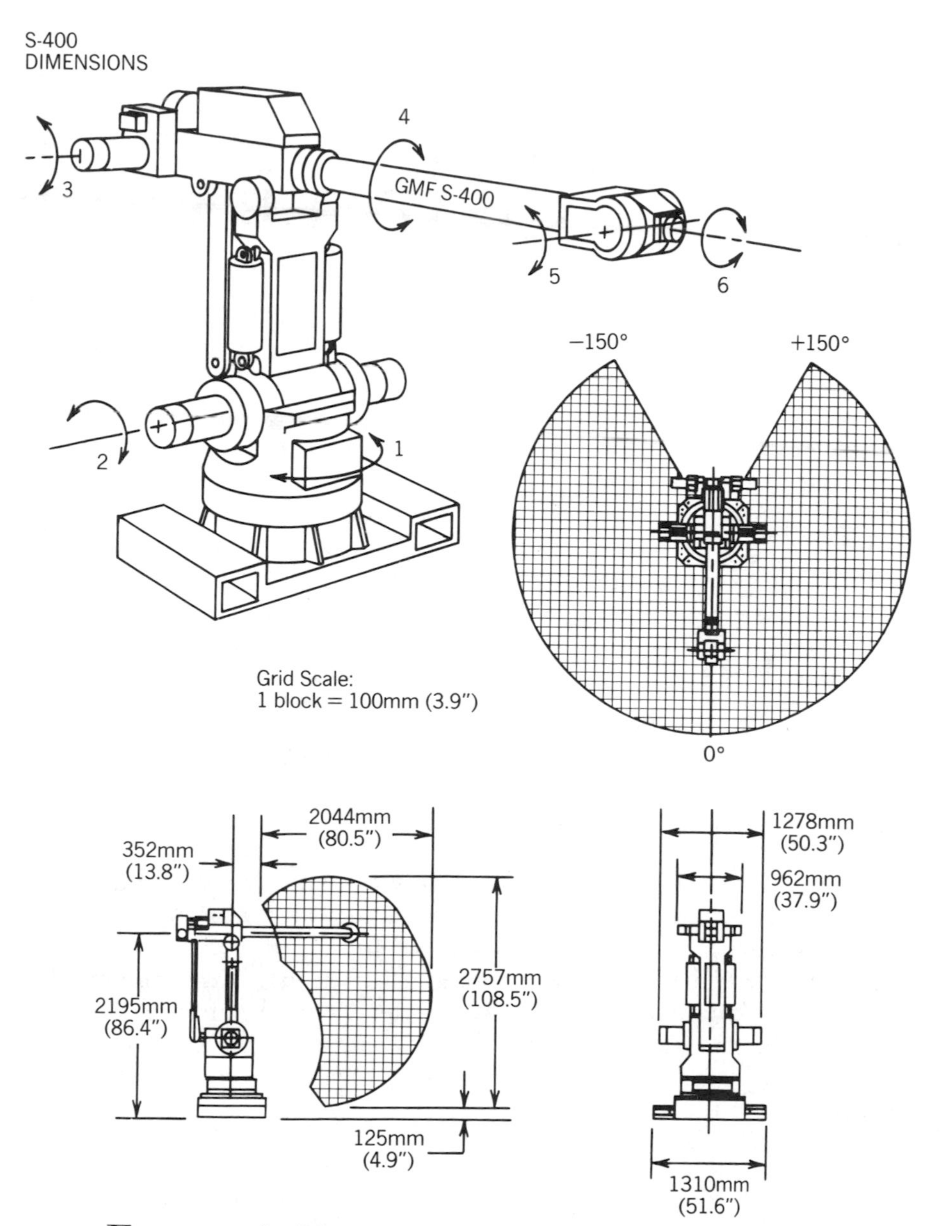

FIGURE 3-23

GMF S-400 robot. (Courtesy GMF Robotics.)

CHAPTER FOUR

INVERSE KINEMATICS

4.1 INTRODUCTION

In the previous chapter we showed how to determine the end-effector position and orientation in terms of the joint variables. This chapter is concerned with the inverse problem of finding the joint variables in terms of the end-effector position and orientation. This is the problem of **inverse kinematics**, and is, in general, more difficult than the forward kinematics problem. The general problem of inverse kinematics can be stated as follows. Given a 4×4 homogeneous transformation

$$H = \begin{bmatrix} R & \mathbf{d} \\ 0 & 1 \end{bmatrix} \in E(3) \tag{4.1.1}$$

with $R \in SO(3)$, find (one or all) solutions of the equation

$$T_0^n(q_1, \ldots, q_n) = H \tag{4.1.2}$$

where

$$T_0^n(q_1, \ldots, q_n) = A_1 \ldots A_n \tag{4.1.3}$$

Equation 4.1.2 results in 12 nonlinear equations in n-unknown variables, which can be written as

$$T_{ij}(q_1, \ldots, q_n) = h_{ij}, \quad i = 1,2,3, \quad j=1,\ldots,4 \tag{4.1.4}$$

where T_{ij}, h_{ij} refer to the 12 nontrivial entries of T_0^n and H, respectively. (Since the bottom row of both T_0^n and H are (0,0,0,1), four of the 16 equations represented by (4.1.2) are trivial.)

(i) *Example 4.1.1*

Recall the Stanford manipulator of Example 3.3.5. Suppose that the position and orientation of the final frame is given as

$$T_0^6 = \begin{bmatrix} r_{11} & r_{12} & r_{13} & d_x \\ r_{21} & r_{22} & r_{23} & d_y \\ r_{31} & r_{32} & r_{33} & d_z \\ 0 & 0 & 0 & 1 \end{bmatrix} \tag{4.1.5}$$

To find the corresponding joint variables θ_1, θ_2, d_3, θ_4, θ_5, and θ_6 we must solve the following simultaneous set of nonlinear trigonometric equations (cf. (3.3.21)).

$$c_1[c_2(c_4c_5c_6-s_4s_6)-s_2s_5c_6]-s_1(s_4c_5c_6+c_4s_6) = r_{11} \tag{4.1.6}$$

$$s_1[c_2(c_4c_5c_6-s_4s_6)-s_2s_5c_6]+c_1(s_4c_5c_6+c_4s_6) = r_{21}$$

$$-s_2(c_4c_5c_6-s_4s_6)-c_2s_5c_6 = r_{31}$$

$$c_1[-c_2(c_4c_5s_6+s_4s_6)+s_2s_5s_6]-s_1(-s_4c_5s_6+c_4c_6) = r_{12}$$

$$s_1[-c_2(c_4c_5s_6+s_4s_6)+s_2s_5s_6]+c_1(-s_4c_5s_6+c_4c_6) = r_{22}$$

$$s_2(c_4c_5s_6+s_4c_6)+c_2s_5s_6 = r_{32}$$

$$c_1(c_2c_4s_5+s_2c_5)-s_1s_4s_5 = r_{13}$$

$$s_1(c_2c_4s_5+s_2c_5)+c_1s_4s_5 = r_{23}$$

$$-s_2c_4s_5+c_2c_5 = r_{33}$$

$$c_1s_2d_3-s_1d_2-d_6(c_1c_2c_4s_5+c_1c_5s_2-s_1s_4s_5) = d_x$$

$$s_1s_2d_3+c_1d_2+d_6(c_1s_4s_5+c_2c_4s_1s_5+c_5s_1s_2) = d_y$$

$$c_2d_3+d_6(c_2c_5-c_4s_2s_5) = d_z$$

These equations are much too difficult to solve directly in closed form and we need to develop efficient and systematic techniques that exploit the particular kinematic structure of the manipulator. Whereas the forward kinematics problem always has a unique solution which can be obtained simply by evaluating the forward equations, the inverse kinematics problem may or may not have a solution. Also, if a solution exists it may or may not be unique. Furthermore, because these forward kinematic equations are in general complicated nonlinear functions of the joint variables, the solutions may be difficult to obtain even when they exist.

In solving the inverse kinematics problem we are most interested in finding a closed form solution of the equations rather than a numerical solution. Finding a closed form solution means finding an explicit relationship:

$$q_k = f_k(h_{11}, \ldots, h_{34}) \quad , \quad k=1, \ldots, n \tag{4.1.7}$$

Closed form solutions are preferable for two reasons. First, in certain applications, such as tracking a welding seam whose location is provided by a vision system, the forward kinematic equations must be solved at a rapid rate, say every 20 milliseconds, and having closed form expressions rather than an iterative search is a practical necessity. Second, the kinematic equations in general have multiple solutions. Having closed form solutions allows one to develop rules for choosing a particular solution among several.

The practical question of the existence of solutions to the inverse kinematics problem depends on engineering as well as mathematical considerations. For example, the motion of the revolute joints may be restricted to less than a full 360 degrees of rotation so that not all mathematical solutions of the kinematic equations will correspond to physically realizable configurations of the manipulator. We will assume that the given position and orientation is such that at least one solution of (4.1.2) exists. Once a solution to the mathematical equations is identified, it must be further checked to see whether or not it satisfies all constraints on the ranges of possible joint motions. For our purposes here we henceforth assume that the given homogeneous matrix H in (4.1.2) corresponds to a configuration within the manipulator's workspace with an attainable orientation. This then guarantees that the mathematical solutions obtained correspond to achievable configurations.

4.2 KINEMATIC DECOUPLING

Although the general problem of inverse kinematics is quite difficult, it turns out that for manipulators having six joints, with the last three joints intersecting at a point (such as the Stanford Manipulator above), it is possible to decouple the inverse kinematics problem into two simpler problems, known respectively, as **inverse position kinematics**, and **inverse orientation kinematics**. To put it another way, for a six-DOF manipulator with a spherical wrist, the inverse kinematics problem may be separated into two simpler problems, namely first finding the position of the intersection of the wrist axes, hereafter called the **wrist center**, and then finding the orientation of the wrist.

For concreteness let us suppose that there are exactly six degrees-of-freedom and that the last three joint axes intersect at a point o. We express (4.1.2) as two sets of equations representing the rotational and positional equations

$$R_0^6(q_1,\ldots,q_6) = R \tag{4.2.1}$$

$$\mathbf{d}_0^6(q_1,\ldots,q_6) = \mathbf{d} \tag{4.2.2}$$

where $\mathbf{d}$ and R are the given position and orientation of the tool frame.

Now assumption of a spherical wrist means that the axes z_4, z_5, and z_6 intersect at o and hence the origins o_4 and o_5 assigned by the D-H convention will always be at the wrist center o. Often o_3 will be at o as well but this is not necessary for our subsequent development. The important point of this assumption for the inverse kinematics is that motion of the final three links about these axes will not change the position of o. The position of the wrist center is thus a function of only the first three joint variables. Since the origin of the tool frame o_6 is simply a translation by a distance d_6 along the z_5 axis from o (see Table 3-3) the vector $\mathbf{o}_6$ in the frame $o_0x_0y_0z_0$ are just

$$\mathbf{o}_6 - \mathbf{o} = -d_6 R\mathbf{k} \tag{4.2.3}$$

Let $\mathbf{p}_c$ denote the vector from the origin of the base frame to the wrist center. Thus in order to have the end-effector of the robot at the point $\mathbf{d}$ with the orientation of the end-effector given by $R = (r_{ij})$, it is necessary and sufficient that the wrist center o be located at the point

$$\mathbf{p}_c = \mathbf{d} - d_6 R\mathbf{k} \tag{4.2.4}$$

and that the orientation of the frame $o_6x_6y_6z_6$ with respect to the base be given by R. If the components of the end-effector position $\mathbf{d}$ are denoted d_x, d_y, d_z and the components of the wrist center $\mathbf{p}_c$ are denoted p_x, p_y, p_z, then (4.2.4) gives the relationship

$$\begin{bmatrix} p_x \\ p_y \\ p_z \end{bmatrix} = \begin{bmatrix} d_x - d_6 r_{13} \\ d_y - d_6 r_{23} \\ d_z - d_6 r_{33} \end{bmatrix} \tag{4.2.5}$$

Using Equation (4.2.5) we may find the values of the first three joint variables. This determines the orientation transformation R_0^3, which depends only on these first three joint variables. We can now determine the orientation of the end-effector relative to the frame $o_3x_3y_3z_3$ from the expression

$$R = R_0^3 R_3^6 \tag{4.2.6}$$

as

$$R_3^6 = (R_0^3)^{-1} R = (R_0^3)^T R \tag{4.2.7}$$

The final three joint angles are then found as a set of Euler angles corresponding to R_3^6. Note that the right hand side of (4.2.7) is completely known since R is given and R_0^3 can be calculated once the first three joint variables are known. The idea of kinematic decoupling is illustrated in Figure 4-1.

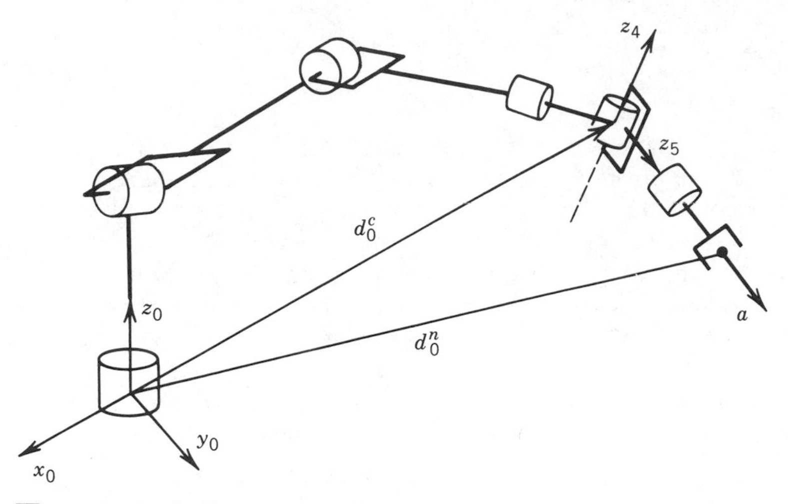

FIGURE 4-1
Kinematic decoupling.

4.2.1 SUMMARY

For this class of manipulators the determination of the inverse kinematics can be summarized by the following algorithm.

Step 1: Find q_1, q_2, q_3 such that the wrist center $\mathbf{p}_c$ is located at

$$\mathbf{p}_c = \mathbf{d} - d_6 R\mathbf{k} \tag{4.2.8}$$

Step 2: Using the joint variables determined in Step 1, evaluate R_0^3.

Step 3: Find a set of Euler angles corresponding to the rotation matrix

$$R_3^6 = (R_0^3)^{-1} R = (R_0^3)^T R \tag{4.2.9}$$

4.3 INVERSE POSITION: A GEOMETRIC APPROACH

For the common kinematic arrangements that we consider, we can use a geometric approach to find the variables, q_1, q_2, q_3, corresponding to $\mathbf{p}_c$ given by (4.2.4). We restrict our treatment to the geometric approach for two reasons. First, as we have said, most present manipulator designs are kinematically simple, usually consisting of one of the five basic configurations of Chapter One with a spherical wrist. Indeed, it is partly due to the difficulty of the general inverse kinematics problem that manipulator designs have evolved to their present state. Second, there are few techniques that can handle the general inverse kinematics problem for arbitrary configurations. Since the reader is most likely to encounter robot configurations of the type considered here, the added difficulty involved in treating the general case seems unjustified. The reader is directed to the references at the end of the chapter for treatment of the general case.

In general the complexity of the inverse kinematics problem increases with the number of nonzero link parameters. For most manipulators, many of the a_i, d_i are zero, the α_i are 0 or $\pm\pi/2$, etc. In these cases especially a geometric approach is the simplest and most natural. We will illustrate this with several important examples.

4.3.1 ARTICULATED CONFIGURATION

Consider the elbow manipulator shown in Figure 4-2. With the components of $\mathbf{p}_c$ denoted by p_x, p_y, p_z, we project $\mathbf{p}_c$ onto the x_0y_0 plane as shown in Figure 4-3. We see from this projection that

$$\theta_1 = Atan(p_x, p_y) \tag{4.3.1}$$

where $Atan(x,y)$ denotes the two argument arctangent function. $Atan(x,y)$ is defined for all $(x,y) \neq 0$ and equals the unique angle θ such that

$$\cos\theta = \frac{x}{(x^2 + y_2)^{1/2}}, \quad \sin\theta = \frac{y}{(x^2 + y^2)^{1/2}} \tag{4.3.2}$$

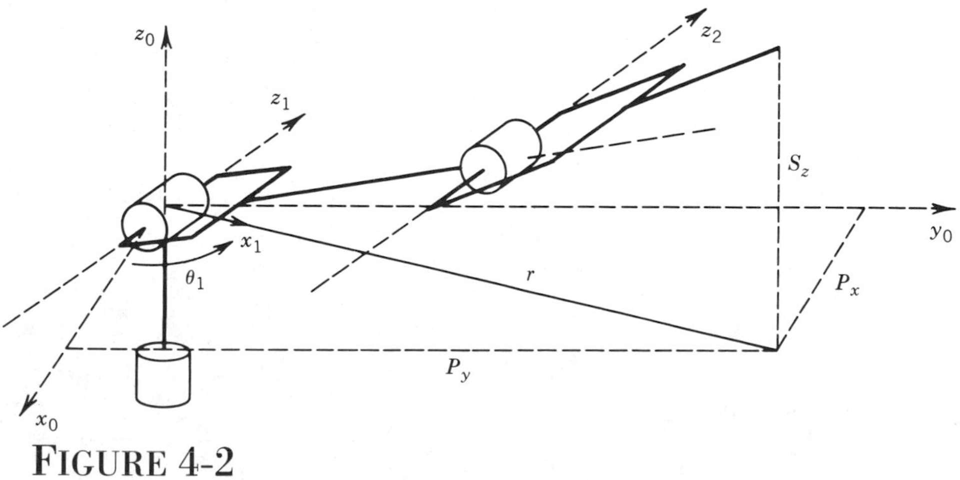

FIGURE 4-2
Elbow manipulator.

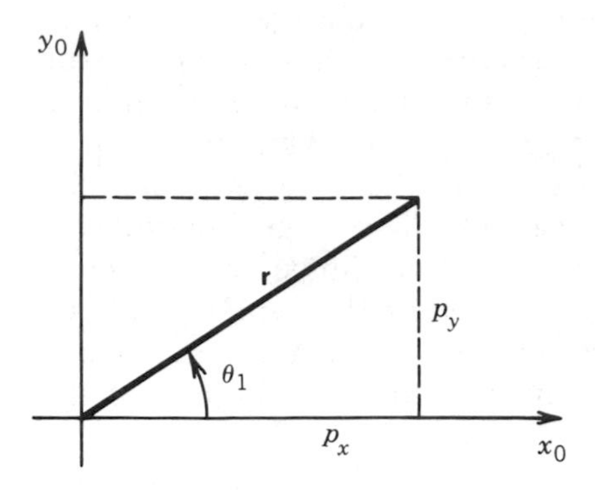

FIGURE 4-3
Projection of the wrist center onto $x_0 - y_0$ plane.

For example, $Atan(1,-1) = -\frac{\pi}{4}$ while $Atan(-1,1) = +\frac{3\pi}{4}$.

Note that a valid second solution for θ_1 is

$$\theta_1 = \pi + Atan(p_x, p_y) \tag{4.3.3}$$

provided that the solution θ_2 corresponding to (4.3.1) is replaced by $\pi - \theta_2$.

These solutions for θ_1 are valid unless $p_x = p_y = 0$. In this case (4.3.1) is undefined and the manipulator is in a singular configuration, shown in Figure 4-4. In this position the wrist center $\mathbf{p}_c$ intersects the z_0 axis; hence any value of θ_1 leaves $\mathbf{p}_c$ fixed. There are thus infinitely many solutions for θ_1 when $\mathbf{p}_c$ intersects z_0.

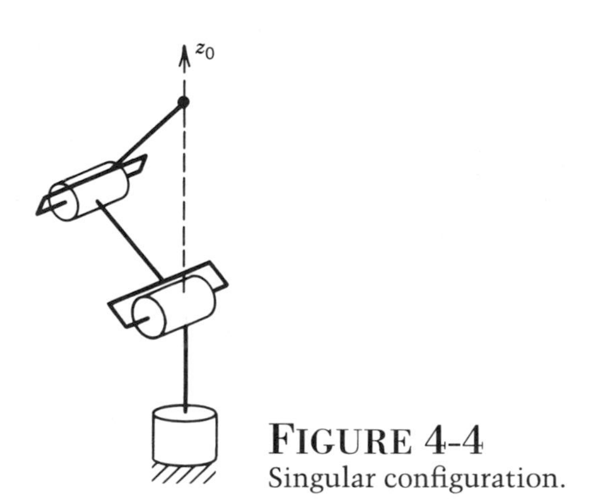

FIGURE 4-4
Singular configuration.

If there is an offset $d_1 \neq 0$ as shown in Figure 4-5 then the wrist center cannot intersect the z_0 axis. In this case there will, in general, be only two solutions for θ_1. These correspond to the so-called **left arm** and **right arm** configurations as shown in Figures 4-6 and 4-7. Figure 4-6 shows the left arm configuration. We see geometrically that

$$\theta_1 = \phi - \alpha \tag{4.3.4}$$

where

$$\phi = Atan(p_x, p_y) \tag{4.3.5}$$

$$\alpha = Atan(d_1, \sqrt{r^2 - d_1^2}) \tag{4.3.6}$$

$$= Atan(d_1, \sqrt{p_x^2 + p_y^2 - d_1^2})$$

The second solution, given by the right arm configuration shown in Figure 4-7 is given by

$$\theta_1 = Atan(p_x, p_y) - Atan(d_1, -\sqrt{p_x^2 + p_y^2 - d_1^2}) \tag{4.3.7}$$

To find the angles θ_2, θ_3 for the elbow manipulator, given θ_1, we consider the plane formed by the second and third links as shown in Figure 4-8. Since the motion of links two and three is planar, the solution is analogous to that of the two-link manipulator of Chapter One. We thus know from our previous derivation (cf. (1.5.8) and (1.5.9)) that

$$\cos\theta_3 = \frac{r^2 + s^2 - a_2^2 - a_3^2}{2a_2a_3} \tag{4.3.8}$$

$$= \frac{p_x^2 + p_y^2 + (p_z^2 - d_1)^2 - a_2^2 - a_3^2}{2a_2a_3} := D$$

and hence, θ_3 is given by

$$\theta_3 = Atan(D, \pm\sqrt{1 - D^2}) \tag{4.3.9}$$

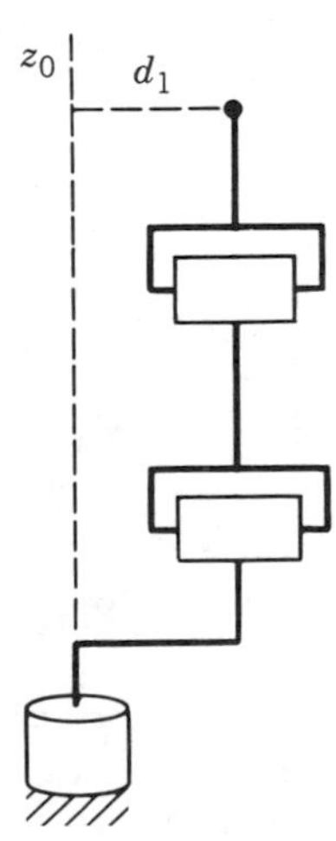

FIGURE 4-5
Elbow manipulator with shoulder offset.

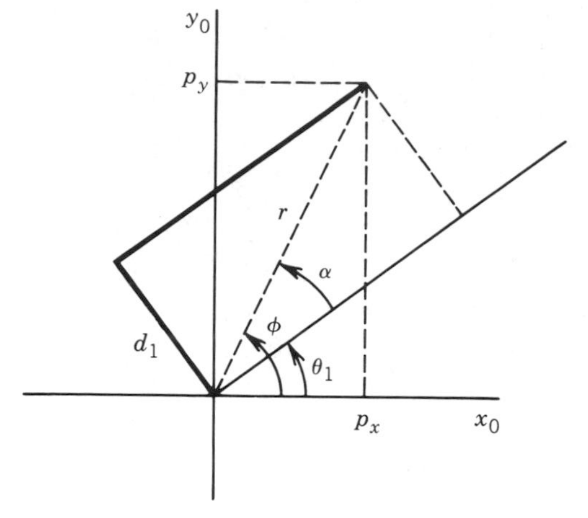

FIGURE 4-6
Left arm configuration.

Similarly θ_2 is given as

$$\theta_2 = Atan(r,s) - Atan(a_2 + a_3c_3, a_3s_3) \tag{4.3.10}$$

$$= Atan(\sqrt{p_x^2 + p_y^2}, p_z - d_1) - Atan(a_2 + a_3c_3, a_3s_3)$$

The two solutions for θ_3 correspond to the elbow-up position and elbow-down position, respectively.

An example of an elbow manipulator with offsets is the PUMA shown in Figure 4-9. There are four solutions to the inverse position kinematics as shown. These correspond to the situations left arm—elbow up, left arm—elbow down, right arm—elbow up and right arm—elbow down. We will see that there are two solutions for the wrist

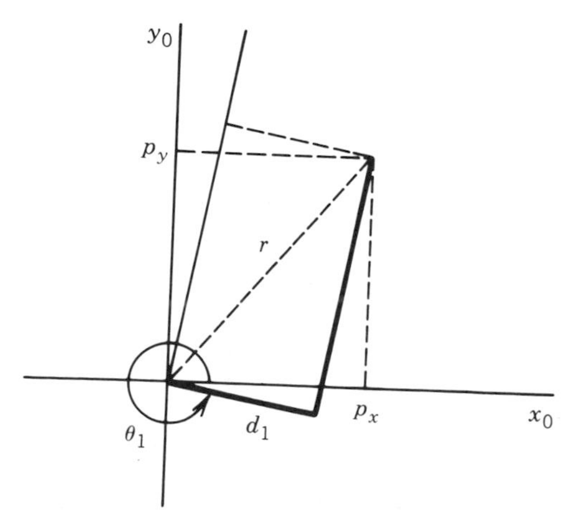

FIGURE 4-7
Right arm configuration.

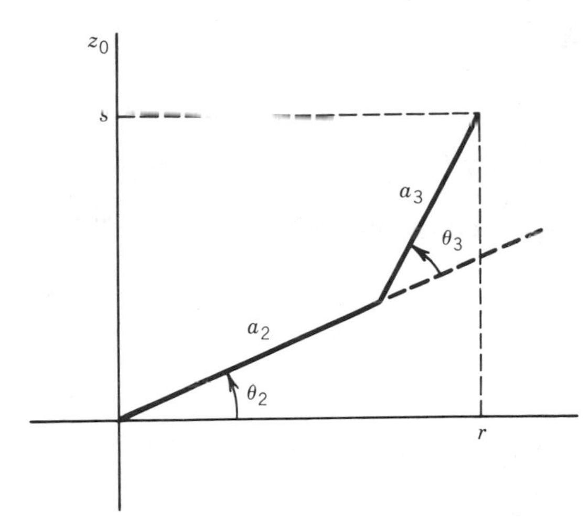

FIGURE 4-8
Projecting onto the plane formed by links 2 and 3.

orientation thus giving a total of eight solutions of the inverse kinematics for the PUMA manipulator.

4.3.2 SPHERICAL CONFIGURATION

We next solve the inverse position kinematics for a three degree of freedom spherical manipulator shown in Figure 4-10. As in the case of the elbow manipulator the first joint variable is the base rotation and a solution is given as

$$\theta_1 = Atan(p_x, p_y) \tag{4.3.11}$$

provided p_x and p_y are not both zero. If both p_x and p_y are zero, the configuration is singular as before and θ_1 may take on any value.

The angle θ_2 is given from Figure 4-10 as

$$\theta_2 = Atan(r, s) \tag{4.3.12}$$

where $r^2 = p_x^2 + p_y^2$, $s = p_z - d_1$. As in the case of the elbow manipulator a second solution for θ_1, θ_2 is given by

$$\theta_1 = \pi + Atan(p_x, p_y); \; \theta_2 = \pi - Atan(r, s) \tag{4.3.13}$$

The linear distance d_3 is found from the expression

$$(d_3 + a_2)^2 = r^2 + s^2 \tag{4.3.14}$$

as

$$d_3 = \sqrt{r^2 + s^2} - a_2 = \sqrt{p_x^2 + p_y^2 + (p_z - d_1)^2} - a_2 \tag{4.3.15}$$

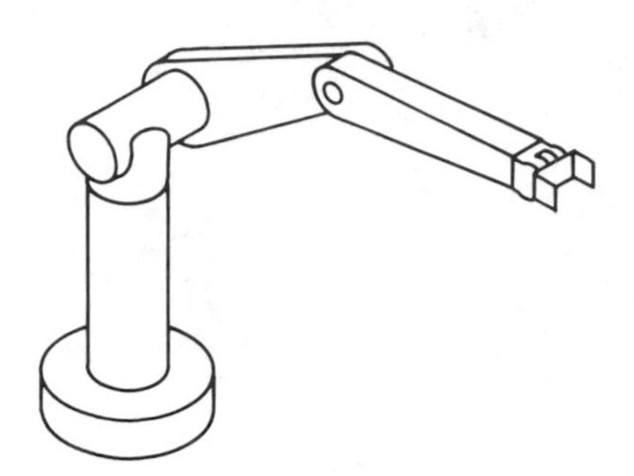

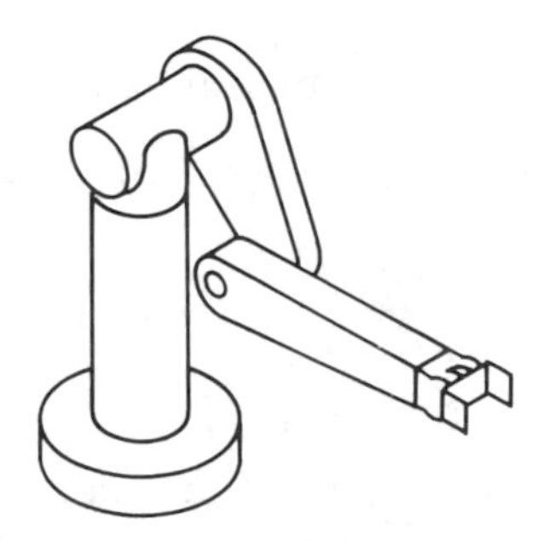

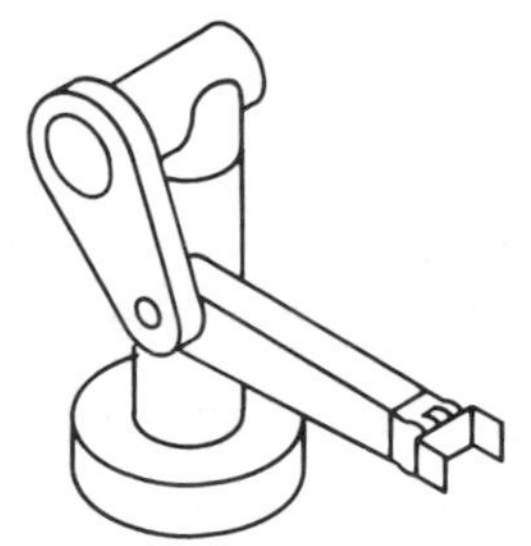

FIGURE 4-9
Four solutions of the inverse position kinematics for the PUMA manipulator.

The negative square root solution for d_3 is disregarded and thus in this case we obtain two solutions to the inverse position kinematics as long as the wrist center does not intersect z_0. If there is an offset then there will be left and right arm configurations as in the case of the elbow manipulator (Problem 4-12).

4.4 INVERSE ORIENTATION

In the previous section we used a geometric approach to solve the inverse position problem. This gives the values of the first three joint variables corresponding to a given position of the wrist origin. The inverse orientation problem is now one on finding the values of the final three joint variables corresponding to a given orientation with respect to the frame $o_3x_3y_3z_3$. For a spherical wrist this can be interpreted as the problem of finding a set of Euler angles corresponding to a given rotation matrix R as we pointed out in Chapter Three.

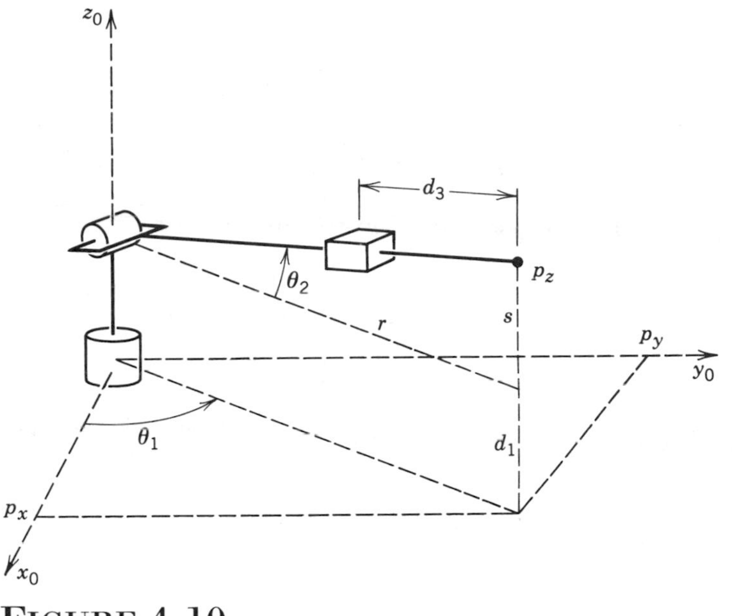

FIGURE 4-10
Spherical manipulator.

4.4.1 EULER ANGLES

Suppose $U = (u_{ij}) \in SO(3)$ is given and R_3^6 is the Euler angle transformation (2.3.9). The problem then is to find the Euler angles ϕ, θ, ψ satisfying the matrix equation

$$\begin{bmatrix} C_\phi C_\theta C_\psi - S_\phi S_\psi & -C_\phi C_\theta S_\psi - S_\phi C_\psi & C_\phi S_\theta \\ S_\phi C_\theta C_\psi + C_\phi S_\psi & -S_\phi C_\theta S_\psi + C_\phi C_\psi & S_\phi S_\theta \\ -S_\theta C_\psi & S_\theta S_\psi & C_\theta \end{bmatrix} = \begin{bmatrix} u_{11} & u_{12} & u_{13} \\ u_{21} & u_{22} & u_{23} \\ u_{31} & u_{32} & u_{33} \end{bmatrix} \tag{4.4.1}$$

Suppose that not both of u_{13}, u_{23} are zero. Then the above equation shows that $s_\theta \neq 0$, and hence that not both of u_{31}, u_{32} are zero. If not both u_{13} and u_{23} are zero, then $u_{33} \neq \pm 1$, and we have $c_\theta = u_{33}$, $s_\theta = \pm\sqrt{1 - u_{33}^2}$ so

$$\theta = Atan(u_{33}, \sqrt{1 - u_{33}^2}) \tag{4.4.2}$$

or

$$\theta = Atan(u_{33}, -\sqrt{1 - u_{33}^2}) \tag{4.4.3}$$

If we choose the first value for θ, then $s_\theta > 0$, and

$$\phi = Atan(u_{13}, u_{23}) \tag{4.4.4}$$

$$\psi = Atan(-u_{31}, u_{32}) \tag{4.4.5}$$

If we choose the second value for θ, then $s_\theta < 0$, and

$$\phi = Atan(-u_{13}, -u_{23}) \tag{4.4.6}$$

$$\psi = Atan(u_{31}, -u_{32}) \tag{4.4.7}$$

Thus there are two solutions depending on the sign chosen for θ.

If $u_{13} = u_{23} = 0$, then the fact that U is orthogonal implies that $u_{33} = \pm 1$, and that $u_{31} = u_{32} = 0$. Thus U has the form

$$U = \begin{bmatrix} u_{11} & u_{12} & 0 \\ u_{21} & u_{22} & 0 \\ 0 & 0 & \pm 1 \end{bmatrix} \tag{4.4.8}$$

If $u_{33} = 1$, then $c_\theta = 1$ and $s_\theta = 0$, so that $\theta = 0$. In this case (4.4.1) becomes

$$\begin{bmatrix} c_\phi c_\psi - s_\phi s_\psi & -c_\phi s_\psi - s_\phi c_\psi & 0 \\ s_\phi c_\psi + c_\phi s_\psi & -s_\phi s_\psi + c_\phi c_\psi & 0 \\ 0 & 0 & 1 \end{bmatrix} = \begin{bmatrix} c_{\phi+\psi} & -s_{\phi+\psi} & 0 \\ s_{\phi+\psi} & c_{\phi+\psi} & 0 \\ 0 & 0 & 1 \end{bmatrix} = \begin{bmatrix} u_{11} & u_{12} & 0 \\ u_{21} & u_{22} & 0 \\ 0 & 0 & 1 \end{bmatrix} \tag{4.4.9}$$

Thus the sum $\phi + \psi$ can be determined as

$$\begin{aligned} \phi + \psi &= Atan(u_{11}, u_{21}) \\ &= Atan(u_{11}, -u_{12}) \end{aligned} \tag{4.4.10}$$

Since only the sum $\phi + \psi$ can be determined in this case there are infinitely many solutions. We may take $\phi = 0$ by convention, and define ψ by (4.4.8). If $u_{33} = -1$, then $c_\theta = -1$ and $s_\theta = 0$, so that $\theta = \pi$. In this case (4.4.1) becomes

$$\begin{bmatrix} -c_{\phi-\psi} & -s_{\phi-\psi} & 0 \\ s_{\phi-\psi} & c_{\phi-\psi} & 0 \\ 0 & 0 & -1 \end{bmatrix} = \begin{bmatrix} u_{11} & u_{12} & 0 \\ u_{21} & u_{22} & 0 \\ 0 & 0 & -1 \end{bmatrix} \tag{4.4.11}$$

The solution is thus

$$\phi - \psi = Atan(-u_{11}, -u_{12}) = Atan(-u_{21}, -u_{22}) \tag{4.4.12}$$

As before there are infinitely many solutions.

(i) Example 4.4.1 Articulated Manipulator

Now the matrix R_0^3 for the articulated or elbow manipulator is easily computed to be

$$R_0^3 = \begin{bmatrix} c_1 c_{23} & -c_1 s_{23} & -s_1 \\ s_1 c_{23} & -s_1 s_{23} & c_1 \\ -s_{23} & -c_{23} & 0 \end{bmatrix} \tag{4.4.13}$$

The matrix $R_3^6 = A_4 A_5 A_6$ is given as

$$R_3^6 = \begin{bmatrix} c_4c_5c_6+s_4s_6 & -c_4c_5s_6+s_4c_6 & c_4s_5 \\ s_4c_5c_6-c_4c_6 & -s_4c_5s_6-c_4c_6 & s_4s_5 \\ s_5c_6 & -s_5s_6 & c_5 \end{bmatrix} \qquad (4.4.14)$$

The equation to be solved now for the final three variables is therefore

$$R_3^6 = (R_0^3)^T R := U \qquad (4.4.15)$$

and the preceding Euler angle solution can be applied to this equation. For example, the (1,3) and (2,3) elements in the above matrix equation are given by

$$c_4s_5 = c_1c_{23}r_{13}+s_1c_{23}r_{23}-s_{23}r_{33} := u_{13} \qquad (4.4.16)$$

$$s_4s_5 = -c_1s_{23}r_{13}-s_1s_{23}r_{23}-c_{23}r_{33} := u_{23} \qquad (4.4.17)$$

Also

$$u_{33} = s_1r_{13}-c_1r_{23} \qquad (4.4.18)$$

Hence, if not both of the expressions (4.4.16), (4.4.17) are zero, then we obtain θ_5 from (4.4.2) and (4.4.3) as

$$\theta_5 = Atan(s_1r_{13}-c_1r_{23}\, ,\ \pm\sqrt{1-(s_1r_{13}-c_1r_{23})^2}\ ) \qquad (4.4.19)$$

If the positive square root is chosen in (4.4.19), then θ_4 and θ_6 are given by (4.4.4) and (4.4.5), respectively, as

$$\theta_4 = Atan(c_1c_{23}r_{13}+s_1c_{23}r_{23}-s_{23}r_{23},\ -c_1s_{23}r_{13}-s_1s_{23}r_{23}-c_{23}r_{33}) \qquad (4.4.20)$$

$$\theta_6 = Atan(s_1r_{11}-c_1r_{21},\ s_1r_{12}+c_1r_{22}) \qquad (4.4.21)$$

The other solutions are obtained analogously. If $s_5 = 0$, then joint axes 3 and 5 are collinear. This is a singular configuration and only the sum $\theta_4+\theta_6$ can be determined. One solution is to choose θ_4 arbitrarily and then determine θ_6 using (4.4.10) or (4.4.12).

4.4.2 SUMMARY OF ELBOW MANIPULATOR SOLUTION

To summarize the preceding development we write down one solution to the inverse kinematics of the six degree-of-freedom elbow manipulator shown in Figure 4-2 which has no joint offsets and a spherical wrist.

Given

$$\mathbf{d} = \begin{bmatrix} d_x \\ d_y \\ d_z \end{bmatrix}; \quad R = \begin{bmatrix} r_{11} & r_{12} & r_{13} \\ r_{21} & r_{22} & r_{23} \\ r_{31} & r_{32} & r_{33} \end{bmatrix} \qquad (4.4.22)$$

then with

$$p_x = d_x - d_6 r_{13} \tag{4.4.23}$$

$$p_y = d_y - d_6 r_{23} \tag{4.4.24}$$

$$p_z = d_z - d_6 r_{33} \tag{4.4.25}$$

a set of D-H joint variables is given by

$$\theta_1 = Atan(p_x, p_y) \tag{4.4.26}$$

$$\theta_2 = Atan(\sqrt{p_x^2 + p_y^2},\ p_z - d_1) - Atan(a_2 + a_3 c_3,\ a_3 s_3) \tag{4.4.27}$$

$$\theta_3 = Atan(D, \pm\sqrt{1-D^2}), \text{ where } D = \frac{p_x^2 + p_y^2 + (p_z - d_1)^2 - a_2^2 - a_3^2}{2a_2 a_3} \tag{4.4.28}$$

$$\theta_4 = Atan(c_1 c_{23} r_{13} + s_1 c_{23} r_{23} - s_{23} r_{23},\ -c_1 s_{23} r_{13} - s_1 s_{23} r_{23} - c_{23} r_{33}) \tag{4.4.29}$$

$$\theta_5 = Atan(s_1 r_{13} - c_1 r_{23},\ +\sqrt{1 - (s_1 r_{13} - c_1 r_{23})^2}) \tag{4.4.30}$$

$$\theta_6 = Atan(s_1 r_{11} - c_1 r_{21},\ s_1 r_{12} + c_1 r_{22}) \tag{4.4.31}$$

The other possible solutions are left as an exercise (Problem 4-11).

(i) *Example 4.4.2 SCARA Manipulator*

As another example, we consider the SCARA manipulator whose forward kinematics is defined by T_0^4 from (3.3.27). The inverse kinematics is then given as the set of solutions of the equation

$$\begin{bmatrix} c_{12}c_4 + s_{12}s_4 & s_{12}c_4 - c_{12}s_4 & 0 & a_1 c_1 + a_2 c_{12} \\ s_{12}c_4 - c_{12}s_4 & -c_{12}c_4 - s_{12}s_4 & 0 & a_1 s_1 + a_2 s_{12} \\ 0 & 0 & -1 & d_3 - d_4 \\ 0 & 0 & 0 & 1 \end{bmatrix} = \begin{bmatrix} R & d \\ 0 & 1 \end{bmatrix} \tag{4.4.32}$$

We first note that, since the SCARA has only four degrees-of-freedom, not every element H of $E(3)$ allows a solution of (4.4.32). In fact we can easily see that there is no solution of (4.4.32) unless R is of the form

$$R = \begin{bmatrix} c_\alpha & s_\alpha & 0 \\ s_\alpha & -c_\alpha & 0 \\ 0 & 0 & -1 \end{bmatrix} \tag{4.4.33}$$

and if this is the case, the sum $\theta_1 + \theta_2 - \theta_4$ is determined by

$$\theta_1 + \theta_2 - \theta_4 = \alpha = Atan(r_{12}, r_{11}) \tag{4.5.34}$$

Projecting the manipulator configuration onto the $x_0 y_0$ plane immediately yields the situation of Figure 4-11.

We see from this that

$$\theta_2 = Atan(\pm\sqrt{1 - r^2},\ r) \tag{4.4.35}$$

where

$$r^2 = \frac{d_x^2 + d_y^2 - a_1^2 - a_2^2}{2a_1 a_2} \tag{4.4.36}$$

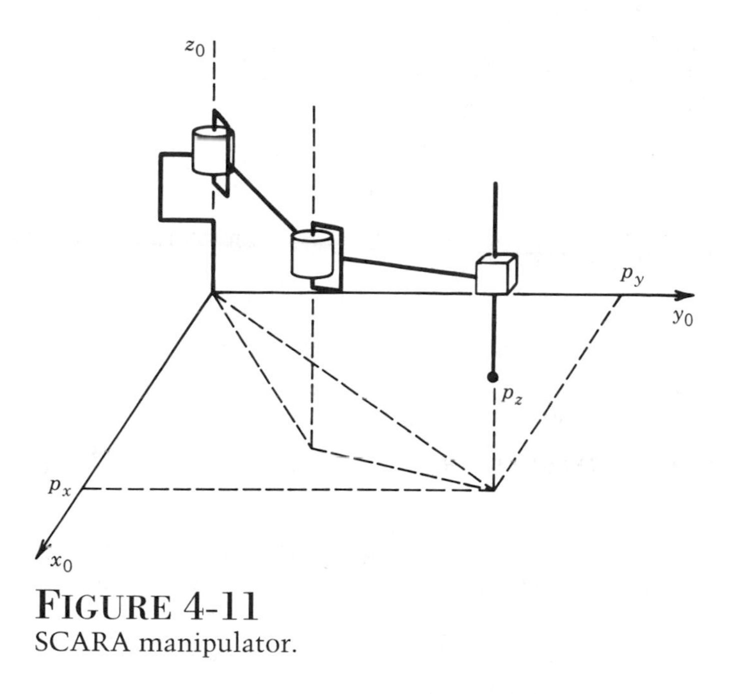

FIGURE 4-11
SCARA manipulator.

$$\theta_1 = Atan(d_x\ ,\ d_y) - Atan(a_1 + a_2c_2\ ,\ a_2s_2) \tag{4.4.37}$$

We may then determine θ_4 from (4.4.34) as

$$\begin{aligned}\theta_4 &= \theta_1 + \theta_2 - \alpha \\ &= \theta_1 + \theta_2 - Atan(r_{12}, r_{11})\end{aligned} \tag{4.4.38}$$

Finally d_3 is given as

$$d_3 = d_z + d_4 \tag{4.4.39}$$

REFERENCES AND SUGGESTED READING

[1] ASADA, H., and CRO-GRANITO, J.A., "Kinematic and Static Characterization of Wrist Joints and Their Optimal Design," *IEEE Conf. on Robotics and Automation*, St. Louis, Mar. 1985.

[2] GOLDENBERG, A.A., BENHABIB, B., and FENTON, R.G., "A Complete Generalized Solution to the Inverse Kinematics of Robots," *IEEE J. Robotics and Automation*, Vol. RA–1, No. 1, pp. 14–20, 1985.

[3] HOLLERBACH, J.M., and GIDEON, S., "Wrist-Partitioned Inverse Kinematic Accelerations and Manipulator Dynamics," *Int. J. Robotics Res.*, Vol. 4, pp. 61–76, 1983.

[4] LEE, C.S.G, GONZALEZ, R.C., and FU, K.S., *Tutorial on Robotics*, IEEE Computer Society Press, Silver Spring, MD, 1983.

[5] LEE, C.S.G., and ZIEGLER, M., "A Geometric Approach in Solving the Inverse Kinematics of PUMA Robots," *IEEE Trans. Aero and Elect Sys*, Vol. AES–20, No. 6, pp. 695–706.

[6] PAUL, R.P., SHIMANO, B., and MAYER, G., "Kinematic Control Equations for Simple Manipulators," *IEEE Trans. Sys., Man., and Cyber.*, Vol. SMC–11, No. 6, 1981.

[7] PIEPER, D.L., "The Kinematics of Manipulators under Computer Control," *Ph.D. Thesis*, Stanford University, 1968.

[8] SHAHINPOOR, M., "The Exact Inverse Kinematics Solutions for the Rhino XR–2 Robot," *Robotics Age*, Vol. 7, No. 8, pp. 6–14, 1985.

[9] TSAI, L., and MORGAN, A., "Solving the Kinematics of the Most General Six- and Five-degree-of-freedom Manipulators by Continuation Methods," *ASME Mechanisms Conference*, Paper 84-DET-20, Boston, Oct. 1984.

PROBLEMS

4-1 Given a desired position of the end-effector, how many solutions are there to the inverse kinematics of the three-link planar arm shown in Figure 4-12? If the orientation of the end-effector is also specified, how many solutions are there? Use the geometric approach to find them.

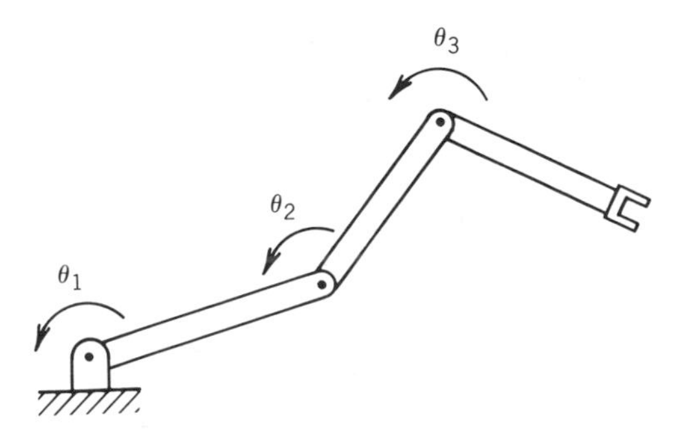

FIGURE 4-12
Three-link planar robot with revolute joints.

4-2 Repeat Problem 4-1 for the three-link planar arm with prismatic joint of Figure 4-13.

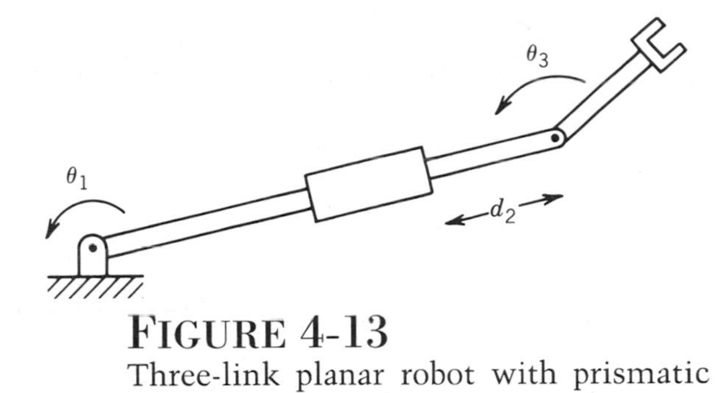

FIGURE 4-13
Three-link planar robot with prismatic joint.

4-3 Solve the inverse position kinematics for the cylindrical manipulator of Figure 4-14.

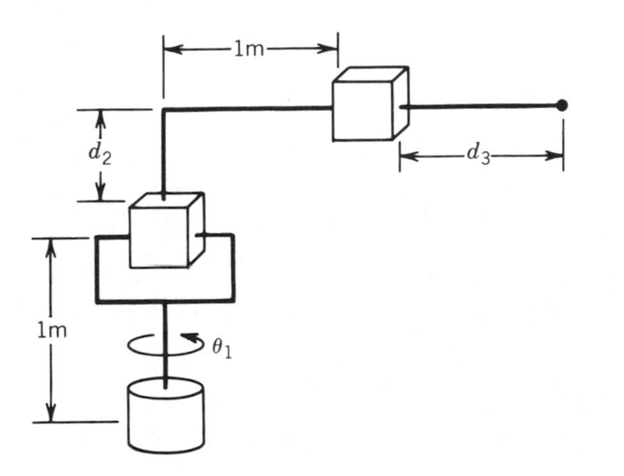

FIGURE 4-14
Cylindrical configuration.

4-4 Solve the inverse position kinematics for the cartesian manipulator of Figure 4-15.

4-5 Add a spherical wrist to the three-link cylindrical arm of Problem 4-3 and write the complete inverse kinematics solution.

4-6 Repeat Problem 4-5 for the cartesian manipulator of Problem 4-4.

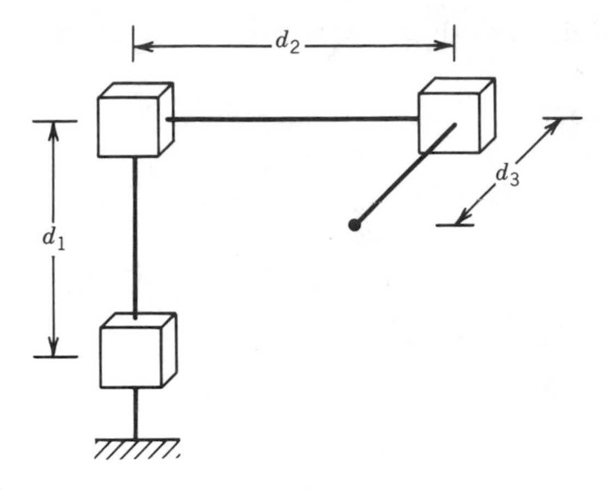

FIGURE 4-15
Cartesian configuration.

4-7 Write a computer program to compute the inverse kinematic equations for the elbow manipulator using Equations 4.4.26–4.4.31. Include procedures for identifying singular configurations and choosing a particular solution when the configuration is singular. Test your routine for various special cases, including singular configurations.

4-8 The Stanford manipulator of Example 3.3.5 has a spherical wrist. Therefore, given a desired position **d** and orientation R of the end-effector,

a) Compute the desired position of the wrist center $\mathbf{p}_c$.

b) Solve the inverse position kinematics, that is, find values of the first three joint variables that will place the wrist center at $\mathbf{p}_c$. Is the solution unique? How many solutions did you find?

c) Compute the rotation matrix R_0^3. Solve the inverse orientation problem for this manipulator by finding a set of Euler angles corresponding to R_3^6 given by (4.4.14).

4-9 Repeat Problem 4-8 for the PUMA 260 manipulator of Problem 3-9, which also has a spherical wrist. How many total solutions did you find?

4-10 Solve the inverse position kinematics for the Rhino robot.

4-11 Find all other solutions to the inverse kinematics of the elbow manipulator of Example 4.4.1.

4-12 Modify the solutions θ_1 and θ_2 for the spherical manipulator given by Equations 4.3.11 and 4.3.12 in the case of a shoulder offset.

CHAPTER FIVE

VELOCITY KINEMATICS—THE MANIPULATOR JACOBIAN

In the previous chapters we derived the forward and inverse position equations relating joint positions and end-effector positions and orientations. In this chapter we derive the velocity relationships, relating the linear and angular velocities of the end-effector (or any other point on the manipulator) to the joint velocities.

Mathematically, the forward kinematic equations define a function between the space of cartesian positions and orientations and the space of joint positions. The velocity relationships are then determined by the **Jacobian** of this function. The Jacobian is a matrix-valued function and can be thought of as the vector version of the ordinary derivative of a scalar function. This Jacobian or Jacobian matrix is one of the most important quantities in the analysis and control of robot motion. It arises in virtually every aspect of robotic manipulation: in the planning and execution of smooth trajectories, in the determination of singular configurations, in the execution of coordinated (anthropomorphic) motion, in the derivation of the dynamic equations of motion, and in the transformation of forces and torques from the end-effector to the manipulator joints.

For an n-link manipulator we first derive the Jacobian representing the instantaneous transformation between the n-vector of joint velocities and the 6-vector consisting of the linear and angular velocities of the end-effector. This Jacobian is then a $6 \times n$ matrix. The same approach is used to determine the transformation between the joint velocities and the linear and angular velocity of *any* point on the manipulator. This will be important when we discuss the derivation of the dynamic equations of motion in Chapter Six. We then discuss the notion of **singular configurations**. These are configurations in which the manipulator loses one or more degrees-of-freedom. We show how the singular configurations are determined geometrically and give several examples. Finally, we discuss the relationship between the accelerations of the joints and of the end-effector.

5.1 DERIVATION OF THE JACOBIAN

Consider an n-link manipulator with joint variables $q_1, \ldots, q_n$. Let

$$T_0^n(\mathbf{q}) = \begin{bmatrix} R_0^n(\mathbf{q}) & \mathbf{d}_0^n(\mathbf{q}) \\ \mathbf{0} & 1 \end{bmatrix} \tag{5.1.1}$$

denote the transformation from the end-effector frame to the base frame, where $\mathbf{q} = (q_1, \ldots, q_n)^T$ is the vector of joint variables. As the robot moves about, both the joint variables q_i and the end-effector position $\mathbf{d}_0^n$ and orientation R_0^n will be functions of time. The objective of this chapter is to relate the linear and angular velocity of the end-effector to the vector of joint velocities $\dot{\mathbf{q}}(t)$. Let

$$S(\boldsymbol{\omega}_0^n) = \dot{R}_0^n (R_0^n)^T \tag{5.1.2}$$

define the angular velocity vector $\boldsymbol{\omega}_0^n$ of the end-effector, and let

$$\mathbf{v}_0^n = \dot{\mathbf{d}}_0^n \tag{5.1.3}$$

denote the linear velocity of the end effector. We seek expressions of the form

$$\mathbf{v}_0^n = J_{\mathbf{v}} \dot{\mathbf{q}} \tag{5.1.4}$$

$$\boldsymbol{\omega}_0^n = J_{\boldsymbol{\omega}} \dot{\mathbf{q}} \tag{5.1.5}$$

where $J_{\mathbf{v}}$ and $J_{\boldsymbol{\omega}}$ are $3 \times n$ matrices. We may write (5.1.4) and (5.1.5) together as

$$\begin{bmatrix} \mathbf{v}_0^n \\ \boldsymbol{\omega}_0^n \end{bmatrix} = J_0^n \dot{\mathbf{q}} \tag{5.1.6}$$

where J_0^n is given by

$$J_0^n = \begin{bmatrix} J_{\mathbf{v}} \\ \hline J_{\boldsymbol{\omega}} \end{bmatrix} \tag{5.1.7}$$

The matrix J_0^n is called the **Manipulator Jacobian** or **Jacobian** for short. Note that J_0^n is a $6 \times n$ matrix where n is the number of links. We next derive a simple expression for the Jacobian of any manipulator.

5.1.1 ANGULAR VELOCITY

Recall from Equation 2.7.14 that angular velocities can be added vectorially provided that they are expressed relative to a common coordinate frame. Thus we can determine the angular velocity of the end-effector relative to the base by expressing the angular velocity of each link in the orientation of the base frame and then summing them. If the i-th joint is revolute, then the i-th joint variable q_i equals θ_i and the axis of rotation is the z_{i-1} axis. Thus the angular velocity of link i expressed in the frame $i-1$ is given by

$$\boldsymbol{\omega}_{i-1}^{i} = \dot{q}_i \mathbf{k} \tag{5.1.8}$$

If the i-th joint is prismatic, then the motion of frame i relative to frame $i-1$ is a translation and

$$\boldsymbol{\omega}_{i-1}^{i} = 0 \tag{5.1.9}$$

Thus, if joint i is prismatic, the angular velocity of the end-effector does not depend on q_i, which now equals d_i.

Therefore, the overall angular velocity of the end-effector, $\boldsymbol{\omega}_0^n$, in the base frame is determined by Equation 2.7.14 as

$$\begin{aligned} \boldsymbol{\omega}_0^n &= \rho_1 \dot{q}_1 \mathbf{k} + \rho_2 \dot{q}_2 R_0^1 \mathbf{k} + \ldots + \rho_n \dot{q}_n R_0^{n-1} \mathbf{k} \\ &= \sum_{i-1}^{n} \rho_i \dot{q}_i z_{i-1} \end{aligned} \tag{5.1.10}$$

where

$$z_{i-1} = R_0^{i-1} \mathbf{k} \tag{5.1.11}$$

denotes the unit vector $\mathbf{k}$ of frame $i-1$ expressed in the orientation of the base frame, and where ρ_i is equal to 1 if joint i is revolute and 0 if joint i is prismatic. Of course $z_0 = \mathbf{k} = (0,0,1)^T$.

The lower half of the Jacobian J_ω in (5.1.7) is thus given as

$$J_\omega = \left[\rho_1 z_0 \ldots \rho_n z_{n-1} \right] \tag{5.1.12}$$

5.1.2 LINEAR VELOCITY

The linear velocity of the end-effector is just $\dot{\mathbf{d}}_0^n$. By the chain rule for differentiation

$$\dot{\mathbf{d}}_0^n = \sum_{i=1}^{n} \frac{\partial \mathbf{d}_0^n}{\partial q_i} \dot{q}_i \tag{5.1.13}$$

Thus we see that the i-th column of J_v is just $\dfrac{\partial \mathbf{d}_0^n}{\partial q_i}$. Furthermore this expression is just the linear velocity of the end-effector that would result if $\dot{q}_i$ is equal to one and the other $\dot{q}_j$ are zero. In other words, the i-th column of the Jacobian is generated by holding all joints fixed but the i-th and actuating the i-th at unit velocity. We now consider two cases separately.

(*i*) *Case 1*

If joint i is prismatic, then R_0^{j-1} is independent of $q_i = d_i$ for all j and

$$\mathbf{d}_{i-1}^{i} = d_i\mathbf{k} + R_{i-1}^{i}a_i\mathbf{i} \tag{5.1.14}$$

If all joints are fixed but the i-th we have that

$$\begin{aligned}\dot{\mathbf{d}}_0^n &= R_0^{i-1}\dot{\mathbf{d}}_{i-1}^{i} \\ &= \dot{d}_i R_0^{i-1}\mathbf{k} \\ &= \dot{d}_i \mathbf{z}_{i-1}\end{aligned} \tag{5.1.15}$$

Thus

$$\frac{\partial \mathbf{d}_0^n}{\partial q_i} = \mathbf{z}_{i-1} \tag{5.1.16}$$

(*ii*) *Case 2*

If joint i is revolute, let $\mathbf{o}_k$ denote the vector $\mathbf{d}_0^k$ from the origin o_0 to the origin o_k for any k, and write

$$\mathbf{d}_0^n = \mathbf{d}_0^{i-1} + R_0^{i-1}\mathbf{d}_{i-1}^{n} \tag{5.1.17}$$

or, in the new notation,

$$\mathbf{o}_n - \mathbf{o}_{i-1} = R_0^{i-1}\mathbf{d}_{i-1}^{n} \tag{5.1.18}$$

Referring to Figure 5-1 we note that both $\mathbf{d}_0^{i-1}$ and R_0^{i-1} are constant if only the i-th joint is actuated. Therefore from (5.1.17)

$$\dot{\mathbf{d}}_0^n = R_0^{i-1}\dot{\mathbf{d}}_{i-1}^{n} \tag{5.1.19}$$

Now, since the motion of link i is a rotation q_i about z_{i-1} we have

$$\dot{\mathbf{d}}_{i-1}^{n} = \dot{q}_i\mathbf{k}\times\mathbf{d}_{i-1}^{n} \tag{5.1.20}$$

and thus

$$\begin{aligned}\dot{\mathbf{d}}_0^n &= R_0^{i-1}(\dot{q}_i\mathbf{k}\times\mathbf{d}_{i-1}^{n}) \\ &= \dot{q}_i R_0^{i-1}\mathbf{k}\times R_0^{i-1}\mathbf{d}_{i-1}^{n} \\ &= \dot{q}_i \mathbf{z}_{i-1}\times(\mathbf{o}_n - \mathbf{o}_{i-1})\end{aligned} \tag{5.1.21}$$

Thus

$$\frac{\partial \mathbf{d}_0^n}{\partial q_i} = \mathbf{z}_{i-1}\times(\mathbf{o}_n - \mathbf{o}_{i-1}) \tag{5.1.22}$$

and the upper half of the Jacobian $J_\mathbf{v}$ is given as

$$J_\mathbf{v} = \begin{bmatrix} J_{v1} & \cdots & J_{vn} \end{bmatrix} \tag{5.1.23}$$

where the i-th column J_{vi} is

$$J_{vi} = \mathbf{z}_{i-1}\times(\mathbf{o}_n - \mathbf{o}_{i-1}) \tag{5.1.24}$$

if joint i is revolute and

$$J_{vi} = \mathbf{z}_{i-1} \tag{5.1.25}$$

if joint i is prismatic.

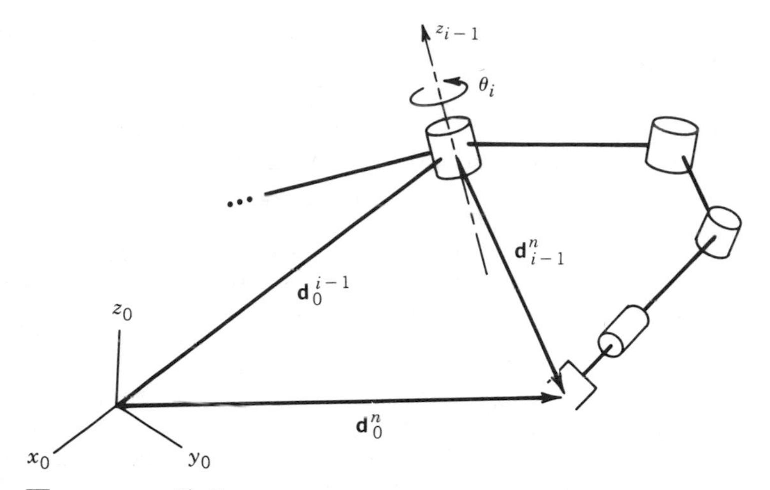

FIGURE 5-1

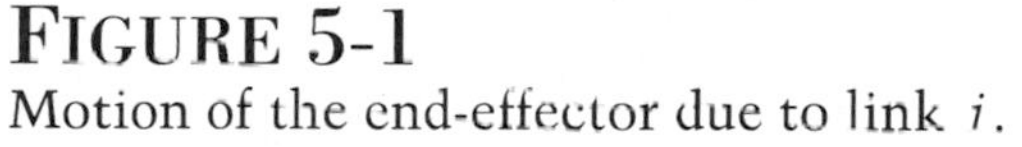
Motion of the end-effector due to link i.

Now putting the upper and lower halves of the Jacobian together we have shown that the Jacobian for an n-link manipulator is of the form

$$J = \left[J_1 J_2 \cdots J_n \right] \tag{5.1.26}$$

where the i–th column J_i is given by

$$J_i = \begin{bmatrix} \mathbf{z}_{i-1} \times (\mathbf{o}_n - \mathbf{o}_{i-1}) \\ \mathbf{z}_{i-1} \end{bmatrix} \tag{5.1.27}$$

if joint i is revolute and

$$J_i = \begin{bmatrix} \mathbf{z}_{i-1} \\ 0 \end{bmatrix} \tag{5.1.28}$$

if joint i is prismatic.

The above formulas make the determination of the Jacobian of any manipulator simple since all of the quantities needed are available once the forward kinematics are worked out. Indeed the only quantities needed to compute the Jacobian are the unit vectors z_i and the vectors from the origin o_0 to the origins $o_1, \ldots, o_n$. A moment's reflection shows that z_i is given by the first three elements in the third column of T_0^i, while $\mathbf{o}_i$ is given by the first three elements of the fourth column of T_0^i. Thus only the third and fourth columns of the T matrices are needed in order to evaluate the Jacobian according to the above formulas.

The above procedure works not only for computing the velocity of the end-effector but also for computing the velocity of any point on the manipulator. This will be important in Chapter Six when we will need to compute the velocity of the center of mass of the various links in order to derive the dynamic equations of motion.

(*iii*) *Example 5.1.1*

Consider the three-link planar manipulator of Figure 5-2. Suppose we wish to compute the linear velocity $\boldsymbol{v}$ and the angular velocity $\boldsymbol{\omega}$ of the center of link 2 as shown. In this case we have that

$$\begin{bmatrix} \boldsymbol{v} \\ \boldsymbol{\omega} \end{bmatrix} = \begin{bmatrix} J_1 & J_2 & J_3 \end{bmatrix} \dot{\mathbf{q}} \tag{5.1.29}$$

where the columns of the Jacobian are determined using the above formula with $\mathbf{o}_c$ in place of $\mathbf{o}_n$. Thus we have

$$J_1 = z_0 \times (\mathbf{o}_c - \mathbf{o}_0) \tag{5.1.30}$$

$$J_2 = z_1 \times (\mathbf{o}_c - \mathbf{o}_1)$$

and

$$J_3 = 0$$

since the velocity of the second link is unaffected by motion of link 3.[1] Note that in this case the vector $\mathbf{o}_c$ must be computed as it is not given directly by the T matrices (Problem 5-1).

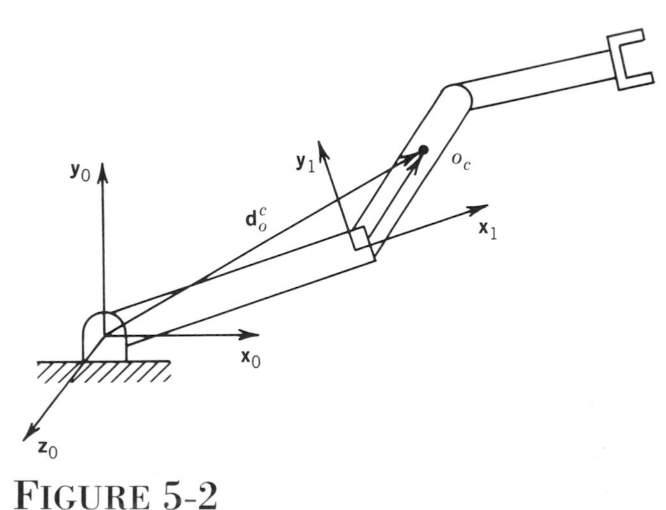

FIGURE 5-2
Finding the velocity of link 2 of a 3-link planar robot.

[1]Note that we are treating only kinematic effects here. Reaction forces on link 2 due to the motion of link 3 will influence the motion of link 2. These dynamic effects are treated by the methods of Chapter Six.

5.2 EXAMPLES

(iv) *Example 5.2.1*

Consider the two-link planar manipulator of Example 3.3.1. Since both joints are revolute the Jacobian matrix, which in this case is 6×2, is of the form

$$J(q) = \begin{bmatrix} z_0 \times (o_2 - o_0) & z_1 \times (o_2 - o_1) \\ z_0 & z_1 \end{bmatrix} \tag{5.2.1}$$

The various quantities above are easily seen to be

$$\mathbf{o}_0 = \begin{bmatrix} 0 \\ 0 \\ 0 \end{bmatrix} \quad \mathbf{o}_1 = \begin{bmatrix} a_1 c_1 \\ a_1 s_1 \\ 0 \end{bmatrix} \quad \mathbf{o}_2 = \begin{bmatrix} a_1 c_1 + a_2 c_{12} \\ a_1 s_1 + a_2 s_{12} \\ 0 \end{bmatrix} \tag{5.2.2}$$

$$z_0 = z_1 = \begin{bmatrix} 0 \\ 0 \\ 1 \end{bmatrix} \tag{5.2.3}$$

Performing the required calculations then yields

$$J = \begin{bmatrix} -a_1 s_1 - a_2 s_{12} & -a_2 s_{12} \\ a_1 c_1 + a_2 c_{12} & a_2 c_{12} \\ 0 & 0 \\ 0 & 0 \\ 0 & 0 \\ 1 & 1 \end{bmatrix} \tag{5.2.4}$$

It is easy to see how the above Jacobian compares with the expression (1.5.11) derived in Chapter One. The first two rows of (5.2.3) are exactly the 2×2 Jacobian of Chapter One and give the linear velocity of the origin $\mathbf{o}_2$ relative to the base. The third row in (5.2.4) is the linear velocity in the z_0 direction, which is of course always zero in this case. The last three rows represent the angular velocity of the final frame, which is simply a rotation about the vertical axis at the rate $\dot{\theta}_1 + \dot{\theta}_2$.

(v) *Example 5.2.2 Stanford Manipulator*

Consider the Stanford manipulator of Example 3.3.5 with its associated Denavit–Hartenberg coordinate frames. Note that joint 3 is prismatic and that $\mathbf{o}_3 = \mathbf{o}_4 = \mathbf{o}_5$ as a consequence of the spherical wrist and the frame assignment. Denoting this common origin by $\mathbf{o}$ we see that the Jacobian is of the form

$$J = \begin{bmatrix} \mathbf{z}_0 \times (\mathbf{o}_6 - \mathbf{o}_0) & \mathbf{z}_1 \times (\mathbf{o}_6 - \mathbf{o}_1) & \mathbf{z}_2 & \mathbf{z}_3 \times (\mathbf{o}_6 - \mathbf{o}) & \mathbf{z}_4 \times (\mathbf{o}_6 - \mathbf{o}) & \mathbf{z}_5 \times (\mathbf{o}_6 - \mathbf{o}) \\ \mathbf{z}_0 & \mathbf{z}_1 & 0 & \mathbf{z}_3 & \mathbf{z}_4 & \mathbf{z}_5 \end{bmatrix} \tag{5.2.5}$$

Now, using the A-matrices given by the expressions (3.3.14)–(3.3.19) and the T-matrices formed as products of the A-matrices, these quantities are easily computed as follows: First, $\mathbf{o}_j$ is given by the first three entries of the last column of $T_0^j = A_1 \ldots A_j$, with $\mathbf{o}_0 = (0,0,0)^T = \mathbf{o}_1$. The vector $\mathbf{z}_j$ is given as

$$\mathbf{z}_j = R_0^j \mathbf{k} \tag{5.2.6}$$

where R_0^j is the rotational part of T_0^j. Thus it is only necessary to compute the matrices T_0^j to calculate the Jacobian. Carrying out these calculations one obtains the following expressions for the Stanford manipulator.

$$\mathbf{o}_6 = (d_x, d_y, d_z)^T = \begin{bmatrix} c_1 s_2 d_3 - s_1 d_2 + d_6(c_1 c_2 c_4 s_5 + c_1 c_5 s_2 - s_1 s_4 s_5) \\ s_1 s_2 d_3 - c_1 d_2 + d_6(c_1 s_4 s_5 + c_2 c_4 s_1 s_5 + c_5 s_1 s_2) \\ c_2 d_3 + d_6(c_2 c_5 - c_4 s_2 s_5) \end{bmatrix} \tag{5.2.7}$$

$$\mathbf{o}_3 = \begin{bmatrix} c_1 s_2 d_3 - s_1 d_2 \\ s_1 s_2 d_3 + c_1 d_2 \\ c_2 d_3 \end{bmatrix} \tag{5.2.8}$$

The vectors $\mathbf{z}_i$ are given as

$$\mathbf{z}_0 = \begin{bmatrix} 0 \\ 0 \\ 1 \end{bmatrix} \qquad \mathbf{z}_1 = \begin{bmatrix} -s_1 \\ c_1 \\ 0 \end{bmatrix} \tag{5.2.9}$$

$$\mathbf{z}_2 = \begin{bmatrix} c_1 s_2 \\ s_1 s_2 \\ c_2 \end{bmatrix} \qquad \mathbf{z}_3 = \begin{bmatrix} c_1 s_2 \\ s_1 s_2 \\ c_2 \end{bmatrix} \tag{5.2.10}$$

$$\mathbf{z}_4 = \begin{bmatrix} -c_1 c_2 s_4 - s_1 c_4 \\ -s_1 c_2 s_4 + c_1 c_4 \\ s_2 s_4 \end{bmatrix} \tag{5.2.11}$$

$$\mathbf{z}_5 = \begin{bmatrix} c_1 c_2 c_4 s_5 - s_1 s_4 s_5 + c_1 s_2 c_5 \\ s_1 c_2 c_4 s_5 + c_1 s_4 s_5 + s_1 s_2 c_5 \\ -s_2 c_4 s_5 + c_2 c_5 \end{bmatrix} \tag{5.2.12}$$

The Jacobian of the Stanford Manipulator is now given by combining these expressions according to the given formulae (Problem 5-7).

(vi) *Example 5.2.3 SCARA Manipulator*

We will now derive the Jacobian of the SCARA manipulator of Example 3.3.6. This Jacobian is a 6×4 matrix since the SCARA has only four degrees-of-freedom. As before we need only compute the matrices $T_0^j = A_1 \dots A_j$, where the A-matrices are given by (3.3.22)–(3.3.25).

Since joints 1,2, and 4 are revolute and joint 3 is prismatic, and since $\mathbf{o}_4 - \mathbf{o}_3$ is parallel to $\mathbf{z}_3$, the Jacobian is of the form

$$J = \begin{bmatrix} \mathbf{z}_0 \times (\mathbf{o}_4 - \mathbf{o}_0) & \mathbf{z}_1 \times (\mathbf{o}_4 - \mathbf{o}_1) & \mathbf{z}_2 & 0 \\ \mathbf{z}_0 & \mathbf{z}_1 & 0 & \mathbf{z}_3 \end{bmatrix} \tag{5.2.13}$$

Performing the indicated calculations, one obtains

$$\mathbf{o}_1 = \begin{bmatrix} a_1 c_1 \\ a_1 s_1 \\ 0 \end{bmatrix} \qquad \mathbf{o}_2 = \begin{bmatrix} a_1 c_1 + a_2 c_{12} \\ a_1 s_1 + a_2 s_{12} \\ 0 \end{bmatrix} \tag{5.2.14}$$

$$\mathbf{o}_4 = \begin{bmatrix} a_1 c_1 + a_2 c_{12} \\ a_1 s_1 + a_2 s_{12} \\ d_3 - d_4 \end{bmatrix} \tag{5.2.15}$$

Similarly $\mathbf{z}_0 = \mathbf{z}_1 = \mathbf{k}$, and $\mathbf{z}_2 = \mathbf{z}_3 = -\mathbf{k}$. Therefore the Jacobian of the SCARA Manipulator is

$$J = \begin{bmatrix} -a_1 s_1 - a_2 s_{12} & -a_2 s_{12} & 0 & 0 \\ a_1 c_1 + a_2 c_{12} & a_2 c_{12} & 0 & 0 \\ 0 & 0 & -1 & 0 \\ 0 & 0 & 0 & 0 \\ 0 & 0 & 0 & 0 \\ 1 & 1 & 0 & -1 \end{bmatrix} \tag{5.2.16}$$

5.3 SINGULARITIES

The $6 \times n$ Jacobian $J(\mathbf{q})$ defines a mapping

$$\dot{\mathbf{X}} = J(\mathbf{q})\dot{\mathbf{q}} \tag{5.3.1}$$

between the vector $\dot{\mathbf{q}}$ of joint velocities and the vector $\dot{\mathbf{X}} := (\mathbf{v}, \boldsymbol{\omega})^T$ of end-effector velocities. Infinitesimally this defines a linear transformation

$$\mathbf{dX} = J(\mathbf{q})\mathbf{dq} \tag{5.3.2}$$

between the differentials $\mathbf{dq}$ and $\mathbf{dX}$. These differentials may be thought of as defining directions in $\mathbb{R}^6$, and $\mathbb{R}^n$, respectively.

Since the Jacobian is a function of the configuration $\mathbf{q}$, those configurations for which the rank of J decreases are of special significance. Such configurations are called **singularities** or **singular configurations**. Identifying manipulator singularities is important for several reasons.

1. Singularities represent configurations from which certain directions of motion may be unattainable.
2. At singularities, bounded gripper velocities may correspond to unbounded joint velocities.
3. At singularities, bounded gripper forces and torques may correspond to unbounded joint torques. (We will see this in Chapter Nine).
4. Singularities usually (but not always) correspond to points on the boundary of the manipulator workspace, that is, to points of maximum reach of the manipulator.
5. Singularities correspond to points in the manipulator workspace that may be unreachable under small perturbations of the link parameters, such as length, offset, etc.
6. Near singularities there will not exist a unique solution to the inverse kinematics problem. In such cases there may be no solution or there may be infinitely many solutions.

(vii) Example 5.3.1

Consider the two-dimensional system of equations

$$\mathbf{dX} = J\mathbf{dq} = \begin{bmatrix} 1 & 1 \\ 0 & 0 \end{bmatrix} \mathbf{dq} \tag{5.3.3}$$

that corresponds to the two equations

$$\begin{aligned} dx &= dq_1 + dq_2 \\ dy &= 0 \end{aligned} \tag{5.3.4}$$

In this case the rank of J is one and we see that for any values of the variables dq_1 and dq_2 there is no change in the variable dy. Thus any vector $\mathbf{dX}$ having a nonzero second component represents an unattainable direction of instantaneous motion.

5.3.1 DECOUPLING OF SINGULARITIES

We saw in Chapter Three that a set of forward kinematic equations can be derived for any manipulator by attaching a coordinate frame rigidly to each link in any manner that we choose, computing a set of homogeneous transformations relating the coordinate frames, and multiplying them together as needed. The D-H convention is merely a systematic way to do this. Although the resulting equations are dependent on the coordinate frames chosen, the manipulator configurations themselves are geometric quantities, independent of the frames used to describe them. Recognizing this fact allows us to decouple the determination of singular configurations, for those manipulators with spherical wrists, into two simpler problems. The first is to determine so-called **arm singularities**, that is, singularities resulting from motion of the arm, which consists of the first three or more links, while the second is to determine the **wrist singularities** resulting from motion of the spherical wrist.

For the sake of argument, suppose that $n = 6$, that is, the manipulator consists of a 3-DOF arm with a 3-DOF spherical wrist. In this case the Jacobian is a 6×6 matrix and a configuration $\mathbf{q}$ is singular if and only if

$$\det J(\mathbf{q}) = 0 \tag{5.3.5}$$

If we now partition the Jacobian J into 3×3 blocks as

$$J = \left[J_P \mid J_O \right] = \left[\begin{array}{c|c} J_{11} & J_{12} \\ \hline J_{21} & J_{22} \end{array} \right] \tag{5.3.6}$$

then, since the final three joints are always revolute

$$J_O = \begin{bmatrix} \mathbf{z}_3 \times (\mathbf{o}_6 - \mathbf{o}_3) & \mathbf{z}_4 \times (\mathbf{o}_6 - \mathbf{o}_4) & \mathbf{z}_5 \times (\mathbf{o}_6 - \mathbf{o}_5) \\ \mathbf{z}_3 & \mathbf{z}_4 & \mathbf{z}_5 \end{bmatrix} \tag{5.3.7}$$

Since the wrist axes intersect at a common point o, if we choose the coordinate frames so that $\mathbf{o}_3 = \mathbf{o}_4 = \mathbf{o}_5 = \mathbf{o}_6 = \mathbf{o}$, then J_O becomes

$$J_O = \begin{bmatrix} 0 & 0 & 0 \\ \mathbf{z}_3 & \mathbf{z}_4 & \mathbf{z}_5 \end{bmatrix} \tag{5.3.8}$$

and the i-th column J_i of J_P is

$$J_i = \begin{bmatrix} \mathbf{z}_{i-1} \times (\mathbf{o} - \mathbf{o}_{i-1}) \\ \mathbf{z}_{i-1} \end{bmatrix} \tag{5.3.9}$$

if joint i is revolute and

$$J_i = \begin{bmatrix} \mathbf{z}_{i-1} \\ 0 \end{bmatrix} \tag{5.3.10}$$

if joint i is prismatic. In this case the Jacobian matrix has the block triangular form

$$J = \begin{bmatrix} J_{11} & 0 \\ J_{21} & J_{22} \end{bmatrix} \tag{5.3.11}$$

with determinant

$$\det J = \det J_{11} \det J_{22} \tag{5.3.12}$$

where J_{11} and J_{22} are each 3×3 matrices. J_{11} has i-th column $\mathbf{z}_{i-1} \times (\mathbf{o} - \mathbf{o}_{i-1})$ if joint i is revolute, and $\mathbf{z}_{i-1}$ if joint i is prismatic, while

$$J_{22} = \begin{bmatrix} z_3 & z_4 & z_5 \end{bmatrix} \tag{5.3.13}$$

Therefore the set of singular configurations of the manipulator is the union of the set of arm configurations satisfying $\det J_{11} = 0$ and the set of wrist configurations satisfying $\det J_{22} = 0$. *Note that this form of the Jacobian does not necessarily give the correct relation between the velocity of the end-effector and the joint velocities.* It is intended only to simplify the determination of singularities.

5.3.2 WRIST SINGULARITIES

We can now see from (5.3.13) that a spherical wrist is in a singular configuration whenever the vectors z_3, z_4 and z_5 are linearly dependent. Referring to Figure 5-3 we see that this happens when the joint axes z_3 and z_5 are collinear. In fact, whenever two revolute joint axes anywhere are collinear, a singularity results since an equal and opposite rotation about the axes results in no net motion of the end-effector. This is the only singularity of the spherical wrist, and is unavoidable without imposing mechanical limits on the wrist design to restrict its motion in such a way that z_3 and z_5 are prevented from lining up.

5.3.3 ARM SINGULARITIES

In order to investigate arm singularities we need only to compute J_{11} according to (5.3.8) and (5.3.9), which is the same formula derived previously with the wrist center $\mathbf{o}$ in place of $\mathbf{o}_6$.

(i) *Example 5.3.2 Elbow Manipulator Singularities*

Consider the three-link articulated manipulator with coordinate frames attached as shown in Figure 5-4. It is left as an exercise (Problem 5-2)

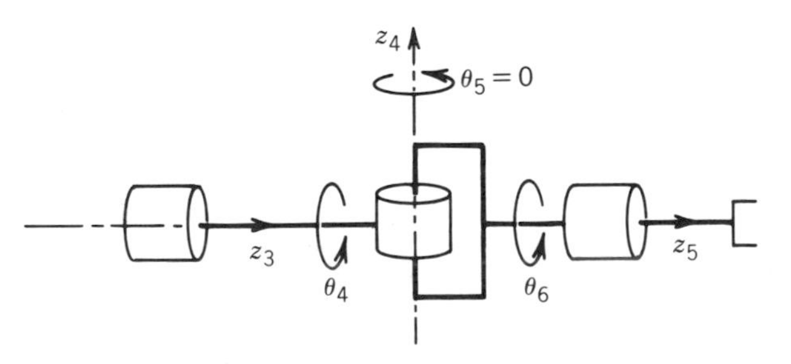

FIGURE 5-3
Spherical wrist singularity.

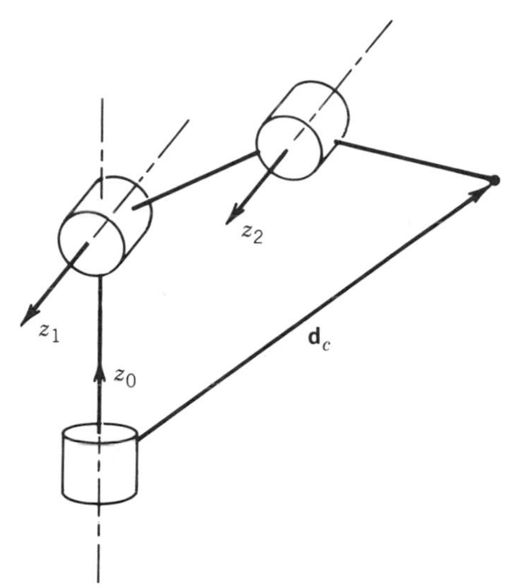

FIGURE 5-4
Elbow manipulator.

to show that

$$J_{11} = \begin{bmatrix} -a_2s_1c_2-a_3s_1c_{23} & -a_2s_2c_1-a_3s_{23}c_1 & -a_3c_1s_{23} \\ a_2c_1c_2+a_3c_1c_{23} & -a_2s_1s_2-a_3s_1s_{23} & -a_3s_1s_{23} \\ 0 & a_2c_2+a_3c_{23} & a_3c_{23} \end{bmatrix} \quad (5.3.14)$$

and that the determinant of J_{11} is

$$\det J_{11} = a_2a_3s_3(a_2c_2+a_3c_{23}) \quad (5.3.15)$$

We see from (5.3.15) that the elbow manipulator is in a singular configuration whenever

$$s_3 = 0, \text{ that is, } \theta_3 = 0 \text{ or } \pi \quad (5.3.16)$$

and whenever

$$a_2c_2+a_3c_{23} = 0 \quad (5.3.17)$$

The situation of (5.3.16) is shown in Figure 5-5 and arises when the elbow is fully extended or fully retracted as shown. The second situation (5.3.17) is shown in Figure 5-6. This configuration occurs when the wrist center intersects the axis of the base rotation z_0. As we saw in Chapter Four, there are infinitely many singular configurations and infinitely many solutions to the inverse position kinematics when the wrist center is along this axis. For an elbow manipulator with an offset, as shown in Figure 5-7, the wrist center cannot intersect z_0, which corroborates our earlier statement that points reachable at singular configurations may not be reachable under arbitrarily small perturbations of the manipulator parameters, in this case an offset in either the elbow or the shoulder.

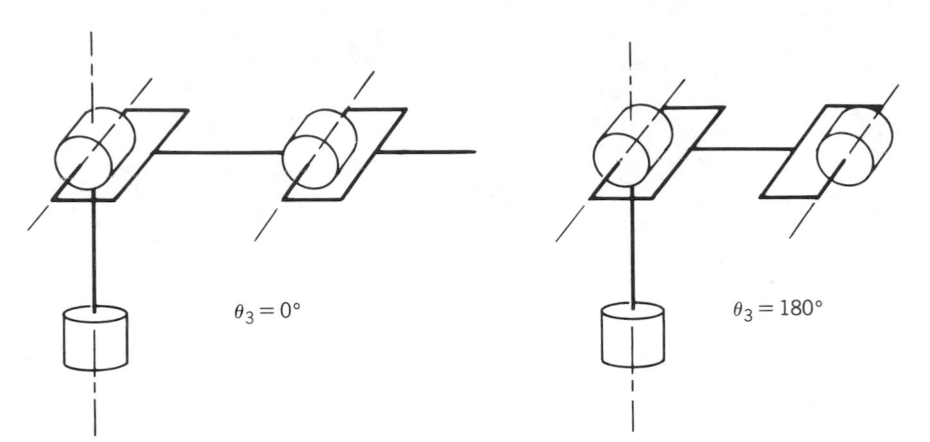

FIGURE 5-5
Elbow singularities of the elbow manipulator.

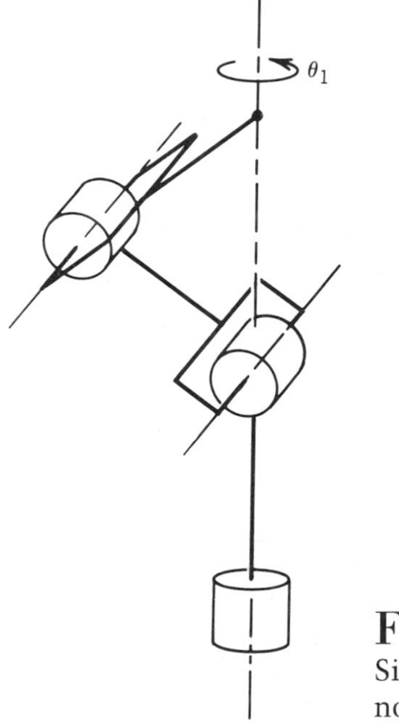

FIGURE 5-6
Singularity of the elbow manipulator with no offsets.

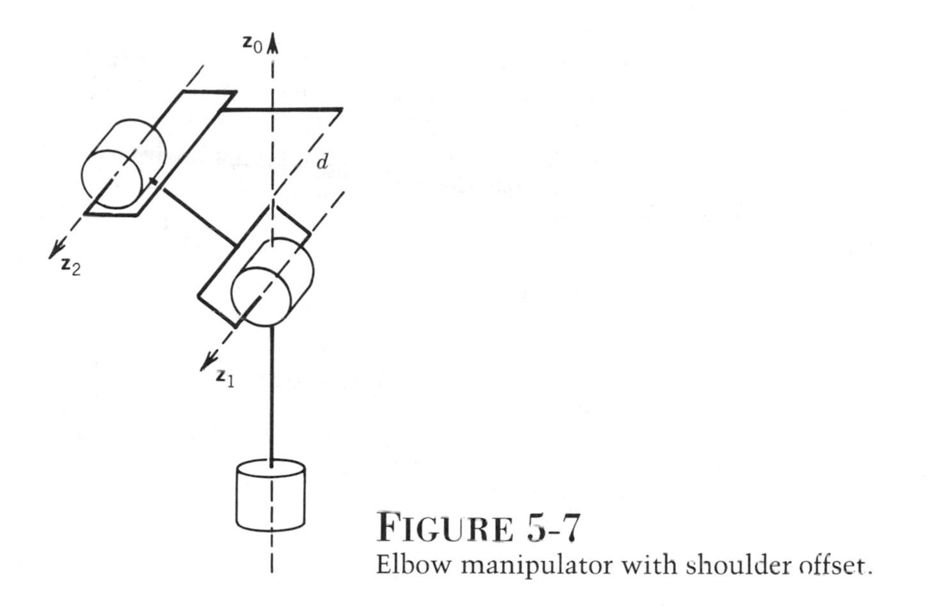

FIGURE 5-7
Elbow manipulator with shoulder offset.

(ii) *Example 5.3.3 Spherical Manipulator*

Consider the spherical arm of Figure 5-8. This manipulator is in a singular configuration when the wrist center intersects z_0 as shown since, as before, any rotation about the base leaves this point fixed.

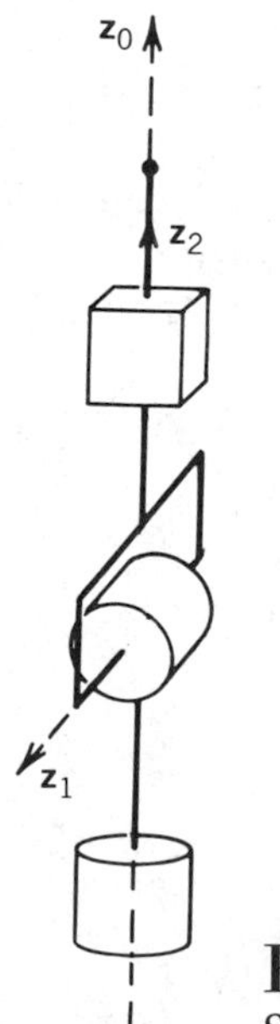

FIGURE 5-8
Singularity of spherical manipulator with no offsets.

(iii) *Example 5.3.4 SCARA Manipulator*

We have already derived the complete Jacobian for the the SCARA manipulator. This Jacobian is simple enough to be used directly rather than deriving the modified Jacobian from this section. Referring to Figure 5-9 we can see geometrically that the only singularity of the SCARA arm is when the elbow is fully extended or fully retracted. Indeed, since the portion of the Jacobian of the SCARA governing arm singularities is given as

$$J_{11} = \begin{bmatrix} \alpha_1 & \alpha_3 & 0 \\ \alpha_2 & \alpha_4 & 0 \\ 0 & 0 & -1 \end{bmatrix} \tag{5.3.18}$$

where

$$\alpha_1 = -a_1 s_1 - a_2 s_{12} \tag{5.3.19}$$

$$\alpha_2 = a_1 c_1 + a_2 c_{12}$$

$$\alpha_3 = -a_1 s_{12}$$

$$\alpha_4 = a_1 c_{12}$$

we see that the rank of J_{11} will be less than three precisely whenever $\alpha_1 \alpha_4 - \alpha_2 \alpha_3 = 0$. It is easy to compute this quantity and show that it is equivalent to (Problem 5-4)

$$s_2 = 0, \quad \text{which implies } \theta_2 = 0, \pi \tag{5.3.20}$$

5.4 INVERSE VELOCITY AND ACCELERATION

It is perhaps a bit surprising that the inverse velocity and acceleration relationships are conceptually simpler than inverse position. Recall from (5.3.1) that the joint velocities and the end-effector velocities are related by the Jacobian as

$$\dot{\mathbf{X}} = J(\mathbf{q})\dot{\mathbf{q}} \tag{5.4.1}$$

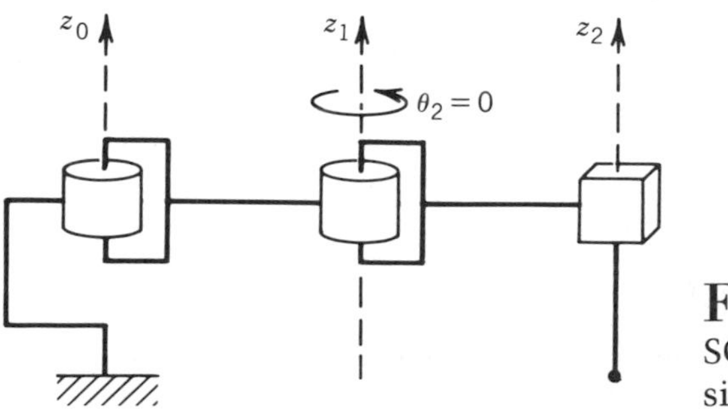

FIGURE 5-9
SCARA manipulator singularity.

Thus the inverse velocity problem becomes one of solving the system of linear equations (5.4.1), which is conceptually simple.

Differentiating (5.4.1) yields the acceleration equations

$$\ddot{\mathbf{X}} = J(\mathbf{q})\ddot{\mathbf{q}} + (\frac{d}{dt}J(\mathbf{q}))\dot{\mathbf{q}} \tag{5.4.2}$$

Thus, given a vector $\ddot{\mathbf{X}}$ of end-effector accelerations, the instantaneous joint acceleration vector $\ddot{\mathbf{q}}$ is given as a solution of

$$\mathbf{b} = J(\mathbf{q})\ddot{\mathbf{q}} \tag{5.4.3}$$

where

$$\mathbf{b} = \ddot{\mathbf{X}} - \frac{d}{dt}J(\mathbf{q})\dot{\mathbf{q}} \tag{5.4.4}$$

For 6-DOF manipulators the inverse velocity and acceleration equations can therefore be written as

$$\dot{\mathbf{q}} = J(\mathbf{q})^{-1}\dot{\mathbf{X}} \tag{5.4.5}$$

and

$$\ddot{\mathbf{q}} = J(\mathbf{q})^{-1}\mathbf{b} \tag{5.4.6}$$

provided $\det J(\mathbf{q}) \neq 0$.

For redundant manipulators or for manipulators with fewer than 6 links the Jacobian can not be inverted. In this case there will be a solution to (5.4.1) or (5.4.4) if and only if the vector on the left-hand side of the equation lies in the range space of the Jacobian. This can be determined by the following simple rank test.

A vector $\mathbf{a}$ belongs to the range of J if and only if

$$rank\, J(\mathbf{q}) = rank\,[J(\mathbf{q}) \mid \mathbf{a}] \tag{5.4.7}$$

In other words, Equation (5.4.1), respectively (5.4.4), may be solved for $\dot{\mathbf{q}} \in \mathbb{R}^n$, respectively $\ddot{\mathbf{q}} \in \mathbb{R}^n$, provided that the rank of the *augmented matrix* $[J(\mathbf{q}) \mid \dot{\mathbf{X}}]$, respectively of $[J(\mathbf{q}) \mid \mathbf{b}]$, is the same as the rank of the Jacobian $J(\mathbf{q})$. This is a standard result from linear algebra, and several algorithms exist, such as Gaussian elimination, for solving such systems of linear equations.

REFERENCES AND SUGGESTED READING

[1] HUNT, K. *Kinematic Geometry of Mechanisms*, Oxford University Press, London, 1978.

[2] ORIN, D., and SCHRADER, W., "Efficient Jacobian Determination for Robot Manipulators," in *Robotics Research: The First International Symposium*, MIT Press, Cambridge, MA, 1984.

PROBLEMS

5-1 For the three-link planar manipulator of Example 5.1.1, compute the vector $\mathbf{o}_c$ and derive the Jacobian (5.1.29).

5-2 Compute the Jacobian J_{11} for the 3-link elbow manipulator of Example 5.3.2 and show that it agrees with (5.3.14). Show that the determinant of this matrix agrees with (5.3.15).

5-3 Compute the Jacobian J_{11} for the three-link spherical manipulator of Example 5.3.3.

5-4 Show from (5.3.19) that the singularities of the SCARA manipulator are given by (5.3.20).

5-5 Find the 6×3 Jacobian for the first three links of the cylindrical manipulator of Figure 3-7. Show that there are no singular configurations for this arm. Thus the only singularities for the cylindrical manipulator must come from the wrist.

5-6 Repeat Problem 5-5 for the cartesian manipulator of Figure 3-17.

5-7 Complete the derivation of the Jacobian for the Stanford manipulator from Example 5.2.2.

CHAPTER SIX

DYNAMICS

The objective of this chapter is to introduce the reader to some methods in analytical mechanics that will enable him or her to derive the dynamical equations describing the time evolution of most common robotic manipulators. The chapter is organized as follows: We begin by deriving so-called **Euler-Lagrange equations**, which describe the evolution of a mechanical system subject to holonomic constraints (this term is defined later on). In order to determine these equations in a specific situation, one has to form the **Lagrangian** of the system, which is the difference of the kinetic and the potential energy of the system; we show how to do this in several commonly encountered situations. We then derive the dynamical equations of several standard types of robotic manipulators, including a two-link cartesian robot, a two-link planar robot, and a five-bar linkage. This chapter is concluded with a derivation of an alternate formulation of the dynamical equations of a robot, known as the Newton–Euler formulation.

6.1 EULER–LAGRANGE EQUATIONS

In this section we derive a general set of differential equations that describe the time evolution of mechanical systems subjected to holonomic constraints, when the constraint forces satisfy the principle of virtual work. These are the **Euler-Lagrange equations** of motion. Note that there are at least two distinct ways of deriving these equations. The method presented here is based on the method of virtual displacements; but it is also possible to derive the same equations based on Hamilton's principle of least action.

Consider a system consisting of k particles, with corresponding coordinates $\mathbf{r}_1, \ldots, \mathbf{r}_k$. If these particles are free to move about without any restrictions, then it is quite an easy matter to describe their motion, by noting merely that the rate of change of the momentum of each mass equals the external force applied to it. However, if the motion of the particles is constrained in some fashion, then one must take into account not only the externally applied forces, but also the so-called constraint forces, that is, the forces needed to make the constraints hold. As a simple illustration of this, suppose the system consists of two particles, which are joined by a massless rigid wire of length ℓ. Then the two coordinates $\mathbf{r}_1$ and $\mathbf{r}_2$ must satisfy the constraint

$$||\mathbf{r}_1 - \mathbf{r}_2|| = \ell, \text{ or } (\mathbf{r}_1 - \mathbf{r}_2)^T(\mathbf{r}_1 - \mathbf{r}_2) = \ell^2 \tag{6.1.1}$$

If one applies some external forces to each particle, then the particles experience not only these external forces but also the force exerted by the wire, which is along the direction $\mathbf{r}_2 - \mathbf{r}_1$ and of appropriate magnitude. Therefore, in order to analyze the motion of the two particles, we can follow one of two options. First, we can compute, under each set of external forces, what the corresponding constraint force must be in order that the equation above continues to hold. Second, we can search for a method of analysis that does not require us to know the constraint force. Clearly, the second alternative is preferable, since it is in general quite an involved task to compute the constraint forces. The contents of this section are aimed at achieving this second objective.

First it is necessary to introduce some terminology. A constraint on the k coordinates $\mathbf{r}_1, \ldots, \mathbf{r}_k$ is called **holonomic** if it is an equality constraint of the form

$$g_i(\mathbf{r}_1, \ldots, \mathbf{r}_k) = 0, \; i = 1, \cdots, \ell \tag{6.1.2}$$

and **nonholonomic** otherwise. The constraint (6.1.1) imposed by connecting two particles by a massless rigid wire is a holonomic constraint. As as example of a nonholonomic constraint, consider a particle moving inside a sphere of radius ρ centered at the origin of the coordinate system. In this case the coordinate vector $\mathbf{r}$ of the particle must satisfy the constraint

$$||\mathbf{r}|| \leq \rho \tag{6.1.3}$$

Note that the motion of the particle is unconstrained so long as the particle remains away from the wall of the sphere; but when the particle comes into contact with the wall, it experiences a constraining force.

If a system is subjected to ℓ holonomic constraints, then one can think in terms of the constrained system having ℓ fewer degrees of freedom than the unconstrained system. In this case it may be possible to express the coordinates of the k particles in terms of n **generalized coordinates** $q_1, \ldots, q_n$. In other words, we assume that the coordinates of the various particles, subjected to the set of constraints (6.1.2), can be expressed in the form

$$\mathbf{r}_i = \mathbf{r}_i(q_1, \ldots, q_n), i = 1, \cdots, k \tag{6.1.4}$$

where $q_1, \ldots, q_n$ are all **independent**. In fact, the idea of generalized coordinates can be used even when there are infinitely many particles. For example, a physical rigid object such as a bar contains an infinity of particles; but since the distance between each pair of particles is fixed throughout the motion of the bar, only *six* coordinates are sufficient to specify completely the coordinates of any particle in the bar. In particular, one could use three position coordinates to specify the location of the center of mass of the bar, and three Euler angles to specify the orientation of the body. To keep the discussion simple, however, we assume in what follows that the number of particles is finite. Typically, generalized coordinates are positions, angles, etc. In fact, in Chapter Three we chose to denote the joint variables by the symbols $q_1, \ldots, q_n$ precisely because these joint variables form a set of generalized coordinates for an n-link robot manipulator.

One can now speak of **virtual displacements**, which are any set $\delta\mathbf{r}_1, \ldots, \delta\mathbf{r}_k$ of infinitesimal displacements that are consistent with the constraints. For example, consider once again the constraint (6.1.1) and suppose $\mathbf{r}_1, \mathbf{r}_2$ are perturbed to $\mathbf{r}_1 + \delta\mathbf{r}_1$, $\mathbf{r}_2 + \delta\mathbf{r}_2$, respectively. Then, in order that the perturbed coordinates continue to satisfy the constraint, we must have

$$(\mathbf{r}_1 + \delta\mathbf{r}_1 - \mathbf{r}_2 - \delta\mathbf{r}_2)^T(\mathbf{r}_1 + \delta\mathbf{r}_1 - \mathbf{r}_2 - \delta\mathbf{r}_2) = \ell^2 \tag{6.1.5}$$

Now let us expand the above product and take advantage of the fact that the original coordinates $\mathbf{r}_1, \mathbf{r}_2$ satisfy the constraint (6.1.1); let us also neglect quadratic terms in $\delta\mathbf{r}_1, \delta\mathbf{r}_2$. This shows that

$$(\mathbf{r}_1 - \mathbf{r}_2)^T(\delta\mathbf{r}_1 - \delta\mathbf{r}_2) = 0 \tag{6.1.6}$$

Thus any perturbations in the positions of the two particles must satisfy the above equation in order that the perturbed positions continue to satisfy the constraint (6.1.1). Any pair of infinitesimal vectors $\delta\mathbf{r}_1, \delta\mathbf{r}_2$ that satisfy (6.1.6) would constitute a set of virtual displacements for this problem.

Now the reason for using generalized coordinates is to avoid dealing with complicated relationships such as (6.1.6) above. If (6.1.4) holds, then one can see that the set of all virtual displacements is precisely

$$\delta\mathbf{r}_i = \sum_{j=1}^{n} \frac{\partial \mathbf{r}_i}{\partial q_j} \delta q_j, i = 1, \ldots, k \tag{6.1.7}$$

where the virtual displacements $\delta q_1, \ldots, \delta q_n$ of the generalized coordinates are unconstrained (that is what makes them generalized coordinates).

Next we begin a discussion of constrained systems in equilibrium. Suppose each particle is in equilibrium. Then the net force on each particle is zero, which in turn implies that the work done by each set of virtual displacements is zero. Hence the sum of the work done by any set of virtual displacements is also zero; that is,

$$\sum_{i=1}^{k} \mathbf{F}_i^T \delta \mathbf{r}_i = 0 \tag{6.1.8}$$

where $\mathbf{F}_i$ is the total force on particle i. As mentioned earlier, the force $\mathbf{F}_i$ is the sum of two quantities, namely (i) the externally applied force $\mathbf{f}_i$, and (ii) the constraint force $\mathbf{f}_i^{(a)}$. Now suppose that the total work done by the constraint forces corresponding to any set of virtual displacements is zero, that is,

$$\sum_{i=1}^{k} (\mathbf{f}_i^{(a)})^T \delta \mathbf{r}_i = 0 \tag{6.1.9}$$

This will be true whenever the constraint force between a pair of particles is directed along the radial vector connecting the two particles (see the discussion in the next paragraph). Substituting (6.1.9) into (6.1.8) results in

$$\sum_{i=1}^{k} \mathbf{f}_i^T \delta \mathbf{r}_i = 0 \tag{6.1.10}$$

The beauty of this equation is that it does *not* involve the unknown constraint forces, but only the known external forces. This equation expresses the **principle of virtual work**, which can be stated in words as follows: The work done by external forces corresponding to any set of virtual displacements is zero. Note that the principle is not universally applicable, but requires that (6.1.9) hold, that is, that the constraint forces do no work. Thus, if the principle of virtual work applies, then one can analyze the dynamics of a system *without* having to evaluate the constraint forces.

It is easy to verify that the principle of virtual work applies whenever the constraint force between a pair of particles acts along the vector connecting the position coordinates of the two particles. In particular, when the constraints are of the form (6.1.1), the principle applies. To see this, consider once again a single constraint of the form (6.1.1). In this case the constraint force, if any, must be exerted by the rigid massless wire, and therefore must be directed along the radial vector connecting the two particles. In other words, the force exerted on particle 1 by the wire must be of the form

$$\mathbf{f}_1^{(a)} = c(\mathbf{r}_1 - \mathbf{r}_2) \tag{6.1.11}$$

for some constant c (which could change as the particles move about). By the law of action and reaction, the force exerted on particle 2 by the wire must be just the negative of the above, that is,

$$\mathbf{f}_2^{(a)} = -c(\mathbf{r}_1 - \mathbf{r}_2) \tag{6.1.12}$$

Now the work done by the constraint forces corresponding to a set of virtual displacements is

$$(\mathbf{f}_1^{(a)})^T\delta\mathbf{r}_1 + (\mathbf{f}_2^{(a)})^T\delta\mathbf{r}_2 = c(\mathbf{r}_1 - \mathbf{r}_2)^T(\delta\mathbf{r}_1 - \delta\mathbf{r}_2) \tag{6.1.13}$$

But (6.1.6) shows that for any set of virtual displacements, the above inner product must be zero. Thus the principle of virtual work applies in a system constrained by (6.1.1). The same reasoning can be applied if the system consists of several particles, which are pairwise connected by rigid massless wires of fixed lengths, in which case the system is subjected to several constraints of the form (6.1.1). Now, the requirement that the motion of a body be rigid can be equivalently expressed as the requirement that the distance between any pair of points on the body remain constant as the body moves, that is, as an infinity of constraints of the form (6.1.1). Thus the principle of virtual work applies whenever rigidity is the only constraint on the motion. There are indeed situations when this principle does not apply, typically in the presence of magnetic fields. However, in all situations encountered in this book, we can safely assume that the principle of virtual work is valid.

In (6.1.10), the virtual displacements $\delta\mathbf{r}_i$ are not independent, so we cannot conclude from this equation that each coefficient $\mathbf{f}_i$ *individually* equals zero. In order to apply such reasoning, we must transform to generalized coordinates. Before doing this, we consider systems that are not necessarily in equilibrium. For such systems, **D'Alembert's principle** states that, if one introduces a fictitious additional force $-\dot{\mathbf{p}}_i$ on particle i for each i, where $\mathbf{p}_i$ is the momentum of particle i, then each particle will be in equilibrium. Thus, if one modifies (6.1.8) by replacing $\mathbf{F}_i$ by $\mathbf{F}_i - \dot{\mathbf{p}}_i$, then the resulting equation is valid for arbitrary systems. One can then remove the constraint forces as before using the principle of virtual work. This results in the equations

$$\sum_{i=1}^{k}\mathbf{f}_i^T\delta\mathbf{r}_i - \sum_{i=1}^{k}\dot{\mathbf{p}}_i^T\delta\mathbf{r}_i = 0 \tag{6.1.14}$$

The above equation does *not* mean that each coefficient of $\delta\mathbf{r}_i$ is zero. For this purpose, express each $\delta\mathbf{r}_i$ in terms of the corresponding virtual displacements of generalized coordinates, as is done in (6.1.7). Then the virtual work done by the forces $\mathbf{f}_i$ is given by

$$\sum_{i=1}^{k}\mathbf{f}_i^T\delta\mathbf{r}_i = \sum_{i=1}^{k}\sum_{j=1}^{n}\mathbf{f}_i^T\frac{\partial\mathbf{r}_i}{\partial q_j}\delta q_j = \sum_{j=1}^{n}\psi_j\delta q_j \tag{6.1.15}$$

where

$$\psi_j = \sum_{i=1}^{k}\mathbf{f}_i^T\frac{\partial\mathbf{r}_i}{\partial q_j} \tag{6.1.16}$$

is called the j-th **generalized force**. Note that ψ_j need not have dimensions of force, just as q_j need not have dimensions of length; however, $\psi_j\delta q_j$ must always have dimensions of work.

Now let us study the second summation in (6.1.14). Since $\mathbf{p}_i = m_i \dot{\mathbf{r}}_i$, it follows that

$$\sum_{i=1}^{k} \dot{\mathbf{p}}_i^T \delta \mathbf{r}_i = \sum_{i=1}^{k} m_i \ddot{\mathbf{r}}_i^T \delta \mathbf{r}_i = \sum_{i=1}^{k} \sum_{j=1}^{n} m_i \ddot{\mathbf{r}}_i^T \frac{\partial \mathbf{r}_i}{\partial q_j} \delta q_j \tag{6.1.17}$$

Next, using the product rule of differentiation, we see that

$$\sum_{i=1}^{k} m_i \ddot{\mathbf{r}}_i^T \frac{\partial \mathbf{r}_i}{\partial q_j} = \sum_{i=1}^{k} \left\{ \frac{d}{dt} \left[m_i \dot{\mathbf{r}}_i^T \frac{\partial \mathbf{r}_i}{\partial q_j} \right] - m_i \dot{\mathbf{r}}_i^T \frac{d}{dt} \left[\frac{\partial \mathbf{r}_i}{\partial q_j} \right] \right\} \tag{6.1.18}$$

Now differentiate (6.1.4) using the chain rule; this gives

$$\mathbf{v}_i = \dot{\mathbf{r}}_i = \sum_{j=1}^{n} \frac{\partial \mathbf{r}_i}{\partial q_j} \dot{q}_j \tag{6.1.19}$$

Observe from the above equation that

$$\frac{\partial \mathbf{v}_i}{\partial \dot{q}_j} = \frac{\partial \mathbf{r}_i}{\partial q_j} \tag{6.1.20}$$

Next,

$$\frac{d}{dt} \left[\frac{\partial \mathbf{r}_i}{\partial q_j} \right] = \sum_{\ell=1}^{n} \frac{\partial^2 \mathbf{r}_i}{\partial q_j \partial q_\ell} \dot{q}_\ell = \frac{\partial \mathbf{v}_i}{\partial q_j} \tag{6.1.21}$$

where the last equality follows from (6.1.19). Substituting from (6.1.20) and (6.1.21) into (6.1.18) and noting that $\dot{\mathbf{r}}_i = \mathbf{v}_i$ gives

$$\sum_{i=1}^{k} m_i \ddot{\mathbf{r}}_i^T \frac{\partial \mathbf{r}_i}{\partial q_j} = \sum_{i=1}^{k} \left\{ \frac{d}{dt} \left[m_i \mathbf{v}_i^T \frac{\partial \mathbf{v}_i}{\partial \dot{q}_j} \right] - m_i \mathbf{v}_i^T \frac{\partial \mathbf{v}_i}{\partial q_j} \right\} \tag{6.1.22}$$

If we define the *kinetic energy* K to be the quantity

$$K = \sum_{i=1}^{k} \frac{1}{2} m_i \mathbf{v}_i^T \mathbf{v}_i \tag{6.1.23}$$

then the sum above can be compactly expressed as

$$\sum_{i=1}^{k} m_i \ddot{\mathbf{r}}_i^T \frac{\partial \mathbf{r}_i}{\partial q_j} = \frac{d}{dt} \frac{\partial K}{\partial \dot{q}_j} - \frac{\partial K}{\partial q_j} \tag{6.1.24}$$

Now, substituting from (6.1.24) into (6.1.17) shows that the second summation in (6.1.14) is

$$\sum_{i=1}^{k} \dot{\mathbf{p}}_i^T \delta \mathbf{r}_i = \sum_{j=1}^{n} \left\{ \frac{d}{dt} \frac{\partial K}{\partial \dot{q}_j} - \frac{\partial K}{\partial q_j} \right\} \delta q_j \tag{6.1.25}$$

Finally, combining (6.1.25) and (6.1.15) gives

$$\sum_{j=1}^{n} \left\{ \frac{d}{dt} \frac{\partial K}{\partial \dot{q}_j} - \frac{\partial K}{\partial q_j} - \psi_j \right\} \delta q_j = 0 \tag{6.1.26}$$

Now, since the virtual displacements δq_j *are* independent, we can conclude that *each coefficient* in (6.1.26) is zero, that is, that

$$\frac{d}{dt}\frac{\partial K}{\partial \dot{q}_j} - \frac{\partial K}{\partial q_j} = \psi_j,\ j = 1, \cdots, n \tag{6.1.27}$$

If the generalized force ψ_j is the sum of an externally applied generalized force and another one due to a potential field, then a further modification is possible. Suppose there exist functions τ_j and $V(q)$ such that

$$\psi_j = -\frac{\partial V}{\partial q_j} + \tau_j \tag{6.1.28}$$

Then (6.1.27) can be written in the form

$$\frac{d}{dt}\frac{\partial L}{\partial \dot{q}_j} - \frac{\partial L}{\partial q_j} = \tau_j \tag{6.1.29}$$

where $L = K - V$ is the **Lagrangian.** The function V is called the **potential energy**.

Note that (6.1.27) and/or (6.1.29) are the **Lagrangian equations** or **Euler-Lagrange equations of motion.**[4]

(*i*) *Example 6.1.1 Single-Link Manipulator*

Consider the single-link arm shown in Figure 6-1, consisting of a rigid link coupled through a gear train to a DC-motor. Let θ_ℓ and θ_m denote the angle of the link and the angle of the motor shaft, respectively. Then $\theta_\ell = \frac{1}{n}\theta_m$ where n:1 is the gear ratio. The kinetic energy of the system is given by

$$\begin{aligned} K &= \tfrac{1}{2} J_m \dot{\theta}_m^2 + \tfrac{1}{2} J_\ell \dot{\theta}_\ell^2 \\ &= \tfrac{1}{2}(J_m + J_\ell / n^2)\dot{\theta}_m^2 \end{aligned} \tag{6.1.30}$$

where J_m, J_ℓ are the rotational inertias of the motor and link, respectively. The potential energy is given as

$$V = Mg\ell(1 - \cos(\theta_\ell)) = Mg\ell(1 - \cos(\theta_m / n)) \tag{6.1.31}$$

where M is the total mass of the link and ℓ is the distance from the joint axis to the link center of mass.

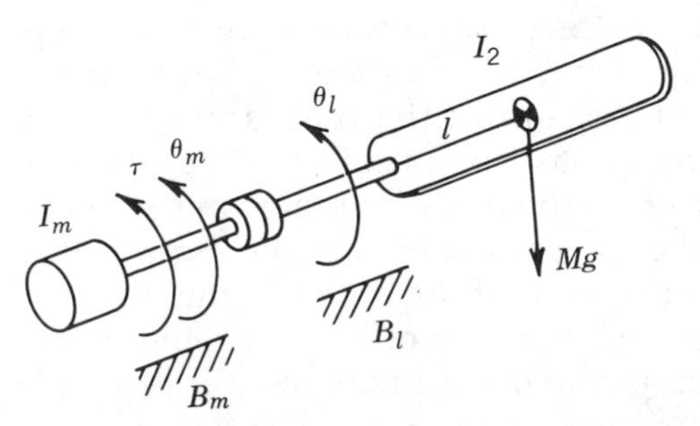

FIGURE 6-1
Single-link robot.

Therefore the Lagrangian L is given by

$$L = \frac{1}{2}(J_m + J_\ell/n^2)\dot{\theta}_m^2 - Mg\ell(1 - \cos(\theta_m/n)) \qquad (6.1.32)$$

Substituting this expression into the Euler–Lagrange equations yields the equation of motion

$$(J_m + J_\ell/n)\ddot{\theta}_m + \frac{Mg\ell}{n} sin(\theta_m/n) = \tau \qquad (6.1.33)$$

The generalized force τ consists of the motor torque input u, and (nonconservative) damping torques $B_m\dot{\theta}_m$, and $B_\ell\dot{\theta}_\ell$. Reflecting the link damping to the motor shaft yields

$$\tau = u - (B_m + B_\ell/n^2)\dot{\theta}_m \qquad (6.1.34)$$

Therefore the complete expression for the dynamics of this system is

$$J\ddot{\theta}_m + B\dot{\theta}_m + C\, sin(\theta_m/n) = u \qquad (6.1.35)$$

where

$$\begin{aligned} J &= J_m + J_\ell/n^2 \\ B &= B_m + B_\ell/n^2 \\ C &= Mg\ell/n \end{aligned} \qquad (6.1.36)$$

In general, for any system of the type considered, an application of the Euler–Lagrange equations leads to a system of second order nonlinear ordinary differential equations in the generalized coordinates.

6.2 EXPRESSIONS FOR KINETIC AND POTENTIAL ENERGY

In the previous section, we derived the Euler–Lagrange equations that can be used to write down the dynamical equations of any configuration of objects in a simple manner, provided one is able to identify the Lagrangian of the system, that is, the difference of the kinetic and potential energy. In order for this result to be useful in a practical context, it is therefore important that one can compute the Lagrangian of a robotic manipulator without too much difficulty. In this section we derive some familiar expressions for the kinetic energy and potential energy of a rigid object. First we show that the kinetic energy of an object is the sum of two terms: the translational energy obtained by concentrating the entire mass of the object at the center of mass, and the rotational energy of the body about the center of mass. Then we show that the potential energy of the body is the same as that obtained by concentrating the entire mass of the object at the center of mass. Once the kinetic and potential energies of each link in a manipulator is known, the Lagrangian of the overall manipulator is simply the sum of the individual Lagrangians.

Suppose we have an object that consists of a continuum of particles, and let ρ denote the density of mass of the object. Let **B** denote the extent of the body itself; in other words,

$$\int_{\mathbf{B}} \rho(x,y,z)dx\, dy\, dz = m \tag{6.2.1}$$

where m is the mass of the object. To reiterate, **B** denotes the region of the three-dimensional space occupied by the body, and is used as a notational device to denote the range of various integrals, as in the above equation. Now the kinetic energy of the object is given by

$$\begin{aligned} K &= \frac{1}{2} \int_{\mathbf{B}} \mathbf{v}^T(x,y,z)\mathbf{v}(x,y,z)\rho(x,y,z)dx\, dy\, dz \\ &= \frac{1}{2} \int_{\mathbf{B}} \mathbf{v}^T(x,y,z)\mathbf{v}(x,y,z)dm \end{aligned} \tag{6.2.2}$$

where dm denotes the infinitesimal mass of the particle located at the coordinates (x,y,z). As the body executes some motion in three-dimensional space, different parts of the body will move with different velocities, so our immediate objective is to bring some order into the above formula. This is accomplished by referring everything to the center of mass of the object. Recall that the center of mass of the object has the coordinates (x_c, y_c, z_c) defined by

$$x_c = \frac{1}{m} \int_{\mathbf{B}} x dm, \quad y_c = \frac{1}{m} \int_{\mathbf{B}} y dm, \quad z_c = \frac{1}{m} \int_{\mathbf{B}} z dm \tag{6.2.3}$$

This equation can be expressed more compactly. Let $\mathbf{r}_c$ denote the three-dimensional coordinate vector of the center of mass, and let $\mathbf{r}$ denote the coordinate vector of a point on the body. Then

$$\mathbf{r}_c = \frac{1}{m} \int_{\mathbf{B}} \mathbf{r} dm \tag{6.2.4}$$

Note that an alternate (and equivalent) definition of the center of mass is

$$\int_{\mathbf{B}} (\mathbf{r}_c - \mathbf{r})dm = \mathbf{0} \tag{6.2.5}$$

Now suppose we attach a coordinate frame rigidly to the center of mass, with its origin at the center of mass. As the body moves, the velocity of a point on the object is given by (cf. (2.6.7))

$$\mathbf{v} = \mathbf{v}_c + \boldsymbol{\omega} \times \mathbf{r} \tag{6.2.6}$$

At this stage it is important to notice the coordinate frame in which the above equation holds. As derived in Section 2.6, the above equation gave the velocity of a particle with respect to an inertial frame, expressed with respect to an inertial frame. However, it is also possible to use this expression to compute the same vector, but expressed in the moving coordinate frame. The transformation is carried out by multiplying the vector by a rotation matrix. As per our custom, let R denote the rotation matrix that transforms free vectors *from* the moving frame *to* the inertial frame. Then the velocity of a particle located at $\mathbf{r}$, expressed with respect to the moving frame, equals

$$R^T(\mathbf{v}_c + \boldsymbol{\omega}\times\mathbf{r}) = R^T\mathbf{v}_c + (R^T\boldsymbol{\omega})\times(R^T\mathbf{r}) \tag{6.2.7}$$

where we use (2.5.8). Alternatively, we can continue to use the same expression (6.2.6), with the understanding that all vectors are referred to the moving frame. Now note that, in computing the kinetic energy according to (6.2.2), it does not make any difference in which frame we express the velocity vector; this is because changing the frame affects the vector but not its length. In this respect, it is better to express all quantities in (6.2.6) with respect to the moving frame attached to the rigid body, for reasons that will become clear shortly. Now, we can express the vector cross product as a matrix multiplying a vector, as follows:

$$\mathbf{v} = \mathbf{v}_c + S(\boldsymbol{\omega})\mathbf{r} \tag{6.2.8}$$

where the skew-symmetric matrix $S(\boldsymbol{\omega})$ is defined in (2.5.4). Substituting from (6.2.8) into the expression (6.2.2) gives the kinetic energy as

$$K = \frac{1}{2}\int_{\mathbf{B}} [\mathbf{v}_c + S(\boldsymbol{\omega})\mathbf{r}]^T[\mathbf{v}_c + S(\boldsymbol{\omega})\mathbf{r}]dm \tag{6.2.9}$$

Let us now expand the product inside the integration; this gives four terms. The first of these is

$$\frac{1}{2}\int_{\mathbf{B}} \mathbf{v}_c^T\mathbf{v}_c\,dm = \frac{1}{2}m\mathbf{v}_c^T\mathbf{v}_c \tag{6.2.10}$$

Here we have exploited the fact that $\mathbf{v}_c$ is independent of the variable of integration and can therefore be moved outside the integral. This quantity is just the kinetic energy of a particle of mass m located at the center of mass and moving with the velocity $\mathbf{v}_c$. This is called the **translational** part of the kinetic energy. The second term is

$$\frac{1}{2}\int_{\mathbf{B}} \mathbf{v}_c^T S(\boldsymbol{\omega})\mathbf{r}\,dm = \frac{1}{2}\mathbf{v}_c^T S(\boldsymbol{\omega})\int_{\mathbf{B}}\mathbf{r}\,dm = 0 \tag{6.2.11}$$

because

$$\int_{\mathbf{B}}\mathbf{r}\,dm = 0$$

Recall that the center of mass is at the origin of the coordinate system. Similarly, the third term, which is

$$\frac{1}{2}\int_{\mathbf{B}} \mathbf{r}^T S^T(\boldsymbol{\omega})\mathbf{v}_c\,dm \tag{6.2.12}$$

also equals zero. Now the final term requires some work. This term is

$$\frac{1}{2}\int_{\mathbf{B}} \mathbf{r}^T S^T(\boldsymbol{\omega})S(\boldsymbol{\omega})\mathbf{r}\,dm := K_4 \tag{6.2.13}$$

Let us first note that, for any two matrices A and B, we have the identity $\mathrm{Tr}\, AB = \mathrm{Tr}\, BA$, where Tr denotes the trace of a matrix, and also that for any two vectors $\mathbf{a}$ and $\mathbf{b}$, $\mathbf{a}^T\mathbf{b} = \mathrm{Tr}\, \mathbf{a}\mathbf{b}^T$. (Appendix A). Using these identities in the above equation gives

$$\begin{aligned} K_4 &= \frac{1}{2} \int_{\mathbf{B}} \mathrm{Tr}\, S(\boldsymbol{\omega})\mathbf{r}\mathbf{r}^T S^T(\boldsymbol{\omega}) dm \\ &= \frac{1}{2} \mathrm{Tr}\, S(\boldsymbol{\omega}) \int_{\mathbf{B}} \mathbf{r}\mathbf{r}^T dm \; S^T(\boldsymbol{\omega}) \\ &= \frac{1}{2} \mathrm{Tr}\, S(\boldsymbol{\omega}) J S^T(\boldsymbol{\omega}) \end{aligned} \tag{6.2.14}$$

where J is a 3×3 matrix defined by

$$J = \int_{\mathbf{B}} \mathbf{r}\mathbf{r}^T dm \tag{6.2.15}$$

We can expand J into its full form, namely

$$J = \begin{bmatrix} \int x^2 dm & \int xy dm & \int xz dm \\ \int xy dm & \int y^2 dm & \int yz dm \\ \int xz dm & \int yz dm & \int z^2 dm \end{bmatrix} \tag{6.2.16}$$

Now let us go back to the angular velocity vector $\boldsymbol{\omega}$ and recall from (2.5.4) that

$$S(\boldsymbol{\omega}) = \begin{bmatrix} 0 & -\omega_3 & \omega_2 \\ \omega_3 & 0 & -\omega_1 \\ -\omega_2 & \omega_1 & 0 \end{bmatrix} \tag{6.2.17}$$

Substituting from the above into (6.2.14) and carrying out the computation of the trace of the product of the three matrices gives the following result:

$$K_4 = \frac{1}{2}\boldsymbol{\omega}^T I \boldsymbol{\omega} \tag{6.2.18}$$

where I is the 3×3 **inertia matrix** defined in the familiar fashion, as follows:

$$I = \begin{bmatrix} \int (y^2 + z^2) dm & -\int xy dm & -\int xz dm \\ -\int xy dm & \int (x^2 + z^2) dm & -\int yz dm \\ -\int xz dm & -\int yz dm & \int (x^2 + y^2) dm \end{bmatrix} \tag{6.2.19}$$

This is known as the **rotational** part of the kinetic energy. Thus the overall kinetic energy of the object is given by

$$K = \frac{1}{2} m \mathbf{v}_c^T \mathbf{v}_c + \frac{1}{2}\boldsymbol{\omega}^T I \boldsymbol{\omega} \tag{6.2.20}$$

This formula is a familiar and useful one, and can be interpreted as follows: The first term is the kinetic energy of a particle of mass m located at the center of mass; if the body were not undergoing any rotational motion, this is all there is to the kinetic energy. The second term is an additional correction term that arises in cases where the body is rotating as well as translating, to account for the fact that different parts of the body are moving at different velocities.

A final word about the coordinate frames in which the various vectors in the above equation are expressed. First, the product $\mathbf{v}_c^T\mathbf{v}_c$ is just the square of the length of the vector $\mathbf{v}_c$; hence it is the same irrespective of the frame in which this vector is expressed, because changing frames merely multiplies this vector by a rotation matrix but does not affect its length. Consider now the rotational part of the kinetic energy. In this case it does matter in which coordinate frame we compute $\boldsymbol{\omega}$ and the inertia matrix I, but it is easy to verify that the **triple product** $\boldsymbol{\omega}^T I\boldsymbol{\omega}$ is the same in all frames. Hence we compute the inertia matrix I with respect to the coordinate frame attached to the object, because in this case I is independent of the motion of the object. In this case, it is necessary to compute the angular velocity also with respect to the same frame; *hence the angular velocity noted previously is different from that introduced in Section 2.7.* In fact, if $\boldsymbol{\omega}_0$ is the angular velocity of the object as measured in an inertial frame, then the angular velocity of the object, expressed in its own frame, is $R^T\boldsymbol{\omega}_0$, where R is the rotation matrix that transforms vectors from the object frame to the inertial frame.

Now consider a manipulator consisting of n links. We have seen in Chapter Five that the linear and angular velocities of any point on any link can be expressed simply in terms of the Jacobian matrix and the derivative of the joint variables. Since in our case the joint variables are indeed the generalized coordinates, it follows that, for appropriate Jacobian matrices $J_{\mathbf{v}_c i}$ and J_{ω_i}, we have that

$$\mathbf{v}_{ci} = J_{\mathbf{v}_{ci}}(\mathbf{q})\dot{\mathbf{q}}, \quad \boldsymbol{\omega}_i = R_i^T(\mathbf{q})J_{\omega_i}(\mathbf{q})\dot{\mathbf{q}} \tag{6.2.21}$$

where the extra matrix $R_i^T(\mathbf{q})$ takes care of the fact that the angular velocity must be expressed in the frame attached to the link. Now suppose the mass of link i is m_i and that the inertia matrix of link i, evaluated around a coordinate frame parallel to frame i but whose origin is at the center of mass, equals I_i. Then from (6.2.20) it follows that the overall kinetic energy of the manipulator equals

$$K = \frac{1}{2}\dot{\mathbf{q}}^T\sum_{i=1}^{n}\left[m_i J_{\mathbf{v}_{ci}}(\mathbf{q})^T J_{\mathbf{v}_{ci}}(\mathbf{q}) + J_{\omega_i}(\mathbf{q})^T R_i(\mathbf{q}) I_i R_i(\mathbf{q})^T J_{\omega_i}(\mathbf{q})\right]\dot{\mathbf{q}} \tag{6.2.22}$$

In other words, the kinetic energy of the manipulator is of the form

$$K = \frac{1}{2}\dot{\mathbf{q}}^T D(\mathbf{q})\dot{\mathbf{q}} \tag{6.2.23}$$

where $D(\mathbf{q})$ is a symmetric positive definite matrix that is in general configuration dependent. The matrix D is called the **inertia matrix**, and in Section 6.4 we will compute this matrix for several commonly occurring manipulator configurations.

Now consider the potential energy term. In the case of rigid dynamics, the only source of potential energy is gravity. Let $\mathbf{g}$ denote the gravity vector expressed in the base frame. Then the potential energy of an infinitesimal particle located at $\mathbf{r}$ on the object is $\mathbf{g}^T\mathbf{r}\,dm$. Hence the overall potential energy is

$$V = \int_{\mathbf{B}} \mathbf{g}^T\mathbf{r}\,dm = \mathbf{g}^T\int_{\mathbf{B}}\mathbf{r}\,dm = \mathbf{g}^T\mathbf{r}_c m \tag{6.2.24}$$

In other words, the potential energy of the object is the same as if the mass of the entire object were concentrated at its center of mass. Thus the potential energy of the manipulator depends only on the vector $\mathbf{q}$ and not on $\dot{\mathbf{q}}$.

6.3 EQUATIONS OF MOTION

In this section, we specialize the Euler–Lagrange equations derived in Section 6.1 to the special case when two conditions hold: First, the kinetic energy is a quadratic function of the vector $\dot{\mathbf{q}}$ of the form

$$K = \frac{1}{2}\sum_{i,j}^{n} d_{ij}(\mathbf{q})\dot{q}_i\dot{q}_j := \frac{1}{2}\dot{\mathbf{q}}^T D(\mathbf{q})\dot{\mathbf{q}} \tag{6.3.1}$$

where the $n \times n$ "inertia matrix" $D(\mathbf{q})$ is symmetric and positive definite for each $\mathbf{q} \in \mathbb{R}^n$. Second, the potential energy $V = V(\mathbf{q})$ is independent of $\dot{\mathbf{q}}$. We have already remarked that robotic manipulators satisfy this condition.

The Euler–Lagrange equations for such a system can be derived as follows. Since

$$L = K - V = \frac{1}{2}\sum_{i,j} d_{ij}(\mathbf{q})\dot{q}_i\dot{q}_j - V(\mathbf{q}) \tag{6.3.2}$$

we have that

$$\frac{\partial L}{\partial \dot{q}_k} = \sum_j d_{kj}(\mathbf{q})\dot{q}_j \tag{6.3.3}$$

and

$$\begin{aligned} \frac{d}{dt}\frac{\partial L}{\partial \dot{q}_k} &= \sum_j d_{kj}(\mathbf{q})\ddot{q}_j + \sum_j \frac{d}{dt} d_{kj}(\mathbf{q})\dot{q}_j \\ &= \sum_j d_{kj}(\mathbf{q})\ddot{q}_j + \sum_{i,j} \frac{\partial d_{kj}}{\partial q_i}\dot{q}_i\dot{q}_j \end{aligned} \tag{6.3.4}$$

Also

$$\frac{\partial L}{\partial q_k} = \frac{1}{2}\sum_{i,j}\frac{\partial d_{ij}}{\partial q_k}\dot{q}_i\dot{q}_j - \frac{\partial V}{\partial q_k} \tag{6.3.5}$$

Thus the Euler–Lagrange equations can be written

$$\sum_j d_{kj}(\mathbf{q})\ddot{q}_j + \sum_{i,j}\{\frac{\partial d_{kj}}{\partial q_i} - \frac{1}{2}\frac{\partial d_{ij}}{\partial q_k}\}\dot{q}_i\dot{q}_j - \frac{\partial V}{\partial q_k} = \tau_k$$

$$k = 1,\cdots,n \tag{6.3.6}$$

By interchanging the order of summation and taking advantage of symmetry, we can show that

$$\sum_{i,j}\{\frac{\partial d_{kj}}{\partial q_i}\}\dot{q}_i\dot{q}_j = \frac{1}{2}\sum_{i,j}\left\{\frac{\partial d_{kj}}{\partial q_i} + \frac{\partial d_{ki}}{\partial q_j}\right\}\dot{q}_i\dot{q}_j \tag{6.3.7}$$

Hence

$$\sum_{i,j}\left\{\frac{\partial d_{kj}}{\partial q_i} - \frac{1}{2}\frac{\partial d_{ij}}{\partial q_k}\right\}\dot{q}_i\dot{q}_j$$

$$= \sum_{i,j}\frac{1}{2}\left\{\frac{\partial d_{kj}}{\partial q_i} + \frac{\partial d_{ki}}{\partial q_j} - \frac{\partial d_{ij}}{\partial q_k}\right\}\dot{q}_i\dot{q}_j \tag{6.3.8}$$

The terms

$$c_{ijk} := \frac{1}{2}\left\{\frac{\partial d_{kj}}{\partial q_i} + \frac{\partial d_{ki}}{\partial q_j} - \frac{\partial d_{ij}}{\partial q_k}\right\} \tag{6.3.9}$$

are known as **Christoffel symbols** (of the first kind). Note that, for a fixed k, we have $c_{ijk} = c_{jik}$, which reduces the effort involved in computing these symbols by a factor of about one half. Finally, if we define

$$\phi_k = \frac{\partial V}{\partial q_k} \tag{6.3.10}$$

then we can write the Euler-Lagrange equations as

$$\sum_j d_{kj}(\mathbf{q})\ddot{q}_j + \sum_{i,j} c_{ijk}(\mathbf{q})\dot{q}_i\dot{q}_j + \phi_k(\mathbf{q}) = \tau_k\,, \quad k = 1,\ldots,n \tag{6.3.11}$$

In the above equation, there are three types of terms. The first involve the second derivative of the generalized coordinates. The second are quadratic terms in the first derivatives of $\mathbf{q}$, where the coefficients may depend on $\mathbf{q}$. These are further classified into two types. Terms involving a product of the type $\dot{q}_i^2$ are called **centrifugal**, while those involving a product of the type $\dot{q}_i\dot{q}_j$ where $i \neq j$ are called **Coriolis** terms. The third type of terms are those involving only $\mathbf{q}$ but not its derivatives. Clearly the latter arise from differentiating the potential energy. It is common to write (6.3.11) in matrix form as

$$D(\mathbf{q})\ddot{\mathbf{q}} + C(\mathbf{q},\dot{\mathbf{q}})\dot{\mathbf{q}} + \mathbf{g}(\mathbf{q}) = \boldsymbol{\tau} \tag{6.3.12}$$

where the k,j-th element of the matrix $C(\mathbf{q},\dot{\mathbf{q}})$ is defined as

$$\begin{aligned} c_{kj} &= \sum_{i=1}^{n} c_{ijk}(\mathbf{q})\dot{q}_i \\ &= \sum_{i=1}^{n} \frac{1}{2}\left\{\frac{\partial d_{kj}}{\partial q_i} + \frac{\partial d_{ki}}{\partial q_j} - \frac{\partial d_{ij}}{\partial q_k}\right\}\dot{q}_i \end{aligned} \qquad (6.3.13)$$

We next derive an important relationship between the inertia matrix $D(\mathbf{q})$ and the matrix $C(\mathbf{q},\dot{\mathbf{q}})$ appearing in (6.3.12) that will be of fundamental importance for the problem of manipulator control considered in later chapters.

(ii) *Theorem 6.3.1*

Define the matrix $N(\mathbf{q},\dot{\mathbf{q}}) = \dot{D}(\mathbf{q}) - 2C(\mathbf{q},\dot{\mathbf{q}})$. Then $N(\mathbf{q},\dot{\mathbf{q}})$ is skew symmetric, that is, the components n_{jk} of N satisfy $n_{jk} = -n_{kj}$.

Proof: Given the inertia matrix $D(\mathbf{q})$, the kj-th component of $\dot{D}(\mathbf{q})$ is given by the chain rule as

$$\dot{d}_{kj} = \sum_{i=1}^{n} \frac{\partial d_{kj}}{\partial q_i}\dot{q}_i \qquad (6.3.14)$$

Therefore, the kj-th component of $N = \dot{D} - 2C$ is given by

$$\begin{aligned} n_{kj} &= \dot{d}_{kj} - 2c_{kj} \\ &= \sum_{i=1}^{n}\left[\frac{\partial d_{kj}}{\partial q_i} - \{\frac{\partial d_{kj}}{\partial q_i} + \frac{\partial d_{ki}}{\partial q_j} - \frac{\partial d_{ij}}{\partial q_k}\}\right]\dot{q}_i \\ &= \sum_{i=1}^{n}\left[\frac{\partial d_{ij}}{\partial q_k} - \frac{\partial d_{ki}}{\partial q_j}\right]\dot{q}_i \end{aligned} \qquad (6.3.15)$$

Since the inertia matrix $D(\mathbf{q})$ is symmetric, that is, $d_{ij} = d_{ji}$, it follows from (6.3.15) by interchanging the indices k and j that

$$n_{jk} = -n_{kj} \qquad (6.3.16)$$

which completes the proof.

Now let us examine an important special case, where *the inertia matrix is diagonal and independent of* $\mathbf{q}$. In this case it follows from (6.3.9) that all of the Christoffel symbols are zero, since each d_{ij} is a constant. Moreover, the quantity d_{kj} is nonzero if and only if $k = j$, so that the Equations 6.3.11 decouple nicely into the form

$$d_{kk}\ddot{\mathbf{q}} - \phi_k(\mathbf{q}) = \tau_k, \quad k = 1, \ldots, n \qquad (6.3.17)$$

In summary, the development in this section is very general and applies to *any* mechanical system whose kinetic energy is of the form (6.3.1) and whose potential energy is independent of $\dot{\mathbf{q}}$. In the next section we apply this discussion to study specific robot configurations.

6.4 SOME COMMON CONFIGURATIONS

In this section we apply the above method of analysis to several manipulator configurations and derive the corresponding equations of motion. The configurations are progressively more complex, beginning with a two-link cartesian manipulator and ending with a five-bar linkage mechanism that has a particularly simple inertia matrix.

(iii) *Example 6.4.1 Two-Link Cartesian Manipulator*

Consider the manipulator shown in Figure 6-2, consisting of two links and two prismatic joints. Denote the masses of the two links by m_1 and m_2, respectively, and denote the displacement of the two prismatic joints by q_1 and q_2, respectively. Then it is easy to see, as mentioned in Section 6.1, that these two quantities serve as generalized coordinates for the manipulator. Since the generalized coordinates have dimensions of distance, the corresponding generalized forces have units of force. In fact, they are just the forces applied at each joint. Let us denote these by $f_i,\ i = 1, 2$.

Since we are using the joint variables as the generalized coordinates, we know that the kinetic energy is of the form (6.3.1) and that the potential energy is only a function of q_1 and q_2. Hence we can use the formulae in Section 6.3 to obtain the dynamical equations.[4] Also, since both joints are prismatic, the angular velocity Jacobian is zero and the kinetic energy of each link consists solely of the translational term.

By (5.1.28) it follows that the velocity of the center of mass of link 1 is given by

$$\mathbf{v}_{c1} = J_{\mathbf{v}_{c1}} \dot{\mathbf{q}} \tag{6.4.1}$$

where

$$J_{\mathbf{v}_{c1}} = \begin{bmatrix} 0 & 0 \\ 0 & 0 \\ 1 & 0 \end{bmatrix}, \quad \dot{\mathbf{q}} = \begin{bmatrix} \dot{q}_1 \\ \dot{q}_2 \end{bmatrix} \tag{6.4.2}$$

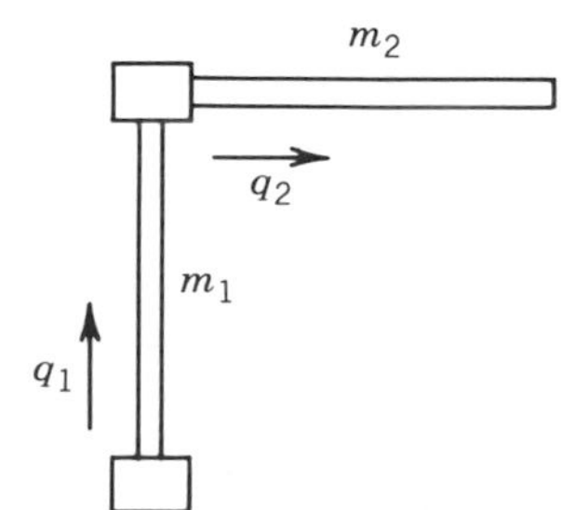

FIGURE 6-2
Two-link cartesian robot.

Similarly,

$$\mathbf{v}_{c2} = J_{\mathbf{v}_{c2}}\dot{\mathbf{q}} \tag{6.4.3}$$

where

$$J_{\mathbf{v}_{c2}} = \begin{bmatrix} 0 & 0 \\ 0 & 1 \\ 1 & 0 \end{bmatrix} \tag{6.4.4}$$

Hence the kinetic energy is given by

$$K = \frac{1}{2}\dot{\mathbf{q}}^T \{m_1 J_{\mathbf{v}_{c1}}^T J_{\mathbf{v}_{c1}} + m_2 J_{\mathbf{v}_{c2}}^T J_{\mathbf{v}_{c2}}\}\dot{\mathbf{q}} \tag{6.4.5}$$

Comparing with (6.3.1), we see that the inertia matrix D is given simply by

$$D = \begin{bmatrix} m_1 + m_2 & 0 \\ 0 & m_2 \end{bmatrix} \tag{6.4.6}$$

Next, the potential energy of link 1 is $m_1 g q_1$, while that of link 2 is $m_2 g q_1$, where g is the acceleration due to gravity. Hence the overall potential energy is

$$V = g(m_1 + m_2)q_1 \tag{6.4.7}$$

Now we are ready to write down the equations of motion. Since the inertia matrix is constant, all Christoffel symbols are zero. Further, the vectors ϕ_k are given by

$$\phi_1 = \frac{\partial V}{\partial q_1} = g(m_1 + m_2), \quad \phi_2 = \frac{\partial V}{\partial q_2} = 0 \tag{6.4.8}$$

Substituting into (6.3.11) gives the dynamical equations as

$$\begin{aligned} (m_1 + m_2)\ddot{q}_1 + g(m_1 + m_2) &= f_1 \\ m_2\ddot{q}_2 &= f_2 \end{aligned} \tag{6.4.9}$$

(iv) *Example 6.4.2 Planar Elbow Manipulator*

Now consider the planar manipulator with two revolute joints shown in Figure 6-3. Let us fix notation as follows: For $i = 1,2$, q_i denotes the joint angle, which also serves as a generalized coordinate; m_i denotes the mass of link i, ℓ_i denotes the length of link i; ℓ_{ci} denotes the distance from the previous joint to the center of mass of link i; and I_i denotes the moment of inertia of link i about an axis coming out of the page, passing through the center of mass of link i.

We will make effective use of the Jacobian expressions in Chapter Five in computing the kinetic energy. Since we are using joint variables as the generalized coordinates, it follows that we can use the contents of Section 6.3. First,

$$\mathbf{v}_{c1} = J_{\mathbf{v}_{c1}}\dot{\mathbf{q}} \tag{6.4.10}$$

where, from (5.1.24),

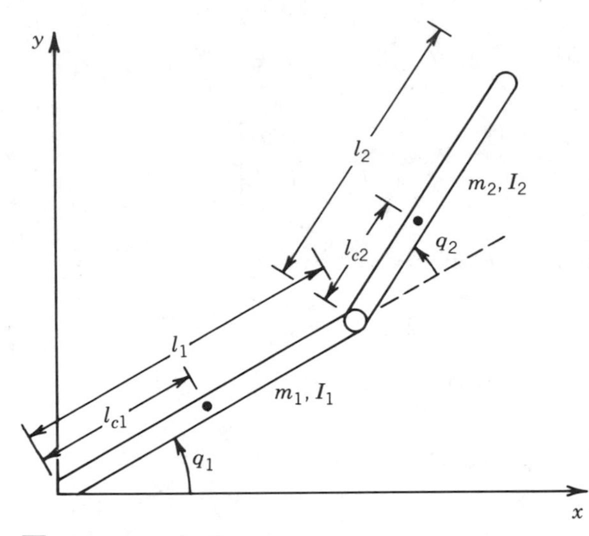

FIGURE 6-3
Two-link revolute joint arm.

$$J_{\mathbf{v}_{c1}} = \begin{bmatrix} -\ell_{c1}\sin q_1 & 0 \\ \ell_{c1}\cos q_1 & 0 \\ 0 & 0 \end{bmatrix} \tag{6.4.11}$$

Similarly,

$$\mathbf{v}_{c2} = J_{\mathbf{v}_{c2}}\dot{\mathbf{q}} \tag{6.4.12}$$

where

$$J_{\mathbf{v}_{c2}} = \begin{bmatrix} -\ell_1\sin q_1 - \ell_{c2}\sin(q_1 + q_2) & -\ell_{c2}\sin(q_1 + q_2) \\ \ell_1\cos q_1 + \ell_{c2}\cos(q_1 + q_2) & \ell_{c2}\cos(q_1 + q_2) \\ 0 & 0 \end{bmatrix} \tag{6.4.13}$$

Hence the translational part of the kinetic energy is

$$\frac{1}{2}m_1\mathbf{v}_{c1}^T\mathbf{v}_{c1} + \frac{1}{2}m_2\mathbf{v}_{c2}^T\mathbf{v}_{c2} = \frac{1}{2}\dot{\mathbf{q}}\left\{m_1 J_{\mathbf{v}_{c1}}^T J_{\mathbf{v}_{c1}} + m_2 J_{\mathbf{v}_{c2}}^T J_{\mathbf{v}_{c2}}\right\}\dot{\mathbf{q}} \tag{6.4.14}$$

Next we deal with the angular velocity terms. Because of the particularly simple nature of this manipulator, many of the potential difficulties do not arise. First, it is clear that

$$\boldsymbol{\omega}_1 = \dot{q}_1\mathbf{k}, \quad \boldsymbol{\omega}_2 = (\dot{q}_1 + \dot{q}_2)\mathbf{k} \tag{6.4.15}$$

when expressed in the base inertial frame. Now we pointed out in Section 6.2 that it is necessary to express these angular velocities in the link-bound coordinate frames. Fortunately, the z-axes of all of these frames are in the same direction, so the above expression is also valid in the link-bound frame. Moreover, since $\boldsymbol{\omega}_i$ is aligned with $\mathbf{k}$, the triple product $\boldsymbol{\omega}_i^T I_i \boldsymbol{\omega}_i$ reduces simply to $(I_{33})_i$ times the square of the magnitude of the angular velocity. This quantity $(I_{33})_i$ is indeed what we have labeled as I_i above. Hence the rotational kinetic energy of the overall system is

$$\frac{1}{2}\dot{\mathbf{q}}^T \left\{ I_1 \begin{bmatrix} 1 & 0 \\ 0 & 0 \end{bmatrix} + I_2 \begin{bmatrix} 1 & 1 \\ 1 & 1 \end{bmatrix} \right\} \dot{\mathbf{q}} \tag{6.4.16}$$

Now we are ready to form the inertia matrix $D(\mathbf{q})$. For this purpose, we merely have to add the two matrices in (6.4.14) and (6.4.16), respectively. Thus

$$D(\mathbf{q}) = m_1 J_{\mathbf{v}_{c1}}^T J_{\mathbf{v}_{c1}} + m_2 J_{\mathbf{v}_{c2}}^T J_{\mathbf{v}_{c2}} + \begin{bmatrix} I_1 + I_2 & I_2 \\ I_2 & I_2 \end{bmatrix} \tag{6.4.17}$$

Carrying out the above multiplications and using the standard trigonometric identities $\cos^2\theta + \sin^2\theta = 1$, $\cos\alpha\cos\beta + \sin\alpha\sin\beta = \cos(\alpha-\beta)$ leads to

$$\begin{aligned} d_{11} &= m_1\ell_{c1}^2 + m_2(\ell_1^2 + \ell_{c2}^2 + 2\ell_1\ell_{c2}\cos q_2) + I_1 + I_2 \\ d_{12} &= d_{21} = m_2(\ell_{c2}^2 + \ell_1\ell_{c2}\cos q_2) + I_2 \\ d_{22} &= m_2\ell_{c2}^2 + I_2 \end{aligned} \tag{6.4.18}$$

Now we can compute the Christoffel symbols using the definition (6.3.9). This gives

$$\begin{aligned} c_{111} &= \frac{1}{2}\frac{\partial d_{11}}{\partial q_1} = 0 \\ c_{121} &= c_{211} = \frac{1}{2}\frac{\partial d_{11}}{\partial q_2} = -m_2\ell_1\ell_{c2}\sin q_2 =: h \\ c_{221} &= \frac{\partial d_{12}}{\partial q_2} - \frac{1}{2}\frac{\partial d_{22}}{\partial q_1} = h \\ c_{112} &= \frac{\partial d_{21}}{\partial q_1} - \frac{1}{2}\frac{\partial d_{11}}{\partial q_2} = -h \\ c_{122} &= c_{212} = \frac{1}{2}\frac{\partial d_{22}}{\partial q_1} = 0 \\ c_{222} &= \frac{1}{2}\frac{\partial d_{22}}{\partial q_2} = 0 \end{aligned} \tag{6.4.19}$$

Next, the potential energy of the manipulator is just the sum of those of the two links. For each link, the potential energy is just its mass multiplied by the gravitational acceleration and the height of its center of mass. Thus

$$V_1 = m_1 g \ell_{c1} \sin q_1$$

$$V_2 = m_2 g(\ell_1 \sin q_1 + \ell_{c2}\sin(q_1 + q_2))$$

$$V = V_1 + V_2 = (m_1\ell_{c1} + m_2\ell_1)g\sin q_1 + m_2\ell_{c2}g\sin(q_1 + q_2) \quad (6.4.20)$$

Hence, the functions ϕ_k defined in (6.3.10) become

$$\phi_1 = \frac{\partial V}{\partial q_1} = (m_1\ell_{c1} + m_2\ell_1)g\cos q_1 + m_2\ell_{c2}g\cos(q_1 + q_2) \quad (6.4.21)$$

$$\phi_2 = \frac{\partial V}{\partial q_2} = m_2\ell_{c2}\cos(q_1 + q_2) \quad (6.4.22)$$

Finally we can write down the dynamical equations of the system as in (6.3.11). Substituting for the various quantities in this equation and omitting zero terms leads to

$$d_{11}\ddot{q}_1 + d_{12}\ddot{q}_2 + c_{121}\dot{q}_1\dot{q}_2 + c_{211}\dot{q}_2\dot{q}_1 + c_{221}\dot{q}_2^2 + \phi_1 = \tau_1$$

$$d_{21}\ddot{q}_1 + d_{22}\ddot{q}_2 + c_{112}\dot{q}_1^2 + \phi_2 = \tau_2 \quad (6.4.23)$$

In this case the matrix $C(\mathbf{q}, \dot{\mathbf{q}})$ is given as

$$C = \begin{bmatrix} h\dot{q}_2 & h\dot{q}_2 + h\dot{q}_1 \\ -h\dot{q}_1 & 0 \end{bmatrix} \quad (6.4.24)$$

(*v*) *Example 6.4.3 Planar Elbow Manipulator with Remotely Driven Link*

Now we illustrate the use of Lagrangian equations in a situation where the generalized coordinates are not the joint variables defined in earlier chapters. Consider again the planar elbow manipulator, but suppose now that both joints are driven by motors mounted at the base. The first joint is turned directly by one of the motors, while the other is turned via a gearing mechanism or a timing belt (see Figure 6-4). In this case one should choose the generalized coordinates as shown in Figure 6-5, because the angle p_2 is determined by driving motor number 2, and is not affected by the angle p_1. We will derive the dynamical equations for this configuration, and show that some simplifications will result.

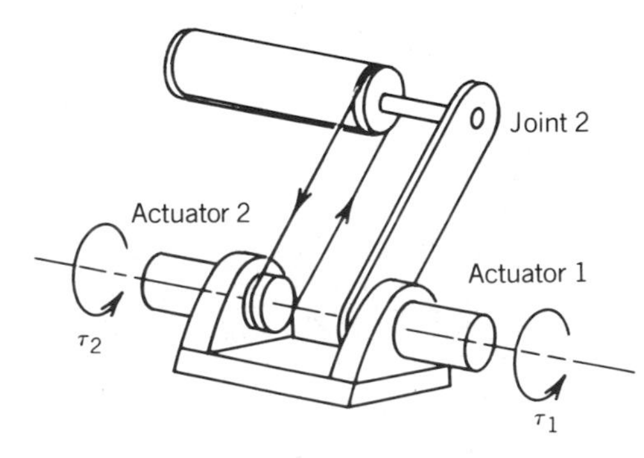

FIGURE 6-4
Two-link revolute joint arm with remotely driven link.

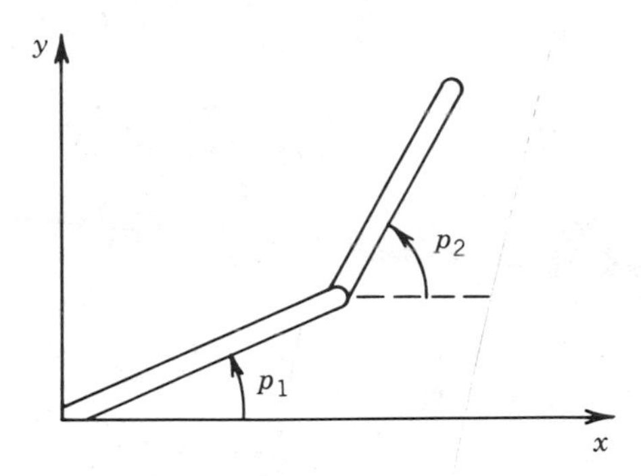

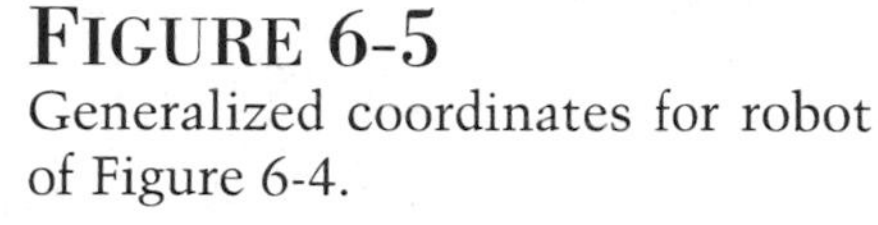

FIGURE 6-5
Generalized coordinates for robot of Figure 6-4.

Since p_1 and p_2 are not the joint angles used earlier, we cannot use the velocity Jacobians derived in Chapter Five in order to find the kinetic energy of each link. Instead, we have to carry out the analysis directly. It is easy to see that

$$\mathbf{v}_{c1} = \begin{bmatrix} -\ell_{c1}\sin p_1 \\ \ell_{c1}\cos p_1 \\ 0 \end{bmatrix} \dot{p}_1, \quad \mathbf{v}_{c2} = \begin{bmatrix} -\ell_1\sin p_1 & -\ell_{c2}\sin p_2 \\ \ell_1\cos p_1 & \ell_{c2}\cos p_2 \\ 0 & 0 \end{bmatrix} \begin{bmatrix} \dot{p}_1 \\ \dot{p}_2 \end{bmatrix} \tag{6.4.25}$$

$$\boldsymbol{\omega}_1 = \dot{p}_1\mathbf{k}, \quad \boldsymbol{\omega}_2 = \dot{p}_2\mathbf{k} \tag{6.4.26}$$

Hence the kinetic energy of the manipulator equals

$$K = \dot{\mathbf{p}}^T D(\mathbf{p})\dot{\mathbf{p}} \tag{6.4.27}$$

where

$$D(\mathbf{p}) = \begin{bmatrix} m_1\ell_{c1}^2 + m_2\ell_1^2 + I_1 & m_2\ell_1\ell_{c2}\cos(p_2-p_1) \\ m_2\ell_1\ell_{c2}\cos(p_2-p_1) & m_2\ell_{c2}^2 + I_2 \end{bmatrix} \tag{6.4.28}$$

Computing the Christoffel symbols as in (6.3.9) gives

$$\begin{aligned}
c_{111} &= \frac{1}{2}\frac{\partial d_{11}}{\partial p_1} = 0 \\
c_{121} = c_{211} &= \frac{1}{2}\frac{\partial d_{11}}{\partial p_2} = 0 \\
c_{221} &= \frac{\partial d_{12}}{\partial p_2} - \frac{1}{2}\frac{\partial d_{22}}{\partial p_1} = -m_2\ell_1\ell_{c2}\sin(p_2-p_1) \\
c_{112} &= \frac{\partial d_{21}}{\partial p_1} - \frac{1}{2}\frac{\partial d_{11}}{\partial p_2} = m_2\ell_1\ell_{c2}\sin(p_2-p_1) \\
c_{212} = c_{122} &= \frac{1}{2}\frac{\partial d_{22}}{\partial p_1} = 0
\end{aligned} \tag{6.4.29}$$

$$c_{222} = \frac{1}{2}\frac{\partial d_{22}}{\partial p_2} = 0$$

Next, the potential energy of the manipulator, in terms of p_1 and p_2, equals

$$V = m_1 g \ell_{c1} \sin p_1 + m_2 g(\ell_1 \sin p_1 + \ell_{c2} \sin p_2) \tag{6.4.30}$$

Hence

$$\phi_1 = (m_1 \ell_{c1} + m_2 \ell_1) g \cos p_1$$

$$\phi_2 = m_2 \ell_{c2} g \cos p_2$$

Finally, the dynamical equations are

$$d_{11}\ddot{p}_1 + d_{12}\ddot{p}_2 + c_{221}\dot{p}_2^2 + \phi_1 = \tau_1$$

$$d_{21}\ddot{p}_1 + d_{22}\ddot{p}_2 + c_{112}\dot{p}_1^2 + \phi_2 = \tau_2 \tag{6.4.31}$$

Comparing (6.4.31) and (6.4.23), we see that by driving the second joint remotely from the base we have eliminated the Coriolis forces, but we still have the centripetal forces coupling the two joints.

6.4.1 FIVE-BAR LINKAGE

Now consider the manipulator shown in Figure 6-6. We will show that, if the parameters of the manipulator satisfy a simple relationship, then the equations of the manipulator are decoupled, so that each quantity q_1 and q_2 can be controlled independently of the other. The mechanism in Figure 6-6 is called a **five-bar linkage**. Clearly there are only four bars in the figure, but in the theory of mechanisms it is a convention to count the ground as an additional linkage, which explains the terminology. In Figure 6-6, it is assumed that the lengths of links 1

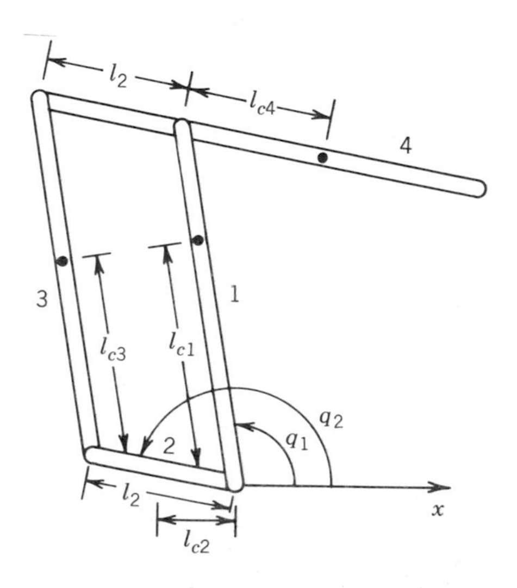

FIGURE 6-6
Five-bar linkage.

and 3 are the same, and that the two lengths marked ℓ_2 are the same; in this way the closed path in the figure is in fact a parallelogram, which greatly simplifies the computations. Notice, however, that the quantities ℓ_{c1} and ℓ_{c3} need not be equal. For example, even though links 1 and 3 have the same length, they need not have the same mass distribution.

It is clear from the figure that, even though there are four links being moved, there are in fact only two degrees-of-freedom, identified as q_1 and q_2. Thus, in contrast to the earlier mechanisms studied in this book, this one is a **closed** kinematic chain (though of a particularly simple kind). As a result, we cannot use the earlier results on Jacobian matrices, and instead have to start from scratch. As a first step we write down the coordinates of the centers of mass of the various links as a function of the generalized coordinates. This gives

$$\begin{bmatrix} x_{c1} \\ y_{c1} \end{bmatrix} = \begin{bmatrix} \ell_{c1}\cos q_1 \\ \ell_{c1}\sin q_1 \end{bmatrix} \tag{6.4.32}$$

$$\begin{bmatrix} x_{c2} \\ y_{c2} \end{bmatrix} = \begin{bmatrix} \ell_{c2}\cos q_2 \\ \ell_{c2}\sin q_2 \end{bmatrix} \tag{6.4.33}$$

$$\begin{bmatrix} x_{c3} \\ y_{c3} \end{bmatrix} = \begin{bmatrix} \ell_2\cos q_1 \\ \ell_2\sin q_2 \end{bmatrix} + \begin{bmatrix} \ell_{c3}\cos q_1 \\ \ell_{c3}\sin q_1 \end{bmatrix} \tag{6.4.34}$$

$$\begin{aligned} \begin{bmatrix} x_{c4} \\ y_{c4} \end{bmatrix} &= \begin{bmatrix} \ell_1\cos q_1 \\ \ell_1\sin q_1 \end{bmatrix} + \begin{bmatrix} \ell_{c4}\cos(q_2-\pi) \\ \ell_{c4}\sin(q_2-\pi) \end{bmatrix} \\ &= \begin{bmatrix} \ell_1\cos q_1 \\ \ell_1\sin q_1 \end{bmatrix} - \begin{bmatrix} \ell_{c4}\cos q_2 \\ \ell_{c4}\sin q_2 \end{bmatrix} \end{aligned} \tag{6.4.35}$$

Next, with the aid of these expressions, we can write down the velocities of the various centers of mass as a function of $\dot{q}_1$ and $\dot{q}_2$. For convenience we drop the third row of each of the following Jacobian matrices as it is always zero. The result is

$$\begin{aligned} \mathbf{v}_{c1} &= \begin{bmatrix} -\ell_{c1}\sin q_1 & 0 \\ \ell_{c1}\cos q_1 & 0 \end{bmatrix} \dot{\mathbf{q}} \\ \mathbf{v}_{c2} &= \begin{bmatrix} 0 & -\ell_{c2}\sin q_2 \\ 0 & \ell_{c2}\cos q_2 \end{bmatrix} \dot{\mathbf{q}} \\ \mathbf{v}_{c3} &= \begin{bmatrix} -\ell_{c3}\sin q_1 & -\ell_2\sin q_2 \\ \ell_{c3}\cos q_1 & \ell_2\cos q_2 \end{bmatrix} \dot{\mathbf{q}} \\ \mathbf{v}_{c4} &= \begin{bmatrix} -\ell_1\sin q_1 & \ell_{c4}\sin q_2 \\ \ell_1\cos q_1 & -\ell_{c4}\cos q_2 \end{bmatrix} \dot{\mathbf{q}} \end{aligned} \tag{6.4.36}$$

Let us define the velocity Jacobians $J_{\mathbf{v}_{ci}}$, $i = 1, \ldots, 4$ in the obvious fashion, that is, as the four matrices appearing in the above equations. Next, it is clear that the angular velocities of the four links are simply given by

$$\omega_1 = \omega_3 = \dot{q}_1\mathbf{k},\ \omega_2 = \omega_4 = \dot{q}_2\mathbf{k} \tag{6.4.37}$$

Thus the inertia matrix is given by

$$D(\mathbf{q}) = \sum_{i=1}^{4} m_i J_{vc}^T J_{vc} + \begin{bmatrix} I_1 + I_3 & 0 \\ 0 & I_2 + I_4 \end{bmatrix} \tag{6.4.38}$$

If we now substitute from (6.4.36) into the above equation and use the standard trigonometric identities, when the dust settles we are left with

$$\begin{aligned} d_{11}(\mathbf{q}) &= m_1\ell_{c1}^2 + m_3\ell_{c3}^2 + m_4\ell_1^2 + I_1 + I_3 \\ d_{12}(\mathbf{q}) &= d_{21}(\mathbf{q}) = (m_3\ell_2\ell_{c3} - m_4\ell_1\ell_{c4})\cos(q_2 - q_1) \\ d_{22}(\mathbf{q}) &= m_2\ell_{c2}^2 + m_3\ell_2^2 + m_4\ell_{c4}^2 + I_2 + I_4 \end{aligned} \tag{6.4.39}$$

Now the thing to note is that if

$$m_3\ell_2\ell_{c3} = m_4\ell_1\ell_{c4} \tag{6.4.40}$$

then the inertia matrix is diagonal and constant, and as a consequence the dynamical equations will contain neither coriolis nor centripetal terms.

Turning now to the potential energy, we have that

$$\begin{aligned} V &= g\sum_{i=1}^{4} \mathbf{y}_{ci} \\ &= g\sin q_1(m_1\ell_{c1} + m_3\ell_{c3} + m_4\ell_1) \\ &\quad + g\sin q_2(m_2\ell_{c2} + m_3\ell_2 - m_4\ell_{c4}) \end{aligned} \tag{6.4.41}$$

Hence

$$\begin{aligned} \phi_1 &= g\cos q_1(m_1\ell_{c1} + m_3\ell_{c3} + m_4\ell_1) \\ \phi_2 &= g\cos q_2(m_2\ell_{c2} + m_3\ell_2 - m_4\ell_{c4}) \end{aligned} \tag{6.4.42}$$

Notice that ϕ_1 depends only on q_1 but not on q_2, and similarly that ϕ_2 depends only on q_2 but not on q_1. Hence, if the relationship (6.4.40) is satisfied, then the rather complex-looking manipulator in Figure 6-6 is described by the **decoupled** set of equations

$$d_{11}\ddot{q}_1 + \phi_1(q_1) = \tau_1,\quad d_{22}\ddot{q}_2 + \phi_2(q_2) = \tau_2 \tag{6.4.43}$$

This discussion helps to explain the increasing popularity of the parallelogram configuration in industrial robots (e.g., P-50). If the relationship (6.4.40) is satisfied, then one can adjust the two angles q_1 and q_2 independently, without worrying about interactions between the two angles. Compare this with the situation in the case of the planar elbow manipulators discussed earlier in this section.

6.5 NEWTON–EULER FORMULATION

In this section, we present a method for analyzing the dynamics of robot manipulators known as the **Newton–Euler formulation.** This method leads to exactly the same final answers as the Lagrangian formulation presented in earlier sections, but the route taken is quite different. In particular, in the Lagrangian formulation we treat the manipulator as a whole and perform the analysis using a Lagrangian function (the difference between the kinetic energy and the potential energy). In contrast, in the Newton–Euler formulation we treat each link of the robot in turn, and write down the equations describing its linear motion and its angular motion. Of course, since each link is coupled to other links, these equations that describe each link contain coupling forces and torques that appear also in the equations that describe neighboring links. By doing a so-called forward–backward recursion, we are able to determine all of these coupling terms and eventually to arrive at a description of the manipulator as a whole. Thus we see that the philosophy of the Newton–Euler formulation is quite different from that of the Lagrangian formulation.

At this stage the reader can justly ask whether there is a need for another formulation, and the answer is not clear. Historically, both formulations were evolved in parallel, and each was perceived as having certain advantages. For instance, it was believed at one time that the Newton–Euler formulation is better suited to **recursive** computation than the Lagrangian formulation. However, the current situation is that both of the formulations are equivalent in almost all respects. Thus at present the main reason for having another method of analysis at our disposal is that it might provide different insights.

In any mechanical system one can identify a set of generalized coordinates (which we introduced in Section 6.1 and labeled $\mathbf{q}$) and corresponding generalized forces (also introduced in Section 6.1 and labeled $\boldsymbol{\tau}$). Analyzing the dynamics of a system means finding the relationship between $\mathbf{q}$ and $\boldsymbol{\tau}$. At this stage we must distinguish between two aspects: First, we might be interested in obtaining **closed-form equations** that describe the time evolution of the generalized coordinates, such as (6.4.24) for example. Second, we might be interested in knowing what generalized forces need to be applied in order to realize a *particular* time evolution of the generalized coordinates. The distinction is that in the latter case we only want to know what time dependent function $\boldsymbol{\tau}(\cdot)$ produces a particular trajectory $\mathbf{q}(\cdot)$ and may not care to know the general functional relationship between the two. It is perhaps fair to say that in the former type of analysis, the Lagrangian formulation is superior while in the latter case the Newton–Euler formulation is superior. Looking ahead to topics beyond the scope of the book, if one wishes to study more advanced mechanical phenomena such as elastic deformations of the links (i.e., if one no longer assumes rigidity of the links), then the Lagrangian formulation is clearly superior.

In this section we present the general equations that describe the Newton–Euler formulation. In the next section we illustrate the method by applying it to the planar elbow manipulator studied in Section 6.4 and show that the resulting equations are the same as (6.4.23).

The facts of Newtonian mechanics that are pertinent to the present discussion can be stated as follows:

1. Every action has an equal and opposite reaction. Thus, if body 1 applies a force $\mathbf{f}$ and torque $\boldsymbol{\tau}$ to body 2, then body 2 applies a force of $-\mathbf{f}$ and torque of $-\boldsymbol{\tau}$ to body 1.
2. The rate of change of the linear momentum equals the total force applied to the body.
3. The rate of change of the angular momentum equals the total torque applied to the body.

Applying the second fact to the linear motion of a body yields the relationship

$$\frac{d(m\mathbf{v})}{dt} = \mathbf{f} \tag{6.5.1}$$

where m is the mass of the body, $\mathbf{v}$ is the velocity of the center of mass with respect to an inertial frame, and $\mathbf{f}$ is the sum of external forces applied to the body. Since in robotic applications the mass is constant as a function of time, (6.5.1) can be simplified to the familiar relationship

$$m\mathbf{a} = \mathbf{f} \tag{6.5.2}$$

where $\mathbf{a} = \dot{\mathbf{v}}$ is the acceleration of the center of mass.

Applying the third fact to the angular motion of a body gives

$$\frac{d(I_0\boldsymbol{\omega}_0)}{dt} = \boldsymbol{\tau}_0 \tag{6.5.3}$$

where I_0 is the moment of inertia of the body about an inertial frame whose origin is at the center of mass, $\boldsymbol{\omega}_0$ is the angular velocity of the body, and $\boldsymbol{\tau}_0$ is the sum of torques applied to the body. Now there is an essential difference between linear motion and angular motion. Whereas the mass of a body is constant in most applications, its moment of inertia with respect an inertial frame may or may not be constant. To see this, suppose we attach a frame rigidly to the body, and let I denote the inertia matrix of the body with respect to this frame. Then I remains the same irrespective of whatever motion the body executes. However, the matrix I_0 is given by

$$I_0 = RIR^T \tag{6.5.4}$$

where R is the rotation matrix that transforms coordinates from the body attached frame to the inertial frame. Thus there is no reason to expect that I_0 is constant as a function of time.

One possible way of overcoming this difficulty is to write the angular motion equation in terms of a frame rigidly attached to the body. This leads to

$$I\dot{\boldsymbol{\omega}} + \boldsymbol{\omega}\times(I\boldsymbol{\omega}) = \boldsymbol{\tau} \tag{6.5.5}$$

where I is the (constant) inertia matrix of the body with respect to the body attached frame, $\boldsymbol{\omega}$ is the angular velocity, but expressed in the body attached frame, and $\boldsymbol{\tau}$ is the total torque on the body, again expressed in the body attached frame. Let us now give a derivation of (6.5.5) to demonstrate clearly where the term $\boldsymbol{\omega}\times(I\boldsymbol{\omega})$ comes from; note that this term is called the **gyroscopic term.**

Let R denote the orientation of the frame rigidly attached to the body w.r.t. the inertial frame; note that it could be a function of time. Then (6.5.4) gives the relation between I and I_0. Now by the definition of the angular velocity given in Section 2.4, we know that

$$\dot{R}R^T = S(\boldsymbol{\omega}_0) \tag{6.5.6}$$

In other words, the angular velocity of the body, *expressed in an inertial frame*, is given by (6.5.6). Of course, the same vector, expressed in the body attached frame, is given by

$$\boldsymbol{\omega}_0 = R\boldsymbol{\omega}, \; \boldsymbol{\omega} = R^T\boldsymbol{\omega}_0 \tag{6.5.7}$$

Hence the angular momentum, expressed in the inertial frame, is

$$\mathbf{h} = RIR^TR\boldsymbol{\omega} = RI\boldsymbol{\omega} \tag{6.5.8}$$

Differentiating and noting that I is constant gives an expression for the rate of change of the angular momentum, expressed as a vector in the inertial frame:

$$\dot{\mathbf{h}} = \dot{R}I\boldsymbol{\omega} + RI\dot{\boldsymbol{\omega}} \tag{6.5.9}$$

Now

$$S(\boldsymbol{\omega}_0) = \dot{R}R^T, \quad \dot{R} = S(\boldsymbol{\omega}_0)R \tag{6.5.10}$$

Hence, with respect to the inertial frame,

$$\dot{\mathbf{h}} = S(\boldsymbol{\omega}_0)RI\boldsymbol{\omega} + RI\dot{\boldsymbol{\omega}} \tag{6.5.11}$$

With respect to the frame rigidly attached to the body, the rate of change of the angular momentum is

$$\begin{aligned} R^T\dot{\mathbf{h}} &= R^TS(\boldsymbol{\omega}_0)RI\boldsymbol{\omega} + I\dot{\boldsymbol{\omega}} \\ &= S(R^T\boldsymbol{\omega}_0)I\boldsymbol{\omega} + I\dot{\boldsymbol{\omega}} \\ &= S(\boldsymbol{\omega})I\boldsymbol{\omega} + I\dot{\boldsymbol{\omega}} = \boldsymbol{\omega}\times(I\boldsymbol{\omega}) + I\dot{\boldsymbol{\omega}} \end{aligned} \tag{6.5.12}$$

This establishes (6.5.5). Of course we can, if we wish, write the same equation in terms of vectors expressed in an inertial frame. But we will see shortly that there is an advantage to writing the force and moment equations with respect to a frame attached to link i, namely that a great many vectors in fact reduce to constant vectors, thus leading to significant simplifications in the equations.

Now we derive the Newton–Euler formulation of the equations of motion of an n-link manipulator. For this purpose, we first choose frames $0, \ldots, n$, where frame 0 is an inertial frame, and frame i is rigidly attached to link i for $i \geq 1$. We also introduce several vectors, *which are all expressed in frame* i. The first set of vectors pertain to the velocities and accelerations of various parts of the manipulator.

$\mathbf{a}_{c,i}$ = the acceleration of the center of mass of link i.

$\mathbf{a}_{e,i}$ = the acceleration of the end of link i (i.e. joint $i+1$).

$\boldsymbol{\omega}_i$ = the angular velocity of frame i w.r.t. frame 0.

$\boldsymbol{\alpha}_i$ = the angular acceleration of frame i w.r.t. frame 0.

The next several vectors pertain to forces and torques.

$\mathbf{g}_i$ = the acceleration due to gravity (expressed in frame i).

$\mathbf{f}_i$ = the force exerted by link $i-1$ on link i.

$\boldsymbol{\tau}_i$ = the torque exerted by link $i-1$ on link i.

R_i^{i+1} = the rotation matrix from frame $i+1$ to frame i.

The final set of vectors pertain to physical features of the manipulator. *Note that each of the following vectors is constant as a function of* $\mathbf{q}$. In other words, each of the vectors listed here is independent of the configuration of the manipulator.

m_i = the mass of link i.

I_i = the inertia matrix of link i about a frame parallel to frame i whose origin is at the center of mass of link i.

$\mathbf{r}_{i,ci}$ = the vector from joint i to the center of mass of link i.

$\mathbf{r}_{i+1,ci}$ = the vector from joint $i+1$ to the center of mass of link i.

$\mathbf{r}_{i,i+1}$ = the vector from joint i to joint $i+1$.

Now consider the free body diagram shown in Figure 6-7; this shows link i together with all forces and torques acting on it. Let us analyze each of the forces and torques shown in the figure. First, $\mathbf{f}_i$ is the force applied by link $i-1$ to link i. Next, by the law of action and reaction, link $i+1$ applies a force of $-\mathbf{f}_{i+1}$ to link i, but this vector is expressed in frame $i+1$ according to our convention. In order to express the same

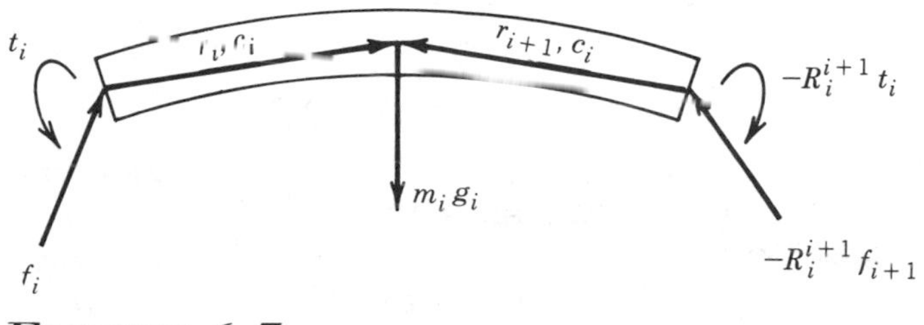

FIGURE 6-7
Forces and moments on link i.

vector in frame i, it is necessary to multiply it by the rotation matrix R_i^{i+1}. Similar explanations apply to the torques $\boldsymbol{\tau}_i$ and $-R_i^{i+1}\boldsymbol{\tau}_{i+1}$. The force $m_i\mathbf{g}_i$ is the gravitational force. Since all vectors in Figure 6-7 are expressed in frame i, the gravity vector $\mathbf{g}_i$ is in general a function of i.

Writing down the force balance equation for link i gives

$$\mathbf{f}_i - R_i^{i+1}\mathbf{f}_{i+1} + m_i\mathbf{g}_i = m_i\mathbf{a}_{c,i} \tag{6.5.13}$$

Next we write down the moment balance equation for link i. For this purpose, it is important to note two things: First, the moment exerted by a force $\mathbf{f}$ about a point is given by $\mathbf{f}\times\mathbf{r}$, where $\mathbf{r}$ is the radial vector *from* the point where the force is applied *to* the point about which we are computing the moment. Second, in the moment equation below, the vector $m_i\mathbf{g}_i$ does not appear, since it is applied directly at the center of mass. Thus we have

$$\boldsymbol{\tau}_i - R_i^{i+1}\boldsymbol{\tau}_{i+1} + \mathbf{f}_i\times\mathbf{r}_{i,ci} - (R_i^{i+1}\mathbf{f}_{i+1})\times\mathbf{r}_{i+1,ci} = \boldsymbol{\alpha}_i + \boldsymbol{\omega}_i\times(I_i\boldsymbol{\omega}_i) \tag{6.5.14}$$

Now we present the heart of the Newton–Euler formulation, which consists of finding the vectors $\mathbf{f}_1, \cdots, \mathbf{f}_n$ and $\boldsymbol{\tau}_1, \cdots, \boldsymbol{\tau}_n$ corresponding to a given set of vectors $\mathbf{q}, \dot{\mathbf{q}}, \ddot{\mathbf{q}}$. In other words, we find the forces and torques in the manipulator that correspond to a given set of generalized coordinates and first two derivatives. This information can be used to perform either type of analysis, as described above. That is, we can either use the equations below to find the $\mathbf{f}$ and $\boldsymbol{\tau}$ corresponding to a **particular trajectory** $\mathbf{q}(\cdot)$, or else to obtain closed-form dynamical equations. The general idea is as follows: Given $\mathbf{q}, \dot{\mathbf{q}}, \ddot{\mathbf{q}}$, suppose we are somehow able to determine all of the velocities and accelerations of various parts of the manipulator, that is, all of the quantities $\mathbf{a}_{c,i}$, $\boldsymbol{\omega}_i$ and $\boldsymbol{\alpha}_i$. Then we can solve (6.5.13) and (6.5.14) recursively to find all the forces and torques, as follows: First, set $\mathbf{f}_{n+1} = 0$ and $\boldsymbol{\tau}_{n+1} = 0$. This expresses the fact that there is no link $n + 1$. Then we can solve (6.5.13) to obtain

$$\mathbf{f}_i = R_i^{i+1}\mathbf{f}_{i+1} + m_i\mathbf{a}_{c,i} - m_i\mathbf{g}_i \tag{6.5.15}$$

By successively substituting $i = n, n-1, \cdots, 1$ we find all forces. Similarly, we can solve (6.5.14) to obtain

$$\boldsymbol{\tau}_i = R_i^{i+1} \boldsymbol{\tau}_{i+1} - \mathbf{f}_i \times \mathbf{r}_{i,ci} + (R_i^{i+1} \mathbf{f}_{i+1}) \times \mathbf{r}_{i+1,ci} + \boldsymbol{\alpha}_i + \boldsymbol{\omega}_i \times (I_i \boldsymbol{\omega}_i) \qquad (6.5.16)$$

By successively substituting $i = n, n-1, \ldots, 1$ we find all torques. Note that the above iteration is running in the direction of decreasing i.

Thus the solution is complete once we find an easily computed relation between $\mathbf{q}$, $\dot{\mathbf{q}}$, $\ddot{\mathbf{q}}$ and $\mathbf{a}_{c,i}$, $\boldsymbol{\omega}_i$ and $\boldsymbol{\alpha}_i$. This can be obtained by a recursive procedure in the direction of *increasing* i. This procedure is given below, for the case of revolute joints; the corresponding relationships for prismatic joints are actually easier to derive.

In order to distinguish between quantities expressed with respect to frame i and the base frame, we use a superscript (0) to denote the latter. Thus, for example, $\boldsymbol{\omega}_i$ denotes the angular velocity of frame i expressed in frame i, while $\boldsymbol{\omega}_i^{(0)}$ denotes the same quantity expressed in an inertial frame.

Now from Section 2.4 we have that

$$\boldsymbol{\omega}_i^{(0)} = \boldsymbol{\omega}_{i-1}^{(0)} + \mathbf{z}_{i-1} \dot{q}_i \qquad (6.5.17)$$

This merely expresses the fact that the angular velocity of frame i equals that of frame $i-1$ plus the added rotation from joint i. To get a relation between $\boldsymbol{\omega}_i$ and $\boldsymbol{\omega}_{i-1}$, we need only express the above equation in frame i rather than the base frame, taking care to account for the fact that $\boldsymbol{\omega}_i$ and $\boldsymbol{\omega}_{i-1}$ are expressed in different frames. This leads to

$$\boldsymbol{\omega}_i = (R_{i-1}^{i})^T \boldsymbol{\omega}_{i-1} + \mathbf{b}_i \dot{q}_i \qquad (6.5.18)$$

where

$$\mathbf{b}_i = (R_0^{i})^T \mathbf{z}_{i-1} \qquad (6.5.19)$$

is the axis of rotation of joint i expressed in frame i.

Next let us work on the angular acceleration $\boldsymbol{\alpha}_i$. It is vitally important to note here that

$$\boldsymbol{\alpha}_i = (R_0^{i})^T \dot{\boldsymbol{\omega}}_i^{(0)} \qquad (6.5.20)$$

In other words, $\boldsymbol{\alpha}_i$ is the derivative of the angular velocity of frame i, but expressed in frame i. It is *not* true that $\boldsymbol{\alpha}_i = \dot{\boldsymbol{\omega}}_i$! We will encounter a similar situation with the velocity and acceleration of the center of mass. Now we see directly from (6.5.17) that

$$\dot{\boldsymbol{\omega}}_i^{(0)} = \dot{\boldsymbol{\omega}}_{i-1}^{(0)} + \mathbf{z}_{i-1} \ddot{q}_i + \boldsymbol{\omega}_i^{(0)} \times \mathbf{z}_{i-1} \dot{q}_i \qquad (6.5.21)$$

Expressing the same equation in frame i gives

$$\boldsymbol{\alpha}_i = (R_{i-1}^{i})^T \boldsymbol{\alpha}_{i-1} + \mathbf{b}_i \ddot{q}_i + \boldsymbol{\omega}_i \times \mathbf{b}_i \dot{q}_i \qquad (6.5.22)$$

Now we come to the linear velocity and acceleration terms. Note that, in contrast to the angular velocity, the linear velocity does not appear anywhere in the dynamic equations; however, an expression for the linear velocity is needed before we can derive an expression for the linear acceleration. From Section 2.5, we get that the velocity of the center of mass of link i is given by

$$\mathbf{v}_{c,i}^{(0)} = \mathbf{v}_{e,i-1}^{(0)} + \boldsymbol{\omega}_i^{(0)} \times \mathbf{r}_{i,ci}^{(0)} \tag{6.5.23}$$

To obtain an expression for the acceleration, we use (2.6.11), and note that the vector $\mathbf{r}_{i,ci}^{(0)}$ is constant in frame i. Thus

$$\mathbf{a}_{c,i}^{(0)} = \mathbf{a}_{e,i-1}^{(0)} + \dot{\boldsymbol{\omega}}_i^{(0)} \times \mathbf{r}_{i,ci}^{(0)} + \boldsymbol{\omega}_i^{(0)} \times (\boldsymbol{\omega}_i^{(0)} \times \mathbf{r}_{i,ci}^{(0)}) \tag{6.5.24}$$

Now

$$\mathbf{a}_{c,i} = (R_0^i)^T \mathbf{a}_{c,i}^{(0)} \tag{6.5.25}$$

Let us carry out the multiplication and use the familiar property

$$R(\mathbf{a} \times \mathbf{b}) = (R\,\mathbf{a}) \times (R\,\mathbf{b}) \tag{6.5.26}$$

We also have to account for the fact that $\mathbf{a}_{e,i-1}$ is expressed in frame $i-1$ and transform it to frame i. This gives

$$\mathbf{a}_{c,i} = (R_{i-1}^i)^T \mathbf{a}_{e,i-1} + \dot{\boldsymbol{\omega}}_i \times \mathbf{r}_{i,ci} + \boldsymbol{\omega}_i \times (\boldsymbol{\omega}_i \times \mathbf{r}_{i,ci}) \tag{6.5.27}$$

Now to find the acceleration of the end of link i, we can use (6.5.27) with $\mathbf{r}_{i,i+1}$ replacing $\mathbf{r}_{i,ci}$. Thus

$$\mathbf{a}_{e,i} = (R_{i-1}^i)^T \mathbf{a}_{e,i-1} + \dot{\boldsymbol{\omega}}_i \times \mathbf{r}_{i,i+1} + \boldsymbol{\omega}_i \times (\boldsymbol{\omega}_i \times \mathbf{r}_{i,i+1}) \tag{6.5.28}$$

Now the recursive formulation is complete. We can now state the Newton–Euler formulation as follows.

1. Start with the initial conditions

$$\boldsymbol{\omega}_0 = 0,\ \boldsymbol{\alpha}_0 = 0,\ \mathbf{a}_{c,0} = 0,\ \mathbf{a}_{e,0} = 0 \tag{6.5.29}$$

   and solve (6.5.18), (6.5.22), (6.5.28) and (6.5.27) (in that order!) to compute $\boldsymbol{\omega}_i$, $\boldsymbol{\alpha}_i$ and $\mathbf{a}_{c,i}$ for i increasing from 1 to n.

2. Start with the terminal conditions

$$\mathbf{f}_{n+1} = 0,\ \boldsymbol{\tau}_{n+1} = 0 \tag{6.5.30}$$

   and use (6.5.15) and (6.5.16) to compute $\mathbf{f}_i$ and $\boldsymbol{\tau}_i$ for i decreasing from n to 1.

6.6 PLANAR ELBOW MANIPULATOR REVISITED

In this section we apply the recursive Newton–Euler formulation derived in Section 6.5 to analyze the dynamics of the planar elbow manipulator of figure 6-3, and show that the Newton–Euler method leads to the same equations as the Lagrangian method, namely (6.4.23).

We begin with the forward recursion to express the various velocities and accelerations in terms of q_1, q_2 and their derivatives. Note that, in this simple case, it is quite easy to see that

$$\boldsymbol{\omega}_1 = \dot{q}_1\mathbf{k},\ \alpha_1 = \ddot{q}_1\mathbf{k},\ \boldsymbol{\omega}_2 = (\dot{q}_1 + \dot{q}_2)\mathbf{k},\ \alpha_2 = (\ddot{q}_1 + \ddot{q}_2)\mathbf{k} \tag{6.6.1}$$

so that there is no need to use (6.5.18) and (6.5.22). Also, the vectors that are independent of the configuration are as follows:

$$\mathbf{r}_{1,c1} = \ell_{c1}\mathbf{i},\ \mathbf{r}_{2,c1} = (\ell_1 - \ell_{c1})\mathbf{i},\ \mathbf{r}_{1,2} = \ell_1\mathbf{i} \tag{6.6.2}$$

$$\mathbf{r}_{2,c2} = \ell_{c2}\mathbf{i},\ \mathbf{r}_{3,c2} = (\ell_2 - \ell_{c2})\mathbf{i},\ \mathbf{r}_{2,3} = \ell_2\mathbf{i} \tag{6.6.3}$$

6.6.1 Forward Recursion: Link 1

Using (6.5.27) with $i = 1$ and noting that $\mathbf{a}_{e,0} = 0$ gives

$$\mathbf{a}_{c,1} = \ddot{q}_1\mathbf{k}\times\ell_{c1}\mathbf{i} + \dot{q}_1\mathbf{k}\times(\dot{q}_1\mathbf{k}\times\ell_{c1}\mathbf{i})$$

$$= \ell_{c1}\ddot{q}_1\mathbf{j} - \ell_{c1}\dot{q}_1^2\mathbf{i} = \begin{bmatrix} -\ell_{c1}\dot{q}_1^2 \\ \ell_{c1}\ddot{q}_1 \\ 0 \end{bmatrix} \tag{6.6.4}$$

Notice how simple this computation is when we do it with respect to frame 1. Compare with the same computation in frame 0! Finally, we have

$$\mathbf{g}_1 = -(R_0^1)^T g\mathbf{j} = g\begin{bmatrix} \sin q_1 \\ -\cos q_1 \\ 0 \end{bmatrix} \tag{6.6.5}$$

where g is the acceleration due to gravity. At this stage we can economize a bit by not displaying the third components of these accelerations, since they are obviously always zero. Similarly, the third component of all forces will be zero while the first two components of all torques will be zero. To complete the computations for link 1, we compute the acceleration of end of link 1. Clearly, this is obtained from (6.6.4) by replacing ℓ_{c1} by ℓ_1. Thus

$$\mathbf{a}_{e,1} = \begin{bmatrix} -\ell_1\dot{q}_1^2 \\ \ell_1\ddot{q}_1 \end{bmatrix} \tag{6.6.6}$$

6.6.2 Forward Recursion: Link 2

Once again we use (6.5.27) and substitute for $\boldsymbol{\omega}_2$ from (6.6.1); this yields

$$a_{c,2} = (R_1^2)^T\mathbf{a}_{e,1} + [(\ddot{q}_1 + \ddot{q}_2)\mathbf{k}]\times\ell_{c2}\mathbf{i} + (\dot{q}_1 + \dot{q}_2)\mathbf{k}\times[(\dot{q}_1 + \dot{q}_2)\mathbf{k}\times\ell_{c2}\mathbf{i}] \tag{6.6.7}$$

The only quantity in the above equation which is configuration-dependent is the first one. This can be computed as

$$(R_1^2)^T \mathbf{a}_{e,1} = \begin{bmatrix} \cos q_2 & \sin q_2 \\ -\sin q_2 & \cos q_2 \end{bmatrix} \begin{bmatrix} -\ell_1 \dot{q}_1^2 \\ \ell_1 \ddot{q}_1 \end{bmatrix}$$

$$= \begin{bmatrix} -\ell_1 \dot{q}_1^2 \cos q_2 + \ell_1 \ddot{q}_1 \sin q_2 \\ \ell_1 \dot{q}_1^2 \sin q_2 + \ell_1 \ddot{q}_1 \cos q_2 \end{bmatrix} \tag{6.6.8}$$

Substituting into (6.6.7) gives

$$\mathbf{a}_{c,2} = \begin{bmatrix} -\ell_1 \dot{q}_1^2 \cos q_2 + \ell_1 \ddot{q}_1 \sin q_2 - \ell_{c2}(\dot{q}_1 + \dot{q}_2)^2 \\ \ell_1 \dot{q}_1^2 \sin q_2 + \ell_1 \ddot{q}_1 \cos q_2 + \ell_{c2}(\ddot{q}_1 + \ddot{q}_2) \end{bmatrix} \tag{6.6.9}$$

The gravitational vector is

$$\mathbf{g}_2 = g \begin{bmatrix} \sin(q_1 + q_2) \\ -\cos(q_1 + q_2) \end{bmatrix} \tag{6.6.10}$$

Since there are only two links, there is no need to compute $\mathbf{a}_{e,2}$. Hence the forward recursions are complete at this point.

6.6.3 BACKWARD RECURSION: LINK 2

Now we carry out the backward recursion to compute the forces and joint torques. Note that, in this instance, the joint torques are the externally applied quantities, and our ultimate objective is to derive dynamical equations involving the joint torques. First we apply (6.5.15) with $i = 2$ and note that $\mathbf{f}_3 = 0$. This results in

$$\mathbf{f}_2 = m_2 \mathbf{a}_{c,2} - m_2 \mathbf{g}_2 \tag{6.6.11}$$

$$\boldsymbol{\tau}_2 = I_2 \boldsymbol{\alpha}_2 + \boldsymbol{\omega}_2 \times (I_2 \boldsymbol{\omega}_2) - \mathbf{f}_2 \times \ell_{c2} \mathbf{i} \tag{6.6.12}$$

Now we can substitute for $\boldsymbol{\omega}_2, \alpha_2$ from (6.6.1), and for $\mathbf{a}_{c,2}$ from (6.6.9). We also note that the gyroscopic term equals zero, since both $\boldsymbol{\omega}_2$ and $I_2 \boldsymbol{\omega}_2$ are aligned with $\mathbf{k}$. Now the cross product $\mathbf{f}_2 \times \ell_{c2} \mathbf{i}$ is clearly aligned with $\mathbf{k}$ and its magnitude is just the second component of $\mathbf{f}_2$. The final result is

$$\boldsymbol{\tau}_2 = I_2(\ddot{q}_1 + \ddot{q}_2)\mathbf{k} + [m_2 \ell_1 \ell_{c2} \sin q_2 \dot{q}_1^2 + m_2 \ell_1 \ell_{c2} \cos q_2 \ddot{q}_1$$
$$+ m_2 \ell_{c2}^2 (\ddot{q}_1 + \ddot{q}_2) + m_2 \ell_{c2} g \cos(q_1 + q_2)]\mathbf{k} \tag{6.6.13}$$

Since $\boldsymbol{\tau}_2 = \tau_2 \mathbf{k}$, we see that the above equation is the same as the second equation in (6.4.24).

6.6.4 BACKWARD RECURSION: LINK 1

To complete the derivation, we apply (6.5.15) and (6.5.16) with $i = 1$. First, the force equation is

$$\mathbf{f}_1 = m_1 \mathbf{a}_{c,1} + R_1^2 \mathbf{f}_2 - m_1 \mathbf{g}_1 \tag{6.6.14}$$

and the torque equation is

$$\boldsymbol{\tau}_1 = R_1^2\boldsymbol{\tau}_2 - \mathbf{f}_1\times\ell_{c1}\mathbf{i} - (R_1^2\mathbf{f}_2)\times(\ell_1 - \ell_{c1})\mathbf{i} + I_1\alpha_1 + \boldsymbol{\omega}_1\times(I_1\boldsymbol{\omega}_1) \quad (6.6.15)$$

Now we can simplify things a bit. First, $R_1^2\boldsymbol{\tau}_2 = \boldsymbol{\tau}_2$, since the rotation matrix does not affect the third components of vectors. Second, the gyroscopic term is once again equal to zero. Finally, when we substitute for $\mathbf{f}_1$ from (6.6.14) into (6.6.15), a little algebra gives

$$\boldsymbol{\tau}_1 = \boldsymbol{\tau}_2 - m_1\mathbf{a}_{c,1}\times\ell_{c1}\mathbf{i} + m_1\mathbf{g}_1\times\ell_{c1}\mathbf{i} - (R_1^2\mathbf{f}_2)\times\ell_1\mathbf{i} + I_1\alpha_1 \quad (6.6.16)$$

Once again, all cross products are quite straightforward, and the only difficult calculation is that of $R_1^2\mathbf{f}_2$. The final result is:

$$\boldsymbol{\tau}_1 = \boldsymbol{\tau}_2 + m_1\ell_{c1}^2 + m_1\ell_{c1}g\cos q_1 + m_2\ell_1 g\cos q_1 + I_1\ddot{q}_1$$
$$+ m_2\ell_1^2\ddot{q}_1 - m_1\ell_1\ell_{c2}(\dot{q}_1 + \dot{q}_2)^2\sin q_2 + m_2\ell_1\ell_{c2}(\ddot{q}_1 + \ddot{q}_2)\cos q_2 \quad (6.6.17)$$

If we now substitute for $\boldsymbol{\tau}_1$ from (6.6.13) and collect terms, we will get the first equation in (6.4.24); the details are routine and are left to the reader.

REFERENCES AND SUGGESTED READING

[1] ARMSTRONG, W. W., "Recursive Solution to the Equations of Motion of an n-link Manipulator," *Fifth World Congress on Theory of Machines and Mechanisms*, Montreal, 1979.

[2] ASADA, H., and YOUCEF-TOUMI, K., "Analysis and Design of A Direct-Drive Arm With a Five-Bar Link Parallel Drive Mechanism," *ASME J. Dyn. Syst. Meas. Contr.*, Vol. 106, No.3, 1984.

[3] BEJCZY, A.K.,"Robot Arm Dynamics and Control,", *Jet Propulsion Laboratory*, Pasadena, CA, TM 33-69, 1974.

[4] GOLDSTEIN, H., *Classical Mechanics*, Addison–Wesley, Reading, MA, 1981.

[5] HEMAMI, H., JASWA, V.C., and MCGHEE, R.B., "Some Alternative Formulations of Manipulator Dynamics for Computer Simulation Studies," *Allerton Conf. on Communication, Control and Computing*, Monticello, IL, Oct., 1975.

[6] HOLLERBACH, J.M., "A Recursive Formulation of Lagrangian Manipulator Dynamics," *IEEE Trans. Sys., Man, and Cyber.*, Vol. 10, No. 11, 1980.

[7] HOLLERBACH, J.M., "Wrist-Partitioned Inverse Kinematic Accelerations and Manipulator Dynamics," *Int. J. Robotics Research*, Vol. 2, No. 4, 1983.

[8] KANADE, T., KHOSLA, P.K., and TANAKA, N., "Real-Time Control of the CMU Direct-Drive Arm II Using Customized Inverse Dynamics," *IEEE Conf. on Decision and Control*, Las Vegas, Dec. 1984.

[9] LEE, C.S.G., "Robot Arm Kinematics, Dynamics, and Control," *Computer*, Vol. 15, No. 12, 1982.

[10] LUH, J.Y.S., WALKER, M.H., and PAUL, R.P., "On-line Computational Scheme for Mechanical Manipulators," *ASME J. Dyn. Sys. Meas. and Contr.*, Vol. 102, 1980.

[11] RAIBERT, M.H., "Analytical Equations vs. Table Look-up for Manipulation: A Unifying Concept," *IEEE Conf. on Decision and Control*, New Orleans, Dec. 1977.

[12] RAIBERT, M.H., and HORN, B.K.P., "Manipulator Control Using the Configuration Space Method," *Industrial Robot*, Vol. 5, No. 2, 1978.

[13] SCHEINMAN, V.C., "Design of a Computer Controlled Manipulator," *Stanford Artificial Intelligence Laboratory*, Vol. 92, 1969.

[14] SILVER, D.B., "On the Equivalence of Lagrangian and Newton–Euler Dynamics for Manipulators," *Int. J. Robotics Res.*, Vol. 1, No. 2, 1982.

[15] YOUCEF-TOUMI, K., and ASADA, H., "The Design of Arm Linkages with Decoupled and Configuration-Invariant Inertia Tensors," *ASME Winter Annual Meeting*, Miami, 1985.

PROBLEMS

6-1 Verify the expression (6.2.18).

6-2 Find the moments of inertia and cross products of inertia of a uniform rectangular solid of sides a, b, c with respect to a coordinate system with origin at the one corner and axes along the edges of the solid.

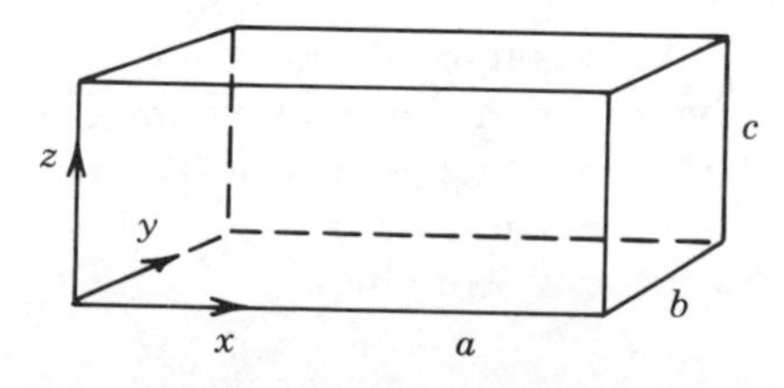

FIGURE 6-8
Diagram for Problem 6-2.

6-3 Given the cylindrical shell shown, show that

$$I_{xx} = \frac{1}{2}mr^2 + \frac{1}{12}m\ell^2$$

$$I_{x_1x_1} = \frac{1}{2}mr^2 + \frac{1}{3}m\ell^2$$

$$I_z = mr^2$$

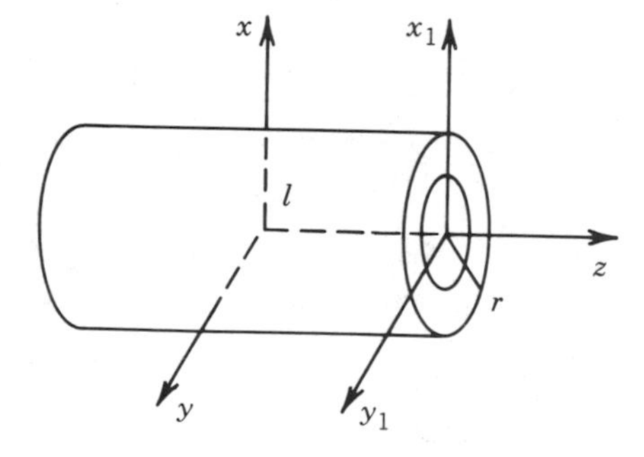

FIGURE 6-9
Diagram for Problem 6-3.

6-4 Given the inertia matrix $D(\mathbf{q})$ defined by (6.4.18) show that $\det D(\mathbf{q}) \neq 0$ for all $\mathbf{q}$.

6-5 Consider a rigid body undergoing a pure rotation with no external forces acting on it. The kinetic energy is then given as

$$K = \frac{1}{2}(I_{xx}\omega_x^2 + I_{yy}\omega_y^2 + I_{zz}\omega_z^2)$$

with respect to a coordinate located at the center of mass and whose coordinate axes are principal axes. Take as generalized coordinates the Euler angles ϕ,θ,ψ. Use the results of Problem 2-28 together with Lagrange's equations to show that the equations of motion of the rotating body are

$$I_{xx}\dot{\omega}_x + (I_{zz}-I_{yy})\omega_y\omega_z = 0$$

$$I_{yy}\dot{\omega}_y + (I_{xx}-I_{zz})\omega_z\omega_x = 0$$

$$I_{zz}\dot{\omega}_z + (I_{yy}-I_{xx})\omega_x\omega_y = 0$$

6-6 Many robots incorporate harmonic drives for speed reduction. A harmonic drive is a special type of gear mechanism. It is known that harmonic drives introduce torsional elasticity into the joints. To model this elastic behavior in the joints, consider the figure shown here. Using Lagrange's equations derive the equations of motion of this system. Assume that there is an input generalized force τ_m acting on the motor shaft.

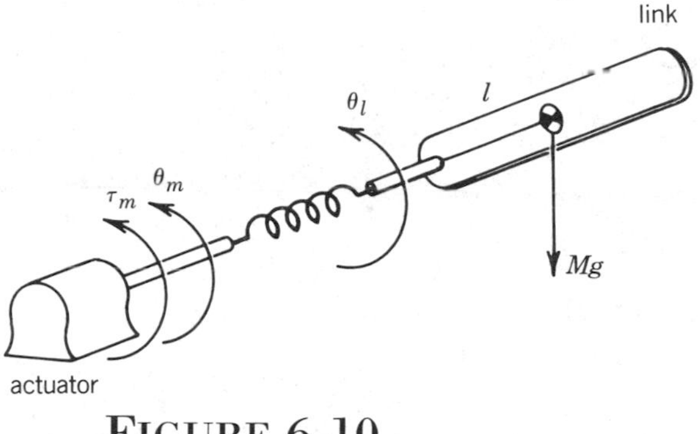

FIGURE 6-10
Diagram for problem 6-6.

6-7 Consider a 3-link cartesian manipulator,

a) Compute the inertia tensor J_i for each link i=1,2,3 assuming that the links are uniform rectangular solids of length 1, width $\frac{1}{4}$, and height $\frac{1}{4}$, and mass 1.

b) Compute the 3×3 inertia matrix $D(\mathbf{q})$ for this manipulator.

c) Show that the Christoffel symbols c_{ijk} are all zero for this robot. Interpret the meaning of this for the dynamic equations of motion.

d) Derive the equations of motion in matrix form:

$$D(\mathbf{q})\ddot{\mathbf{q}} + C(\mathbf{q},\dot{\mathbf{q}})\dot{\mathbf{q}} + \mathbf{g}(\mathbf{q}) = \mathbf{u}$$

6-8 Recall for a particle with kinetic energy $K = \frac{1}{2}m\dot{x}^2$, the **momentum** is defined as

$$p = m\dot{x} = \frac{dK}{d\dot{x}}$$

Therefore for a mechanical system with generalized coordinates $q_1, \ldots, q_n$ we define the **generalized momentum** p_k as

$$p_k = \frac{\partial L}{\partial \dot{q}_k}$$

where L is the Lagrangian of the system. With $K = \frac{1}{2}\dot{\mathbf{q}}^T D(\mathbf{q})\dot{\mathbf{q}}$, and $L = K - V$, prove that

$$\sum_{k=1}^{n} \dot{q}_k p_k = 2K$$

6-9 There is another formulation of the equations of motion of a mechanical system that is useful, the so-called **Hamiltonian** formulation: Define the Hamiltonian function H by

$$H = \sum_{k-1}^{n} \dot{q}_k p_k - L$$

a) Show that $H = K + V$.

b) Using Lagrange's equations, derive Hamilton's equations

$$\dot{q}_k = \frac{\partial H}{\partial p_k}$$

$$\dot{p}_k = -\frac{\partial H}{\partial q_k} + \tau_k$$

where τ_k is the input generalized force.

c) For two-link manipulator of Figure 6-3 compute Hamiltonian equations in matrix form. Note that Hamilton's equations are a system of first order differential equations as opposed to second order system given by Lagrange's equations.

6-10 Given the Hamiltonian H for a rigid robot, show that

$$\frac{dH}{dt} = \dot{\mathbf{q}}^T \tau$$

where τ is the external force applied at the joints. [Hint: Use Theorem 6.3.1.] What are the units of $\frac{dH}{dt}$?

CHAPTER SEVEN

INDEPENDENT JOINT CONTROL

7.1 INTRODUCTION

The control problem for robot manipulators is the problem of determining the time history of joint inputs required to cause the end-effector to execute a commanded motion. The joint inputs may be joint forces and torques, or they may be inputs to the actuators, for example, voltage inputs to the motors, depending on the model used for controller design. The commanded motion is typically specified either as a sequence of end-effector positions and orientations, or as a continuous path.

There are many control techniques and methodologies that can be applied to the control of manipulators. The particular control method chosen as well as the manner in which it is implemented can have a significant impact on the performance of the manipulator and consequently on the range of its possible applications. For example, continuous path tracking requires a different implementation in terms of the hardware and software of the computer interface than does point-to-point control.

In addition, the mechanical design of the manipulator itself will influence the type of control scheme needed. For example, the control problems encountered with a cartesian manipulator are fundamentally different from those encountered with an elbow type manipulator.

This creates a so-called **hardware/software trade-off** between the mechanical structure of the system and the architecture/programming of the controller.

Technological improvements are continually being made in the mechanical design of robots[13], which in turn improves their performance potential and broadens their range of applications. Realizing this increased performance, however, requires more sophisticated approaches to control. One can draw an analogy to the aerospace industry. Early aircraft were relatively easy to fly but possessed limited performance capabilities. As performance increased with technological advances so did the problems of control to the extent that the latest vehicles, such as the space shuttle or forward swept wing fighter aircraft, cannot be flown without sophisticated computer control.

As an illustration of the effect of the mechanical design on the control problem, compare a robot actuated by permanent magnet DC motors with gear reduction to a direct-drive robot using high-torque motors with no gear reduction. In the first case, the motor dynamics are linear and well understood and the effect of the gear reduction is largely to decouple the system by reducing the nonlinear coupling among the joints. However, the presence of the gears introduces friction, drive train compliance and backlash. In order to achieve high performance, the control designer would likely have to pay more attention to these latter effects than to the nonlinear inertia, coriolis forces, etc.

In the case of a direct-drive robot, the problems of backlash, friction, and compliance due to the gears are eliminated. However, the nonlinear coupling among the links is now significant, and the dynamics of the motors themselves may be much more complex. The result is that in order to achieve high performance from this type of manipulator a different set of control problems must be addressed.

We begin our discussion of control by considering the simplest type of control strategy, namely, independent joint control. In this type of control each axis of the manipulator is controlled as a single-input/single-output system. Any coupling effects due to the motion of the other links is either ignored or treated as a disturbance. We assume, in this chapter, that the reader has had an introduction to the theory of feedback control systems up to the level of say, Kuo [10], with some exposure to the state space theory of linear systems.

The basic structure of a single-input/single-output feedback control system is shown in Figure 7-1. The design objective is to choose the compensator in such a way that the plant output "tracks" or follows a desired output, given by the reference signal. The control signal, however, is not the only input acting on the system. Disturbances, which are really inputs that we do not control, also influence the behavior of the output. Therefore, the controller must be designed, in addition, so that the effects of the disturbances on the plant output are

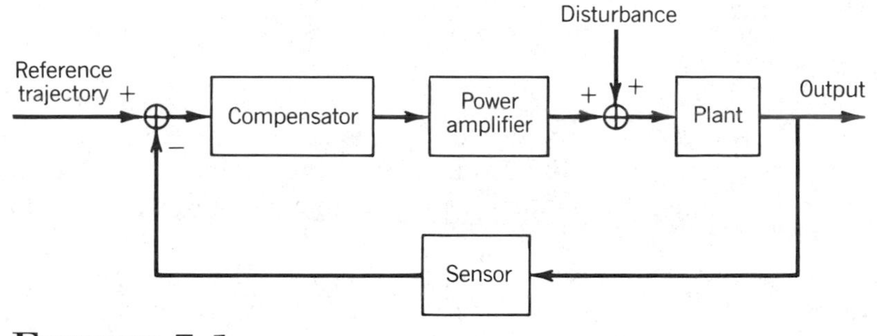

FIGURE 7-1
Basic structure of a feedback control system.

reduced. If this is accomplished, the plant is said to "reject" the disturbances. The twin objectives of **tracking** and **disturbance rejection** are central to any control methodology.

7.2 ACTUATOR DYNAMICS

In Chapter Six we obtained the following set of differential equations describing the motion of an n degree of freedom robot (cf. equation (6.3.12))

$$D(\mathbf{q})\ddot{\mathbf{q}} + C(\mathbf{q}, \dot{\mathbf{q}})\dot{\mathbf{q}} + \mathbf{g}(\mathbf{q}) = \tau \tag{7.2.1}$$

It is important to understand exactly what this equation represents. Equation (7.2.1) represents the dynamics of an interconnected chain of ideal rigid bodies, supposing that there is a generalized force τ acting at the joints. We can assume that the k-th component τ_k of the generalized force vector $\boldsymbol{\tau}$ is a torque about the axis z_{k-1} if joint k is revolute and is a force along z_{k-1} if joint k is prismatic. This generalized force is produced by an actuator, which may be electric, hydraulic or pneumatic. Although (7.2.1) is extremely complicated for all but the simplest manipulators, it nevertheless is an idealization, and there are a number of dynamic effects that are *not* included in (7.2.1). For example, friction at the joints is not accounted for in these equations and may be significant for some manipulators. Also, no physical body is completely rigid. A more detailed analysis of robot dynamics would include various sources of flexibility, such as elastic deformation of bearings and gears, deflection of the links under load, and vibrations. In this section we are interested mainly in the dynamics of the actuators producing the generalized force $\boldsymbol{\tau}$. We treat only the dynamics of permanent magnet DC-motors, as these are increasingly common for use in present-day robots.

A DC-motor basically works on the principle that a current carrying conductor in a magnetic field experiences a force $\mathbf{F} = \boldsymbol{\phi} \times \mathbf{i}$, where $\boldsymbol{\phi}$ is the magnetic flux and $\mathbf{i}$ is the current in the conductor. The motor itself consists of a fixed **stator** and a movable **rotor** that rotates inside the stator, as shown in Figure 7-2. If the stator produces a radial magnetic flux $\boldsymbol{\phi}$ and the current in the rotor (also called the **armature**) is $\mathbf{i}$ then there will be a torque on the rotor causing it to rotate. The magnitude of this torque is

$$\tau_m = K_1 \phi i_a \tag{7.2.2}$$

where τ_m is the motor torque (N-m), ϕ is the magnetic flux (webers), i_a is the armature current (amperes), and K_1 is a physical constant. In addition, whenever a conductor moves in a magnetic field, a voltage V_b is generated across its terminals that is proportional to the velocity of the conductor in the field. This voltage, called the **back emf**, will tend to oppose the current flow in the conductor.

Thus, in addition to the torque τ_m in (7.2.2), we have the back emf relation

$$V_b = K_2 \phi \omega_m \tag{7.2.3}$$

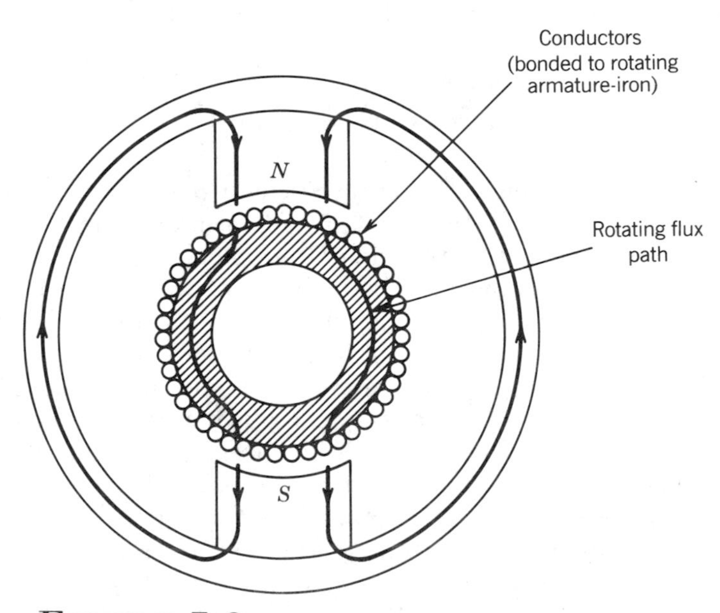

FIGURE 7-2
Cross-sectional view of a surface-wound permanent magnet DC motor. (Source: *Automatic Control Systems, 4th Ed.*, by Benjamin C. Kuo. Copyright 1982, Prentice–Hall, Inc.)

where V_b denotes the back emf in Volts, ω_m is the angular velocity of the rotor (rad/sec), and K_2 is a proportionality constant.

DC-motors can be classified according to the way in which the magnetic field is produced and the armature is designed. Here we discuss only the so-called **permanent magnet** motors whose stator consists of a permanent magnet. In this case we can take the flux ϕ to be a constant. The torque on the rotor is then controlled by controlling the armature current i_a.

Consider the schematic diagram of Figure 7-3 where

$$\begin{aligned} V(t) &= \text{armature voltage} \\ L &= \text{armature inductance} \\ R &= \text{armature resistance} \\ V_b &= \text{back emf} \\ i_a &= \text{armature current} \\ \theta_m &= \text{rotor position (radians)} \\ \tau_m &= \text{generated torque} \\ \tau_\ell &= \text{load torque} \\ \phi &= \text{magnetic flux due to stator.} \end{aligned}$$

The differential equation for the armature current is then

$$L\frac{di_a}{dt} + Ri_a = V - V_b \tag{7.2.4}$$

Since the flux ϕ is constant the torque developed by the motor is

$$\tau_m = K_1\phi i_a = K_m i_a \tag{7.2.5}$$

where K_m is the torque constant in N-m/amp. From (7.2.3) we have

$$V_b = K_2\phi\omega_m = K_b\omega_m = K_b\frac{d\theta_m}{dt} \tag{7.2.6}$$

where K_b is the back emf constant.

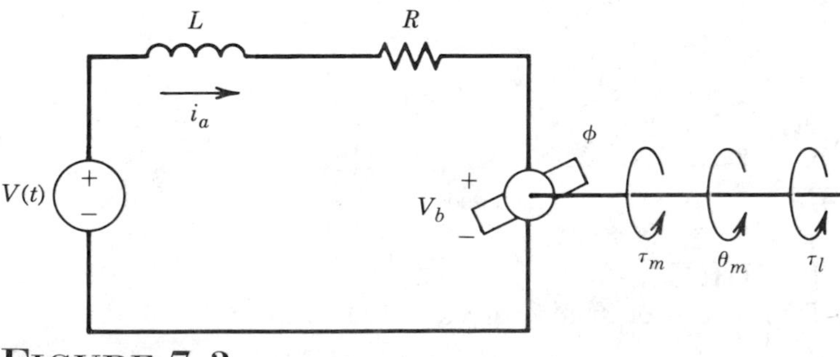

FIGURE 7-3
Circuit diagram for armature controlled DC motor.

We can determine the torque constant of the DC motor using a set of torque-speed curves as shown in Figure 7-4 for various values of the applied voltage V. When the motor is stalled, the blocked-rotor torque at the rated voltage is denoted τ_0. Using equation (7.2.4) with $V_b = 0$ and $di_a/dt = 0$ we have

$$\begin{aligned} V_r &= Ri_a \\ &= \frac{R\tau_0}{K_m} \end{aligned} \tag{7.2.7}$$

Therefore the torque constant is

$$K_m = \frac{R\tau_0}{V_r} \tag{7.2.8}$$

The remainder of the discussion refers to Figure 7-5 consisting of the DC-motor in series with a gear train with gear ratio 1:r and connected to a link of the manipulator. The gear ratio r typically has values in the range 0.05 to 0.005 representing a gear reduction from 20 to 1 up to 200 to 1. Referring to Figure 7-5, we set $J_m = J_a + J_g$, the sum of the actuator and gear inertias. The equation of motion of this system is then

$$\begin{aligned} J_m \frac{d^2\theta_m}{dt^2} + B_m \frac{d\theta_m}{dt} &= \tau_m - r\tau_\ell \\ &= K_m i_a - r\tau_\ell \end{aligned} \tag{7.2.9}$$

the latter equality coming from (7.2.5). In the Laplace domain the three equations (7.2.4), (7.2.6) and (7.2.9) may be combined and written as

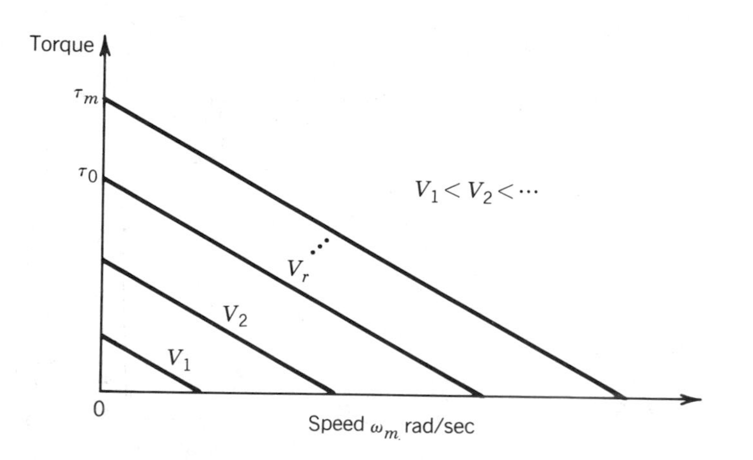

FIGURE 7-4
Typical torque-speed curces of a DC motor.

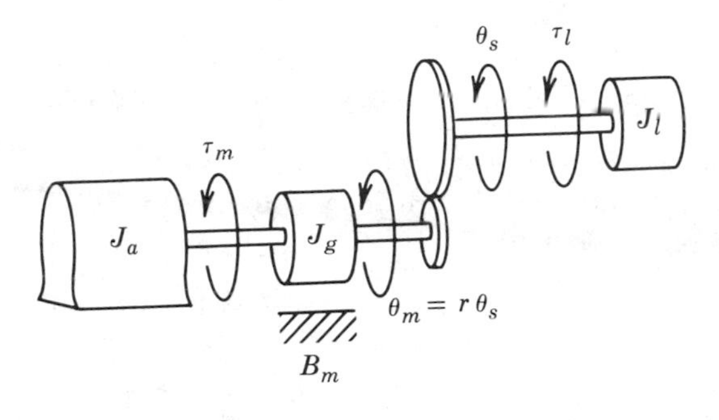

FIGURE 7-5
Lumped model of a single link with actuator/gear train.

$$(Ls + R)I_a(s) = V(s) - K_b s\Theta_m(s) \tag{7.2.10}$$

$$(J_m s^2 + B_m s)\Theta_m(s) = K_i I_a(s) - r\tau_\ell(s) \tag{7.2.11}$$

The block diagram of the above system is shown in Figure 7-6. The transfer function from $V(s)$ to $\Theta_m(s)$ is then given by (with $\tau_\ell = 0$)

$$\frac{\Theta_m(s)}{V(s)} = \frac{K_m}{s\left[(Ls + R)(J_m s + B_m) + K_b K_m\right]} \tag{7.2.12}$$

The transfer function from the load torque τ_ℓ to Θ_m is given by (with $V = 0$)

$$\frac{\Theta_m(s)}{\tau_\ell(s)} = \frac{-r(Ls + R)}{s\left[(Ls + R)(J_m s + B_m) + K_b K_m\right]} \tag{7.2.13}$$

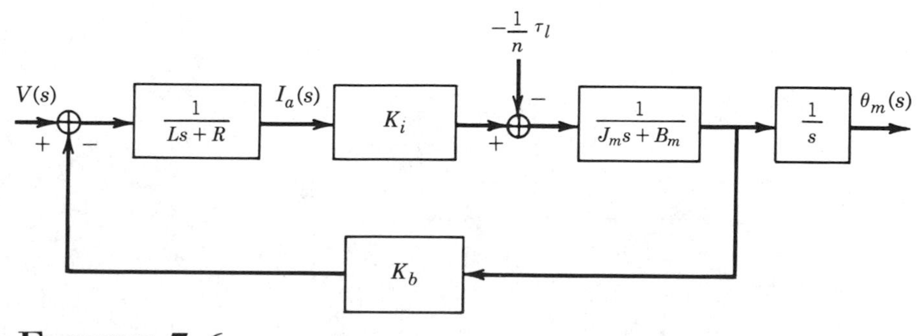

FIGURE 7-6
Block diagram for a DC motor system.

Frequently it is assumed that the "electrical time constant" $\frac{L}{R}$ is much smaller than the "mechanical time constant" $\frac{J_m}{B_m}$. This is a reasonable assumption for many electro-mechanical systems and leads to a reduced order model of the actuator dynamics. If we now divide numerator and denominator of (7.2.10) by R and neglect the electrical time constant by setting $\frac{L}{R}$ equal to zero, the transfer function between Θ_m and V becomes (again, with $\tau_\ell = 0$)

$$\frac{\Theta_m(s)}{V(s)} = \frac{K_m/R}{s(J_m s + B_m + K_b K_m/R)} \tag{7.2.14}$$

Similarly the transfer function between Θ_m and τ_ℓ is

$$\frac{\Theta_m(s)}{\tau_\ell(s)} = -\frac{r}{s(J_m s + B_m + K_b K_m/R)} \tag{7.2.15}$$

In the time domain equations (7.2.14) and (7.2.15) represent, by superposition, the second order differential equation

$$J_m \ddot{\theta}_m(t) + (B_m + K_b K_m/R)\dot{\theta}_m(t) = (K_m/R)V(t) - r\tau_\ell(t) \tag{7.2.16}$$

The block diagram corresponding to the reduced order system (7.2.16) is shown in Figure 7-7.

If the output side of the gear train is directly coupled to the link, then the joint variables and the motor variables are related by

$$q_k = r_k \theta_{m_k} \quad ; \quad k=1,\ldots,n \tag{7.2.17}$$

where r_k is the k-th gear ratio. Similarly, the joint torques τ_k given by (7.2.1) and the actuator load torques τ_{ℓ_k} are related by

$$\tau_{\ell_k} = \tau_k \quad ; \quad k=1,\ldots,n \tag{7.2.18}$$

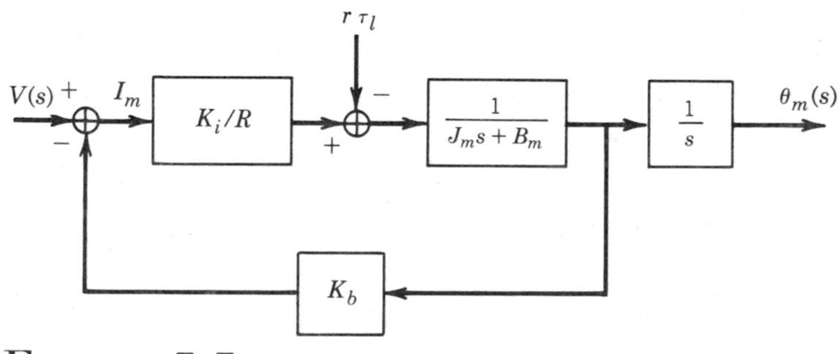

FIGURE 7-7
Block diagram for reduced order system.

However, in manipulators incorporating other types of drive mechanisms such as belts, pulleys, chains, etc., $r_k \theta_{m_k}$ need not equal q_k. In general one must incorporate into the dynamics a transformation between joint space variables and actuator variables of the form

$$q_k = f_k(\theta_{s_1}, \ldots, \theta_{s_n}) \quad ; \quad \tau_{\ell_k} = f_k(\tau_1, \ldots, \tau_n) \tag{7.2.19}$$

where $\theta_{s_k} = r_k \theta_{m_k}$.

(i) *Example 7.2.1*

Consider the two link planar manipulator shown in Figure 7-8, whose actuators are both located on link 1. In this case we have,

$$q_1 = \theta_{s_1} \quad ; \quad q_2 = \theta_{s_1} + \theta_{s_2} \tag{7.2.20}$$

Similarly, the joint torques τ_i and the actuator load torques τ_{ℓ_i} are related by

$$\tau_{\ell_1} = \tau_1 \quad ; \quad \tau_{\ell_2} = \tau_1 + \tau_2 \tag{7.2.21}$$

The inverse transformation is then

$$\theta_{s_1} = q_1 \quad ; \quad \theta_{s_2} = q_2 - q_1 \tag{7.2.22}$$

and

$$\tau_1 = \tau_{\ell_1} \quad ; \quad \tau_2 = \tau_{\ell_2} - \tau_{\ell_1} \tag{7.2.23}$$

7.3 SET-POINT TRACKING

In this section we discuss set-point tracking using a PD or PID compensator. This type of control is adequate for applications not involving very fast motion, especially in robots with large gear reduction between the actuators and the links. The analysis in this section follows typical engineering practice rather than complete mathematical rigor.

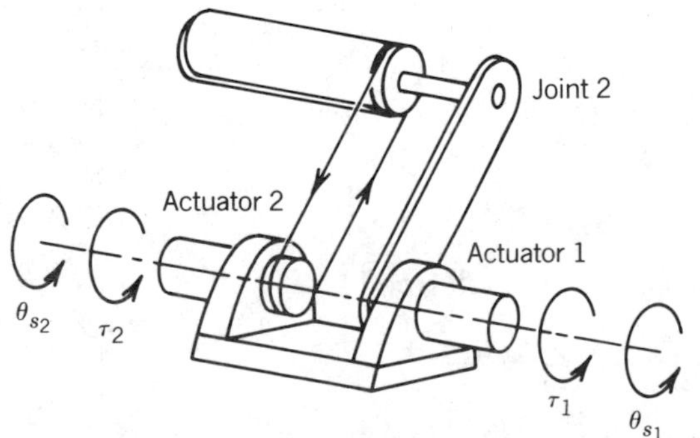

FIGURE 7-8
Two-link manipulator with remotely driven link.

For the following discussion, assume for simplicity that

$$q_k = \theta_{s_k} = r\theta_{m_k} \text{ and} \tag{7.3.1}$$

$$\tau_{\ell_k} = \tau_k$$

Then the equations of motion of the manipulator can be written as

$$\sum_{j=1}^{n} d_{jk}(\mathbf{q})\ddot{q}_j + \sum_{i,j=1}^{n} c_{ijk}(\mathbf{q})\dot{q}_i\dot{q}_j + g_k(\mathbf{q}) = \tau_k \tag{7.3.2}$$

$$J_m\ddot{\theta}_{m_k} + (B_m + K_b K_m/R)\dot{\theta}_{m_k} = K_m/RV_k - r_k\tau_k\ , \ k=1, \ldots, n \tag{7.3.3}$$

Equation (7.3.3) represents the actuator dynamics and (7.3.2) represents the nonlinear inertial, centripetal, coriolis, and gravitational coupling effects due to the motion of the manipulator. The simplest approach to the control of the above system is to consider the nonlinear term τ_k entering (7.3.3) and defined by (7.3.2) as an input disturbance to the motor and design an independent controller for each joint according to the model (7.3.3). The advantage of this approach is its simplicity since the motor dynamics represented by (7.3.3) are linear. Notice that τ_k in (7.3.3) is proportional to the gear reduction r_k. This is an important observation. The effect of the gear ratio is to reduce the coupling nonlinearities represented by (7.3.2), which adds to the validity of the above approach to control. However, for high speed motion, or for manipulators without gear reduction at the joints, the coupling nonlinearities have a much larger effect on the performance of the system and treating the nonlinear coupling terms represented by τ_k simply as a disturbance will generally cause large tracking errors.

Now, since $q_k = r_k\theta_{m_k}$, the actual coefficient of $\ddot{\theta}_{m_k}$ in (7.3.3) includes the term $r_k^2 d_{kk}(\mathbf{q})$ from (7.3.2), and is thus given by

$$J_m + r_k^2 d_{kk}(\mathbf{q}) \tag{7.3.4}$$

which is, of course, configuration dependent. In the present approach one approximates this inertia coefficient by a constant **average**, or **effective inertia** J_{eff}. It should be noted that even with the gear reduction, the inertia in (7.3.4) may vary over a large range, perhaps an order of magnitude. For example, Table 7-1 gives the effective joint inertias for the Stanford manipulator [6].

Setting

$$B_{eff} = B_m + K_b K_m/R \text{ and } K = K_m/R \tag{7.3.5}$$

we write (7.3.3) as

$$J_{eff}\ddot{\theta}_{m_k} + B_{eff}\dot{\theta}_{m_k} = KV_k - r_k d_k \tag{7.3.6}$$

where d_k is treated as a disturbance and is defined by

$$d_k := \sum_{j \neq k} d_{jk}\ddot{q}_j + \sum_{i,j} c_{ijk}\dot{q}_i\dot{q}_j + g_k \tag{7.3.7}$$

TABLE 7-1
Actuator and Effective Link Inertias for the Stanford Manipulator [5]

Link_i	$J_{g_i}^{2}$	J_{ii} no load, min.	J_{ii} no load max.	J_{ii} full load, max.
1	0.953	1.417	6.176	9.570
2	2.193	3.590	6.950	10.300
3	0.782	7.257	7.257	9.057
4	0.106	0.108	0.123	0.234
5	0.097	0.114	0.114	0.225
6	0.040	0.040	0.040	0.040

In the Laplace domain we represent (7.3.6) by the block diagram of Figure 7-9. Henceforth we drop the subscript k representing the k-th joint. The **set-point tracking problem** is now the problem of tracking a constant or step reference command θ^d.

7.3.1 PD COMPENSATOR

As a first illustration we choose a so-called PD-compensator. The resulting closed loop system is shown in Figure 7-10. The input $V(s)$ is given by

$$V(s) = K_P(\Theta^d(s) - \Theta(s)) - K_D s \Theta(s) \tag{7.3.8}$$

where K_P, K_D are the proportional (P) and derivative (D) gains, respectively.

Taking Laplace transforms of both sides of (7.3.6) and using the expression (7.3.8) for the feedback control $V(s)$, leads to the closed loop system

$$\Theta_m(s) = \frac{KK_P}{\Omega(s)}\Theta^d(s) - \frac{r}{\Omega(s)}D(s) \tag{7.3.9}$$

where $\Omega(s)$ is the closed loop characteristic polynomial

$$\Omega(s) = J_{eff}\, s^2 + (B_{eff} + KK_D)s + KK_P \tag{7.3.10}$$

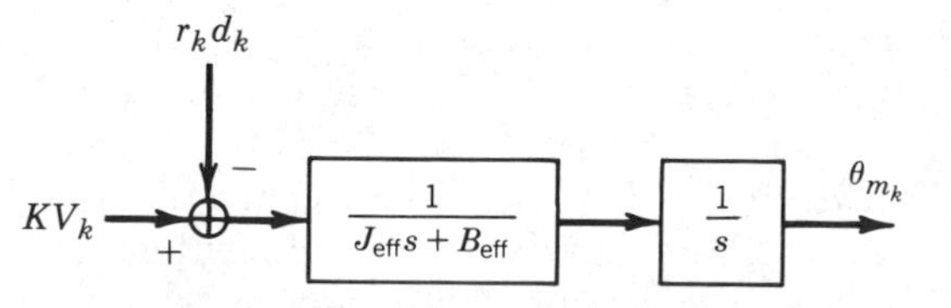

FIGURE 7-9
Block diagram of simplified open loop system with effective inertia and damping.

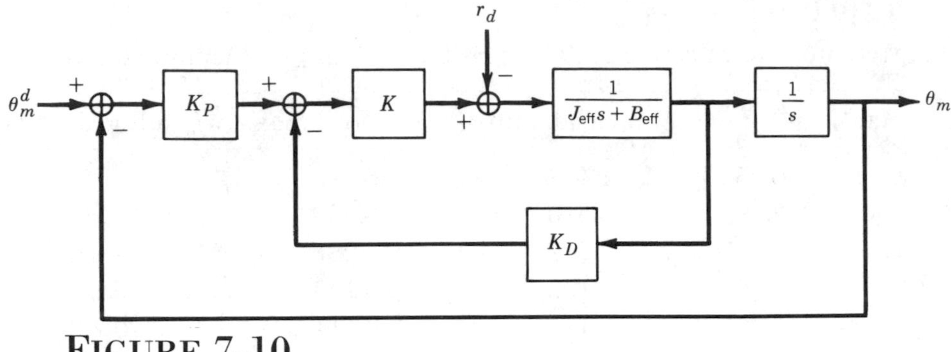

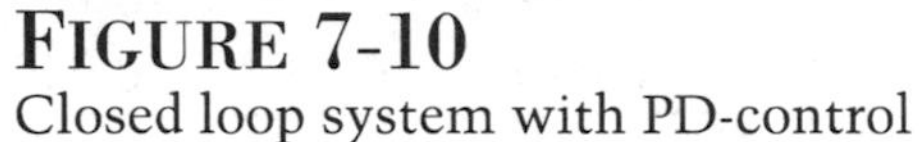

FIGURE 7-10
Closed loop system with PD-control

The closed loop system will be stable for all positive values of K_P and K_D and bounded disturbances, and the tracking error is given by

$$E(s) = \Theta^d(s) - \Theta_m(s) \tag{7.3.11}$$

$$= \frac{J_{eff}\, s^2 + (B_{eff} + KK_D)s}{\Omega(s)} \Theta^d(s) + \frac{r}{\Omega(s)} D(s)$$

For a step reference input

$$\Theta^d(s) = \frac{\Theta^d}{s} \tag{7.3.12}$$

and a constant disturbance

$$D(s) = \frac{D}{s} \tag{7.3.13}$$

it now follows directly from the final value theorem[4] that the steady state error e_{ss} satisfies

$$e_{ss} = \lim_{s \to 0} sE(s) \tag{7.3.14}$$

$$= \frac{-rD}{KK_P}$$

Thus we see that the steady state error due to a constant disturbance is smaller for larger gear reduction and can be made arbitrarily small by making the position gain K_P large, which is to be expected since the system is Type 1.

We know, of course, from (7.3.7) that the disturbance term $D(s)$ in (7.3.11) is not constant. However, in the steady state this disturbance term is just the gravitational force acting on the robot, which is constant. The above analysis therefore, while only approximate, nevertheless gives a good description of the actual steady state error using a PD compensator assuming stability of the closed loop system.

7.3.2 PERFORMANCE OF PD COMPENSATORS

For the PD-compensator given by (7.3.8) the closed loop system is second order and hence the step response is determined by the closed loop natural frequency ω and damping ratio ζ. Given a desired value for these quantities, the gains K_D and K_P can be found from the expression

$$s^2 + \frac{(B_{eff} + KK_D)}{J_{eff}}s + \frac{KK_P}{J_{eff}} = s^2 + 2\zeta\omega s + \omega^2 \tag{7.3.15}$$

as

$$K_P = \frac{\omega^2 J_{eff}}{K}, \quad K_D = \frac{2\zeta\omega J_{eff} - B_{eff}}{K} \tag{7.3.16}$$

It is customary in robotics applications to take $\zeta = 1$ so that the response is critically damped. This produces the fastest non-oscillatory response. In this context ω determines the speed of response.

(i) *Example 7.3.1*

Consider the second order system of Figure 7-11. The closed loop characteristic polynomial is

$$p(s) = s^2 + (1 + K_D)s + K_P \tag{7.3.17}$$

Suppose $\theta^d = 10$ and there is no disturbance $(d = 0)$. With $\zeta = 1$, the required PD gains for various values of ω are shown in Table 7-2. The corresponding step responses are shown in Figure 7-12.

Now suppose that there is a constant disturbance $d = 40$ acting on the system. The response of the system with the PD gains of Table 7-2 are shown in Figure 7-13. We see that the steady state error due to the disturbance is smaller for large gains as expected.

In theory, we could achieve arbitrarily fast response and arbitrarily small steady state error to a constant disturbance by simply increasing

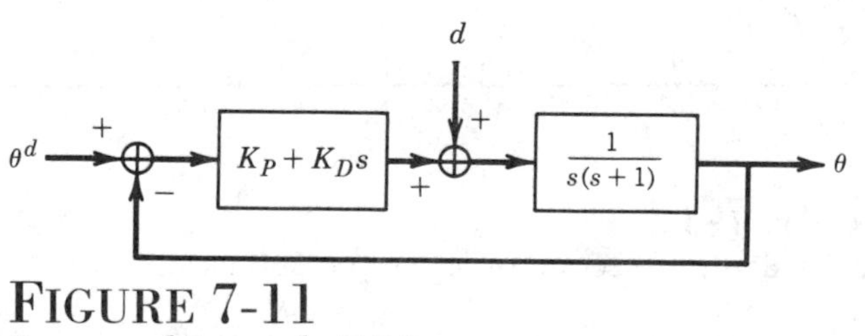

FIGURE 7-11
System of Example 7.3.1.

TABLE 7-2

ω	K_P	K_D
4	16	7
8	64	15
12	144	23

the PD gains in the compensator. In practice, however, there is a maximum speed of response achievable from the system. Two major factors, heretofore neglected, limit the value of ω that may be chosen. The first is due to bounds on the maximum torque (or current) input. Many manipulators, in fact, incorporate current limiters in the servo-system to prevent damage that might result from overdrawing current. The second effect is flexibility in the motor shaft and/or drive train.

(*ii*) ***Example*** *7.3.2*

Consider the block diagram of Figure 7-14, where the saturation function represents the maximum allowable input. The step response of this system with $\omega = 8$ is shown in Figure 7-15. The limits on the input effectively limit the maximum velocity and acceleration of the motor. Increasing the gains further would not result in a faster response but would increase the overshoot.

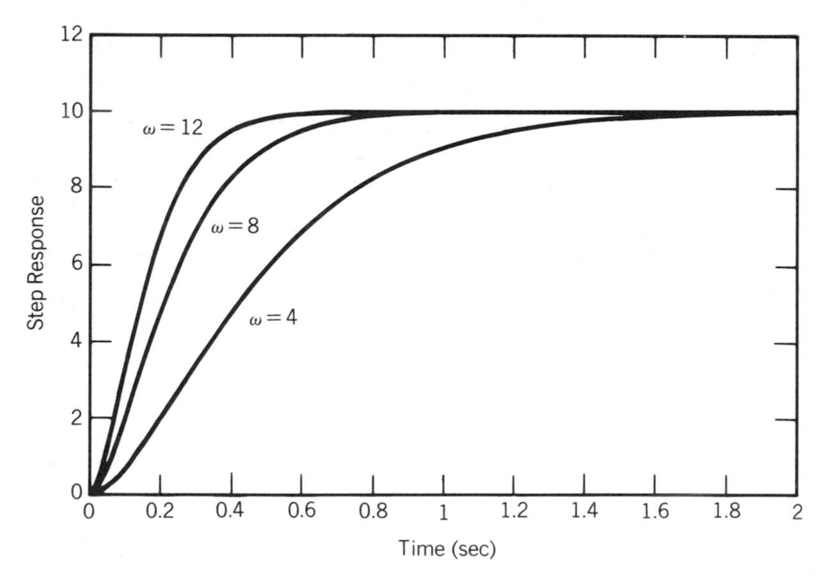

FIGURE 7-12
Critically damped second order step responses.

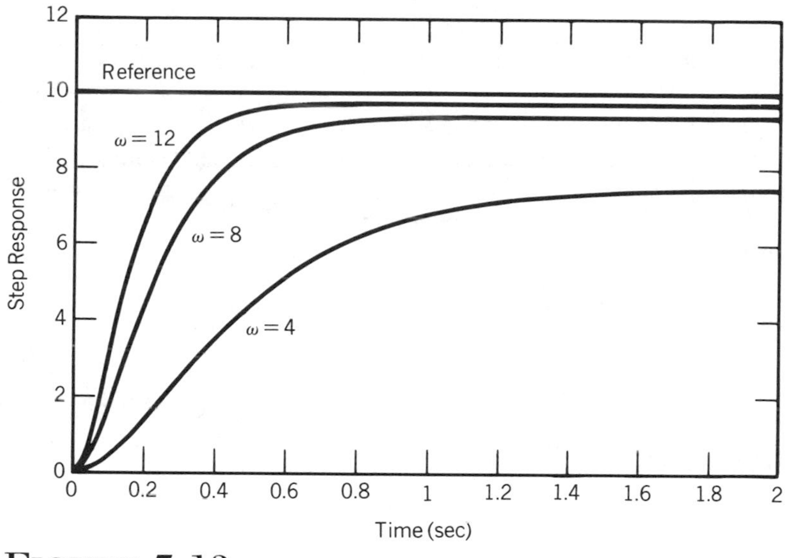

FIGURE 7-13
Second order system response with disturbance added.

The second effect to consider is the joint flexibility. Let k_r be the effective stiffness at the joint. The joint resonant frequency is then $\omega_r = \sqrt{k_r / J_{eff}}$. It is common engineering practice to limit ω in (7.3.16) to no more than half of ω_r to avoid excitation of the joint resonance. We will discuss the effects of the joint flexibility in more detail in the next section.

These examples clearly show the limitations of PD-control when additional effects such as input saturation, disturbances, and unmodeled dynamics must be considered.

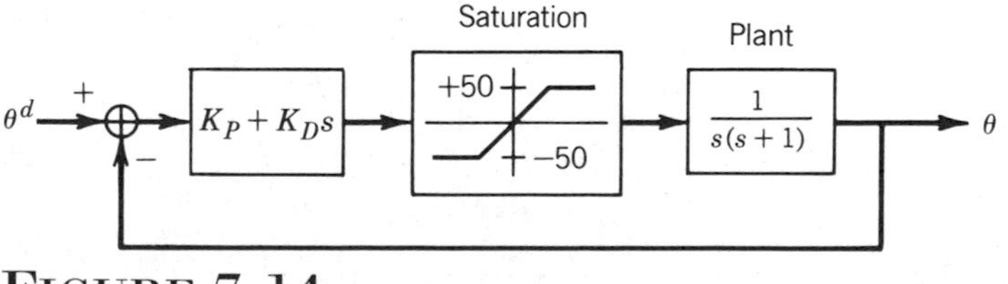

FIGURE 7-14
Second order system with input saturation.

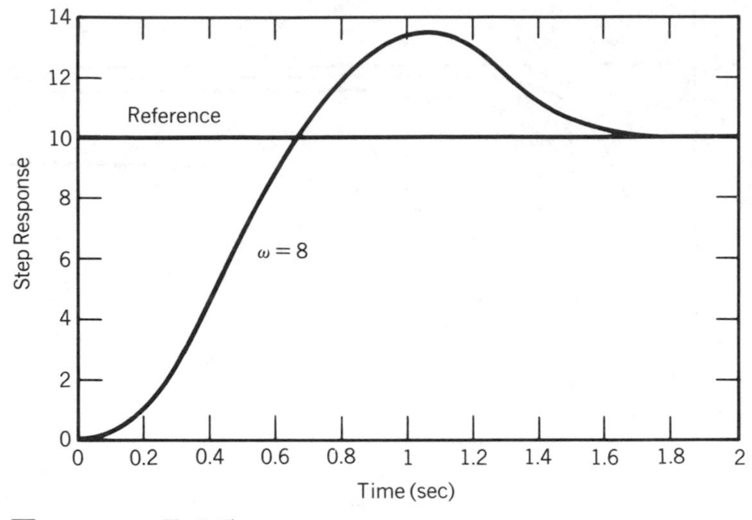

FIGURE 7-15
Overshoot caused by saturating input.

7.3.3 PID COMPENSATOR

In order to reject a constant disturbance using PD control we have seen that large gains are required. By using integral control we may achieve zero steady state error while keeping the gains small. Thus, let us add an integral term $\frac{K_I}{s}$ to the above PD compensator. This leads to the so-called PID control law, as shown in Figure 7-16. The system is now Type 2 and the PID control achieves exact steady tracking of step (and ramp) inputs while rejecting step disturbances, provided of course that the closed loop system is stable.

With the PID compensator

$$C(s) = K_P + K_D s + \frac{K_I}{s} \tag{7.3.18}$$

the closed loop system is now the third order system

$$\Theta_m(s) = \frac{(K_D s^2 + K_P s + K_I)}{\Omega_2(s)} \Theta^d(s) - \frac{rs}{\Omega_2(s)} D(s) \tag{7.3.19}$$

where

$$\Omega_2 = J_{eff} s^3 + (B_{eff} + KK_D)s^2 + KK_P s + KK_I \tag{7.3.20}$$

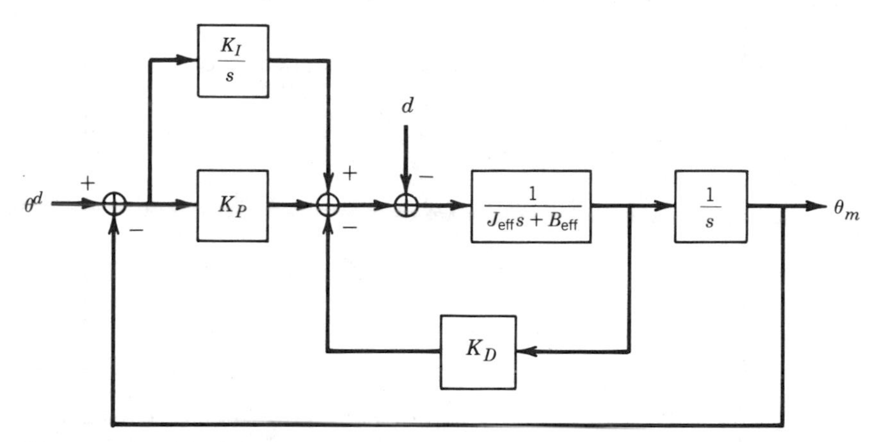

FIGURE 7-16
Closed loop system with PID control.

Applying the Routh criterion[10] to this polynomial, it follows that the closed loop system is stable if the gains are positive, and in addition,

$$K_I < \frac{(B_{eff} + KK_D)K_P}{J_{eff}} \tag{7.3.23}$$

(i) *Example 7.3.3*

To the system of Example 7.3.2 we have added a disturbance and an integral control term in the compensator. The step responses are shown in Figure 7-17. We see that the steady state error due to the disturbance is removed.

7.4 DRIVE TRAIN DYNAMICS

In this section we discuss in more detail the problem of joint flexibility. For many manipulators, particularly those using harmonic drives[1] for torque transmission, the joint flexibility is significant. In addition to torsional flexibility in the gears, joint flexibility is caused by effects such as shaft windup, bearing deformation, and compressibility of the hydraulic fluid in hydraulic robots.

[1]Harmonic drives are a type of gear mechanism that are very popular for use in robots due to their low backlash, high torque transmission and compact size. However, they also introduce unwanted friction and flexibility at the joints.

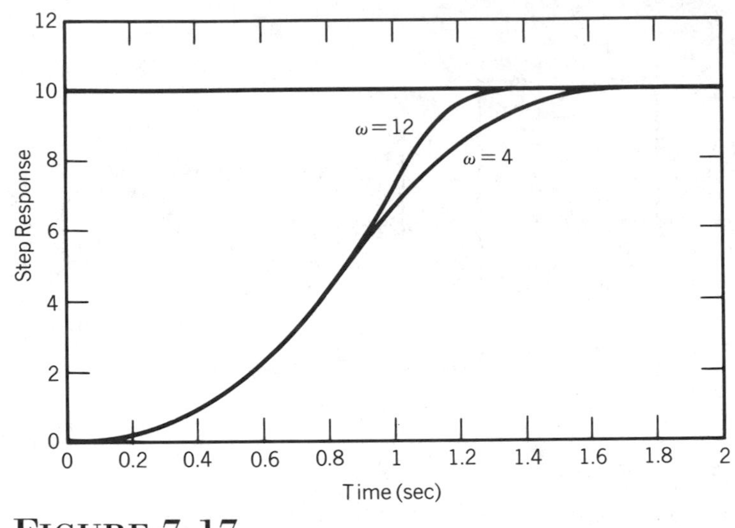

FIGURE 7-17
Response with integral control action.

Consider the idealized situation of Figure 7-18 consisting of an actuator connected to a load through a torsional spring which represents the joint flexibility. For simplicity we take the motor torque u as input. The equations of motion are easily derived using the techniques of Chapter Six, with generalized coordinates θ_ℓ and θ_m, the link angle, and the motor angle, respectively, as

$$J_\ell \ddot{\theta}_\ell + B_\ell \dot{\theta}_\ell + k(\theta_\ell - \theta_m) = 0 \tag{7.4.1}$$

$$J_m \ddot{\theta}_m + B_m \dot{\theta}_m - k(\theta_\ell - \theta_m) = u \tag{7.4.2}$$

where J_ℓ , J_m are the load and motor inertias, B_ℓ and B_m are the load and motor damping constants, and u is the input torque applied to the motor shaft. In the Laplace domain we can write this as

$$p_\ell(s)\Theta_\ell(s) = k\Theta_m(s) \tag{7.4.3}$$

$$p_m(s)\Theta_m(s) = k\Theta_\ell(s) + U(s) \tag{7.4.4}$$

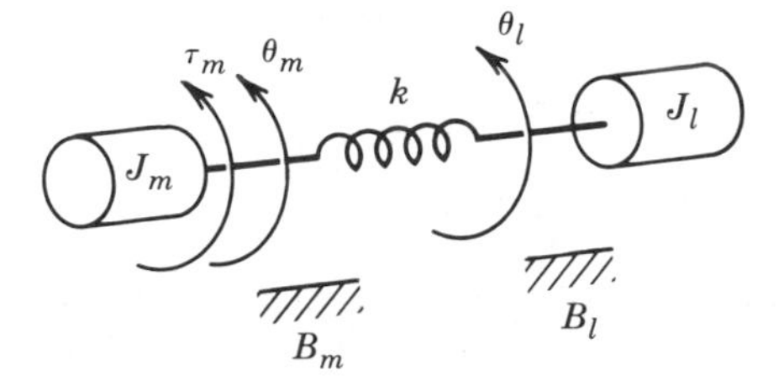

FIGURE 7-18
Idealized model to represent joint flexibility.

where

$$p_\ell(s) = J_\ell s^2 + B_\ell s + k \tag{7.4.5}$$

$$p_m(s) = J_m s^2 + B_m s + k \tag{7.4.6}$$

This system is represented by the block diagram of Figure 7-19.

The output to be controlled is, of course, the load angle θ_ℓ. The open loop transfer function between U and Θ_ℓ is given by

$$\frac{\Theta_\ell(s)}{U(s)} = \frac{k}{p_\ell(s)p_m(s) - k^2} \tag{7.4.7}$$

The open loop characteristic polynomial is

$$J_\ell J_m s^4 + (J_\ell B_m + J_m B_\ell)s^3 + (k(J_\ell + J_m) + B_\ell B_m)s^2 + k(B_\ell + B_m)s \tag{7.4.8}$$

If the damping constants B_ℓ and B_m are neglected, the open loop characteristic polynomial is

$$J_\ell J_m s^4 + k(J_\ell + J_m)s^2 \tag{7.4.9}$$

which has a double pole at the origin and a pair of complex conjugate poles at $s = \pm j\omega$ where $\omega^2 = k(\frac{1}{J_\ell} + \frac{1}{J_m})$. Assuming that the open loop damping constants B_ℓ and B_m are small, then the open loop poles of the system (7.4.3)–(7.4.4) will be in the left half plane near the poles of the undamped system.

Suppose we implement a PD compensator $C(s) = K_P + K_D s$. At this point the analysis depends on whether the position/velocity sensors are placed on the motor shaft or on the load shaft, that is, whether the PD-compensator is a function of the motor variables or the load variables. If the motor variables are measured then the closed loop system is given by the block diagram of Figure 7-20. Set $K_P + K_D s = K_D(s + a)$; $a = K_P/K_D$. The root locus for the closed loop system in terms of K_D is shown in Figure 7-21.

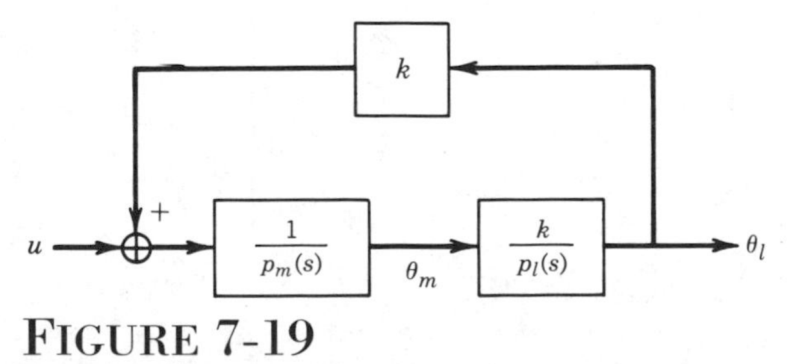

FIGURE 7-19
Block diagram for the system (7.4.3)–(7.4.4).

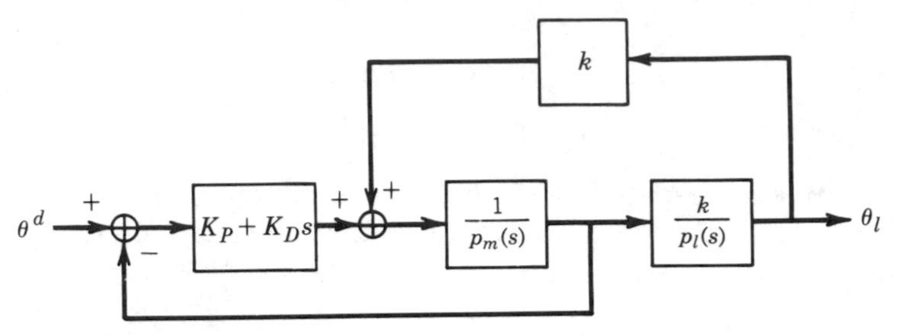

FIGURE 7-20
PD-control with motor angle feedback.

We see that the system is stable for all values of the gain K_D but that the presence of the open loop zeros near the $j\omega$ axis may result in poor overall performance, for example, undesirable oscillations with poor settling time. Also the poor relative stability means that disturbances and other unmodeled dynamics could render the system unstable.

If we measure instead the load angle θ_ℓ, the system with PD control is represented by the block diagram of Figure 7-22. The corresponding root locus is shown in Figure 7-23. In this case the system is unstable for large K_D. The critical value of K_D, that is, the value of K_D for which the system becomes unstable, can be found from the Routh criterion. The best that one can do in this case is to limit the gain K_D so that the closed loop poles remain within the left half plane with a reasonable stability margin.

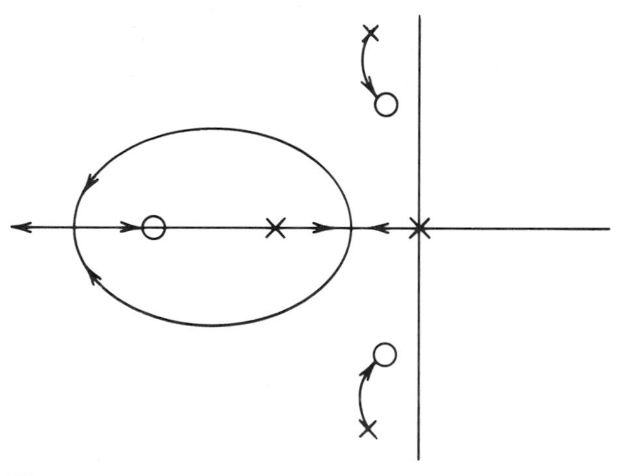

FIGURE 7-21
Root locus for the system of Figure 7-20.

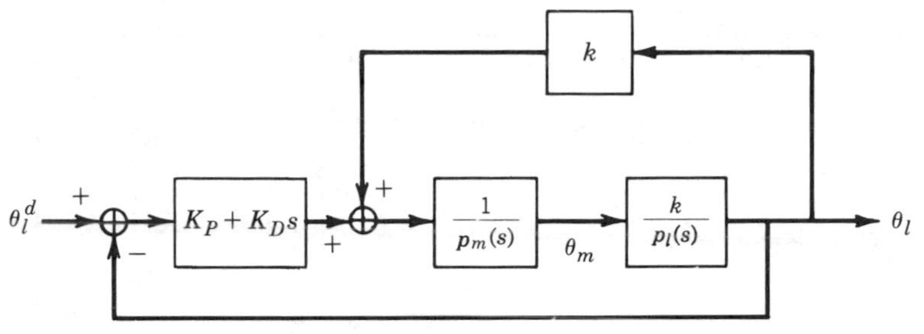

FIGURE 7-22
PD-control with load angle feedback.

(*ii*) *Example 7.4.1*

Suppose that the system (7.4.1)-(7.4.2) has the following parameters (see [1])

$$k = 0.8 Nm/rad \qquad B_m = 0.015 Nms/rad \qquad (7.4.10)$$

$$J_m = 0.0004 Nms^2/rad \qquad B_\ell = 0.0 Nms/rad$$

$$J_\ell = 0.0004 Nm^2/rad$$

If we implement a PD controller $K_D(s + a)$ then the response of the system with motor (respectively, load) feedback is shown in Figure 7-24 (respectively, Figure 7-25).

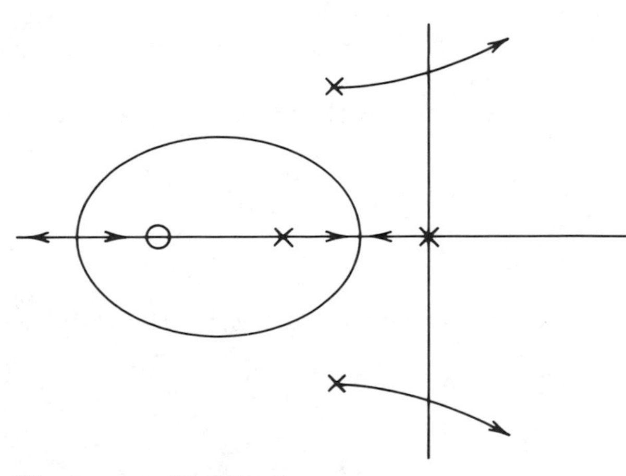

FIGURE 7-23
Root locus for the system of Figure 7-22.

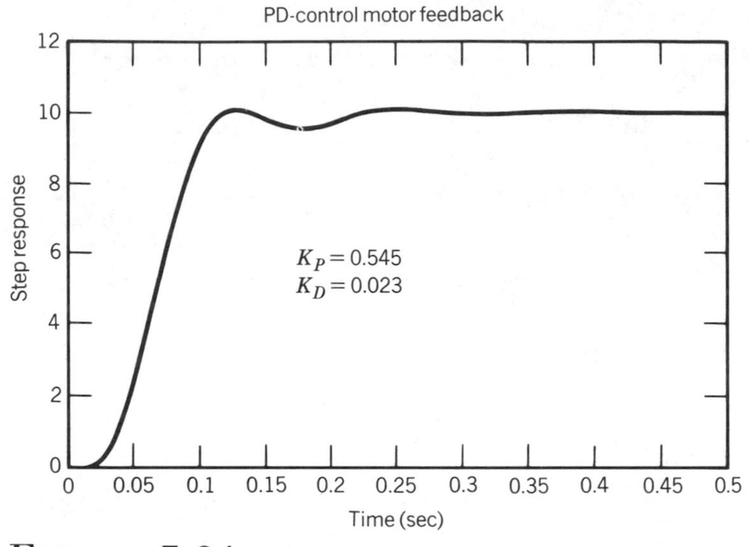

FIGURE 7-24
Step response—PD-control with motor angle feedback.

7.4.1 STATE SPACE DESIGN[2]

The previous analysis has shown that PD control is inadequate for robot control unless the joint flexibility is negligible or unless one is content with relatively slow response of the manipulator. Not only does the joint flexibility limit the magnitude of the gain for stability reasons, it also introduces lightly damped poles into the closed loop system that result in unacceptable oscillation of the transient response. We next investigate the application of more advanced techniques for controlling the above system.

We can write the system (7.4.1)–(7.4.2) in state space by choosing state variables

$$\begin{aligned} x_1 &= \theta_\ell & x_2 &= \dot{\theta}_\ell \\ x_3 &= \theta_m & x_4 &= \dot{\theta}_m \end{aligned} \tag{7.4.11}$$

In terms of these state variables the system (7.4.1)-(7.4.2) becomes

$$\dot{x}_1 = x_2 \tag{7.4.12}$$

$$\dot{x}_2 = -\frac{k}{J_\ell}x_1 - \frac{B_\ell}{J_\ell}x_2 + \frac{k}{J_\ell}x_3$$

[2]This section assumes more knowledge of control theory than previous sections.

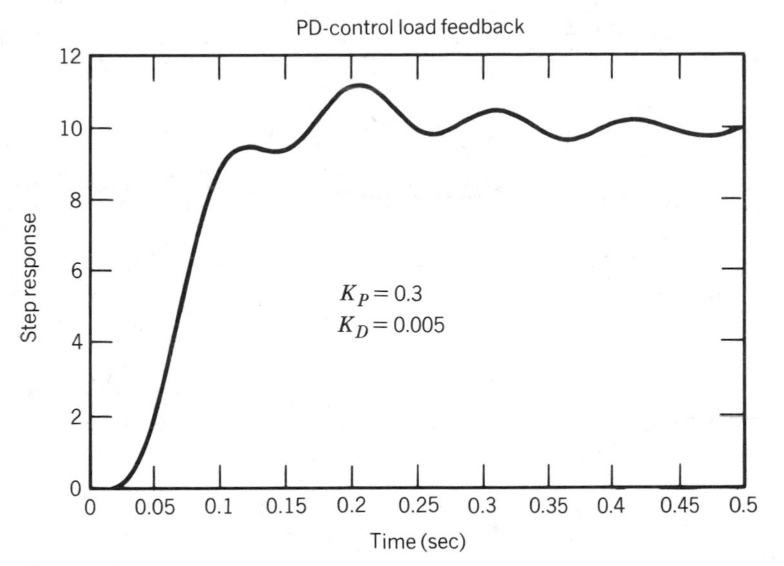

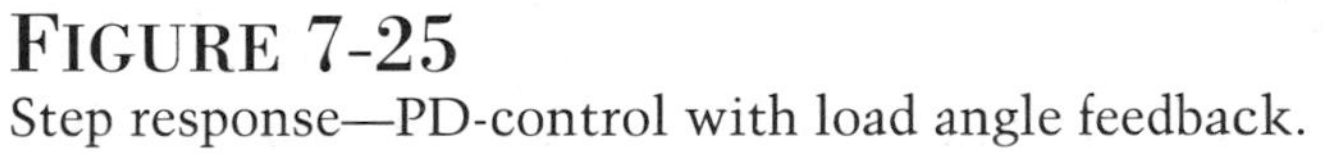

FIGURE 7-25
Step response—PD-control with load angle feedback.

$$\dot{x}_3 = x_4$$

$$\dot{x}_4 = \frac{k}{J_m} x_1 - \frac{B_\ell}{J_m} x_4 - \frac{k}{J_m} x_3 + \frac{1}{J_m} u$$

which, in matrix form, can be written as

$$\dot{\mathbf{x}} = A\mathbf{x} + \mathbf{b}u \tag{7.4.13}$$

where

$$A = \begin{bmatrix} 0 & 1 & 0 & 0 \\ -\frac{k}{J_\ell} & -\frac{B_\ell}{J_\ell} & \frac{k}{J_\ell} & 0 \\ 0 & 0 & 0 & 1 \\ \frac{k}{J_m} & 0 & -\frac{k}{J_m} & -\frac{B_m}{J_m} \end{bmatrix} ; \quad \mathbf{b} = \begin{bmatrix} 0 \\ 0 \\ 0 \\ \frac{1}{J_m} \end{bmatrix} \tag{7.4.14}$$

If we choose an output $y(t)$, say the measured load angle $\theta_\ell(t)$, then we have an output equation

$$y = x_1 = \mathbf{c}^T \mathbf{x} \tag{7.4.15}$$

where

$$\mathbf{c}^T = \begin{bmatrix} 1,0,0,0 \end{bmatrix} \tag{7.4.16}$$

The relationship between the state space form (7.4.13)-(7.4.15) and the transfer function defined by (7.4.7) is found by taking Laplace transforms of (7.4.13)–(7.4.15) with initial conditions set to zero. This yields

$$G(s) = \frac{\Theta_\ell(s)}{U(s)} = \frac{Y(s)}{U(s)} = \mathbf{c}^T(sI - A)^{-1}\mathbf{b} \tag{7.4.17}$$

where I is the $n \times n$ identity matrix. The poles of $G(s)$ are eigenvalues of the matrix A. In the system (7.4.13)-(7.4.15) the converse holds as well, that is, all of the eigenvalues of A are poles of $G(s)$. This is always true if the state space system is defined using a minimal number of state variables [8].

7.4.2 STATE FEEDBACK COMPENSATOR

Given a linear system in state space form, such as (7.4.13), a **linear state feedback control law** is an input u of the form

$$\begin{aligned} u(t) &= -\mathbf{k}^T\mathbf{x} + r \qquad (7.4.18)\\ &= -\sum_{i=1}^{4} k_i x_i + r \end{aligned}$$

where k_i are constants and r is a reference input. In other words, the control is determined as a linear combination of the system states which, in this case, are the motor and load positions and velocities. Compare this to the previous PD-control, which was a function either of the motor position and velocity or of the load position and velocity, but not both. The coefficients k_i in (7.4.18) are the gains to be determined. If we substitute the control law (7.4.18) into (7.4.13) we obtain

$$\dot{\mathbf{x}} = (A - \mathbf{b}\mathbf{k}^T)\mathbf{x} + \mathbf{b}r \tag{7.4.19}$$

Thus we see that the linear feedback control has the effect of changing the poles of the system from those determined by A to those determined by $A - \mathbf{b}\mathbf{k}^T$.

In the previous PD-design the closed loop pole locations were restricted to lie on the root locus shown in Figure 7-21 or 7-23. Since there are more free parameters to choose in (7.4.18) than in the PD controller, it may be possible to achieve a much larger range of closed loop poles. This turns out to be the case if the system (7.4.13) satisfies a property known as **controllability**.

(i) *Definition 7.4.2*

A linear system is said to be **completely state-controllable**, or **controllable** for short, if for each initial state $\mathbf{x}(t_0)$ and each final state $\mathbf{x}(t_f)$ there is a control input $t \to u(t)$ that transfers the system from $\mathbf{x}(t_0)$ at time t_0 to $\mathbf{x}(t_f)$ at time t_f.

The above definition says, in essence, that if a system is controllable we can achieve any state whatsoever in finite time starting from an arbitrary initial state. To check whether a system is controllable we have the following simple test.

(ii) *Lemma 7.4.3*

A linear system of the form (7.4.13) is controllable if and only if

$$\det[\mathbf{b}, A\,\mathbf{b}, A^2\mathbf{b}, \ldots, A^{n-1}\mathbf{b}] \neq 0 \qquad (7.4.20)$$

The $n \times n$ matrix $[\mathbf{b}, A\mathbf{b}, \ldots, A^{n-1}\mathbf{b}]$ Bs called the **controllability matrix** for the linear system defined by the pair $(A, \mathbf{b})$. The fundamental importance of controllability of a linear system is shown by the following

(iii) *Theorem 7.4.4*

Let $\alpha(s) = s^n + \alpha_n s^{n-1} + \cdots + \alpha_2 s + \alpha_1$ be an arbitrary polynomial of degree n. Then there exists a state feedback control law (7.4.18) such that

$$\det(sI - A + \mathbf{b}\mathbf{k}^T) = \alpha(s) \qquad (7.4.21)$$

if and only if the system (7.4.13) is controllable.

This fundamental result says that, for a controllable linear system, we may achieve **arbitrary**[3] closed loop poles using state feedback. Returning to the specific fourth-order system (7.4.14) we see that the system is indeed controllable since

$$\det[\mathbf{b}, A\mathbf{b}, A^2\mathbf{b}, A^3\mathbf{b}] = \frac{k^2}{J_m^4 J_\ell^2} \qquad (7.4.22)$$

which is never zero since $k > 0$. Thus we can achieve any desired set of closed loop poles that we wish, which is much more than was possible using the previous PD compensator.

There are many algorithms that can be used to determine the feedback gains in (7.4.18) to achieve a desired set of closed loop poles. This is known as the **pole assignment problem**. In this case most of the difficulty lies in choosing an appropriate set of closed loop poles based on the desired performance, the limits on the available torque, etc. We would like to achieve a fast response from the system without requiring too much torque from the motor. One way to design the feedback gains is through an optimization procedure. This takes us into the realm of optimal control theory. For example, we may choose as our goal the minimization of the performance criterion

$$J = \int_0^\infty (\mathbf{x}^T Q\,\mathbf{x} + Ru^2)\,dt \qquad (7.4.23)$$

where Q is a given symmetric, positive definite matrix and $R > 0$.

---

[3]Since the coefficients of $\alpha(s)$ are real, the only restriction on the pole locations is that they occur in complex conjugate pairs.

Choosing a control law to minimize (7.4.23) frees us from having to decide beforehand what the closed loop poles should be as they are automatically dictated by the weighting matrices Q and R in (7.4.23). It is shown in optimal control texts that the optimum linear control law that minimizes (7.4.23) is given as

$$u = -\mathbf{k}_*^T \mathbf{x} \tag{7.4.24}$$

where

$$\mathbf{k}_* = R^{-1} \mathbf{b}^T P \tag{7.4.25}$$

and P is the (unique) symmetric, positive definite $n \times n$ matrix satisfying the so-called **matrix Algebraic Riccatti equation**

$$A^T P + PA - P\mathbf{b}R^{-1}\mathbf{b}^T P + Q = 0 \tag{7.4.26}$$

The control law (7.4.24) is referred to as a **Linear Quadratic (LQ) Optimal Control**, since the performance index is quadratic and the control system is linear.

(iv) *Example 7.4.5*

For illustration purposes, let Q and R in (7.4.23) be given as $Q = diag\{100,0.1,100,0.1\}$ and $R = 100$. This puts a relatively large weighting on the position variables and control input with smaller weighting on the velocities of the motor and load. Figure 7-26 shows the optimal gains that result and the response of this LQ-optimal control for the system (7.4.12) with a unit step reference input r.

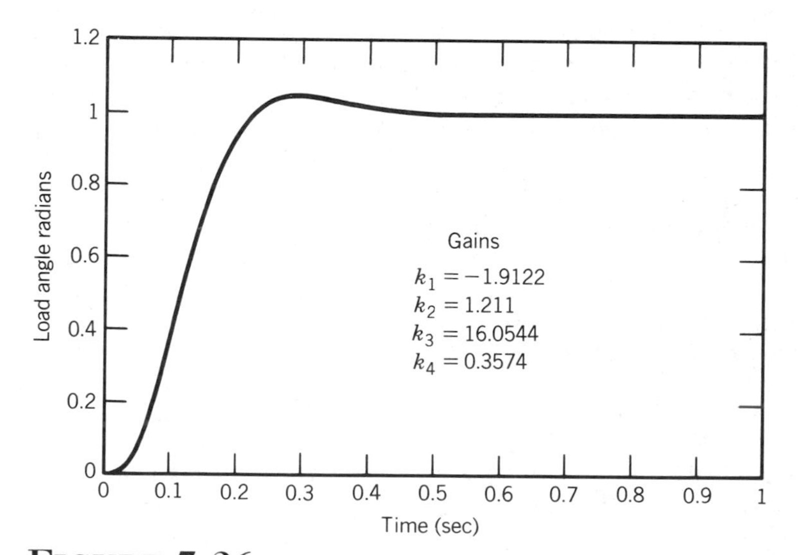

FIGURE 7-26
Step response—Linear, Quadratic–Optimal (LQ) state feedback control.

7.4.3 OBSERVERS

The above result is remarkable; however, in order to achieve it, we have had to pay a price, namely, the control law must be a function of *all* of the states. In order to build a compensator that requires only the measured output, in this case θ_ℓ, we need to introduce the concept of an **observer**. An observer is a state estimator. It is a dynamical system (constructed in software) that attempts to estimate the full state $\mathbf{x}(t)$ using only the system model (7.4.13)–(7.4.15) and the measured output $y(t)$. A complete discussion of observers is beyond the scope of the present text. We give here only a brief introduction to the main idea of observers for linear systems.

Assuming that we know the parameters of the system (7.4.13) we could simulate the response of the system in software and recover the value of the state $\mathbf{x}(t)$ at time t from the simulation. We could then use this simulated or estimated state, call in $\hat{\mathbf{x}}(t)$, in place of the true state in (7.4.24). However, since the true initial condition $\mathbf{x}(t_0)$ for (7.4.13) will generally be unknown, this idea is not feasible. However the idea of using the model of the system (7.4.13) is a good starting point to construct a state estimator in software. Let us, therefore, consider an estimate $\hat{\mathbf{x}}(t)$ satisfying the system

$$\dot{\hat{\mathbf{x}}} = A\hat{\mathbf{x}} + \mathbf{b}u + \boldsymbol{\ell}(y - \mathbf{c}^T\hat{\mathbf{x}}) \tag{7.4.27}$$

Equation (7.4.27) is called an **observer** for (7.4.13) and represents a model of the system (7.4.13) with an additional term $\boldsymbol{\ell}(y - \mathbf{c}^T\hat{\mathbf{x}})$. This additional term is a measure of the error between the output $y(t) = \mathbf{c}^T\mathbf{x}(t)$ of the plant and the estimate of the output, $\mathbf{c}^T\hat{\mathbf{x}}(t)$. Since we know the coefficient matrices in (7.4.27) and can measure y directly, we can solve the above system for $\hat{\mathbf{x}}(t)$ starting from any initial condition, and use this $\hat{\mathbf{x}}$ in place of the true state $\mathbf{x}$ in the feedback law (7.4.24). The additional term $\boldsymbol{\ell}$ in (7.4.27) is to be designed so that $\hat{\mathbf{x}} \to \mathbf{x}$ as $t \to \infty$, that is, so that the estimated state converges to the true (unknown) state independent of the initial condition $\mathbf{x}(t_0)$. Let us see how this is done.

Define $\mathbf{e(t)} = \mathbf{x} - \hat{\mathbf{x}}$ as the **estimation error**. Combining (7.4.13) and (7.4.27), since $y = \mathbf{c}^T\mathbf{x}$, we see that the estimation error satisfies the system

$$\dot{\mathbf{e}} = (A - \boldsymbol{\ell}\mathbf{c}^T)\mathbf{e} \tag{7.4.28}$$

From (7.4.28) we see that the dynamics of the estimation error are determined by the eigenvalues of $A - \boldsymbol{\ell}\mathbf{c}^T$. Since $\boldsymbol{\ell}$ is a design quantity we can attempt to choose it so that $\mathbf{e}(t) \to 0$ as $t \to \infty$, in which case the estimate $\hat{\mathbf{x}}$ converges to the true state $\mathbf{x}$. In order to do this we obviously want to choose $\boldsymbol{\ell}$ so that the eigenvalues of $A - \boldsymbol{\ell}\mathbf{c}^T$ are in the left half plane. This is similar to the pole assignment problem considered previously. In fact it is dual, in a mathematical sense, to the pole assignment problem. It turns out that the eigenvalues of $A - \boldsymbol{\ell}\mathbf{c}^T$

can be assigned arbitrarily if and only if the pair $(A, \mathbf{c})$ satisfies the property known as **observability**. Observability is defined by the following:

(*i*) *Definition 7.4.6*

A linear system is **completely observable**, or **observable** for short, if every initial state $\mathbf{x}(t_0)$ can be exactly determined from measurements of the output $y(t)$ and the input $u(t)$ in a finite time interval $t_0 \le t \le t_f$.

To check whether a system is observable we have the following

(*ii*) *Theorem 7.4.7*

The pair $(A, \mathbf{c})$ is **observable** if and only if

$$\det\left[\mathbf{c}, A^T\,\mathbf{c}, \ldots, A^{T^{n-1}}\,\mathbf{c}\right] \neq 0 \qquad (7.4.29)$$

The $n \times n$ matrix $[\mathbf{c}^T, \mathbf{c}^T A^T, \ldots, \mathbf{c}^T A^{T^{n-1}}]$ is called the **observability matrix** for the pair $(A, \mathbf{c}^T)$. In the system 7.4.13–7.4.15 above we have that

$$\det\left[\mathbf{c}, A^T\,\mathbf{c}, A^{T^2}\mathbf{c}, A^{T^3}\mathbf{c}\right] = \frac{k^2}{J_\ell^2} \qquad (7.4.30)$$

and hence the system is observable. A result known as the **Separation Principle** says that if we use the estimated state in place of the true state in (7.4.24), then the set of closed loop poles of the system will consist of the union of the eigenvalues of $A - \boldsymbol{\ell}\,\mathbf{c}^T$ and the eigenvalues of $A - \mathbf{bk}^T$. As the name suggests the Separation Principle allows us to separate the design of the state feedback control law (7.4.24) from the design of the state estimator (7.4.27). A typical procedure is to place the observer poles to the left of the desired pole locations of $A - \mathbf{bk}^T$. This results in rapid convergence of the estimated state to the true state, after which the response of the system is nearly the same as if the true state were being used in (7.4.24).

The result that the closed loop poles of the system may be placed arbitrarily, under the assumption of controllability and observability, is a powerful theoretical result. There are always practical considerations to be taken into account, however. The most serious factor to be considered in observer design is noise in the measurement of the output. To place the poles of the observer very far to the left of the imaginary axis in the complex plane requires that the observer gains be large. Large gains can amplify noise in the output measurement and result in poor overall performance. Large gains in the state feedback control law (7.4.24) can result in saturation of the input, again resulting in poor performance. Also uncertainties in the system parameters, nonlinearities such as a nonlinear spring characteristic or backlash, will reduce the achievable performance from the above design. Therefore, the above ideas are intended only to illustrate what may be possible by using more advanced concepts from control theory. Virtually all present day industrial robot controllers still use the simple PD or PID design.

7.5 TRAJECTORY INTERPOLATION

The simplest type of robot motion is **point to point** motion. In this approach the robot is commanded to go from an initial configuration T^6_{0init} to a final configuration T^6_{0final} without regard to the intermediate path followed by the end-effector. This type of motion is suitable for materials transfer tasks when the workspace is clear of obstacles and is common in so-called **teach and playback mode** where the robot is taught a sequence of moves with a teach pendant. The configurations of the robot are recorded and can be executed in the playback mode. The motion of the joints in this scheme is typically uncoordinated although the joints may be actuated in such a way that they reach their final positions simultaneously.

For robots that are physically led through the desired motion with a teach pendant, there is no need for calculation of the forward or inverse kinematics. The desired motion is simply recorded as a set of joint angles (actually as a set of encoder values) and the robot can be controlled entirely in joint space. For off-line programming, given the desired initial and final positions and orientations of the end-effector, the inverse kinematic solution must be evaluated to find the required initial and final joint variables.

Suppose that the initial and final configurations of the robot are specified as T^6_{0init} and T^6_{0final}. For a manipulator with a spherical wrist we may decouple the position and orientation as before. The motion of the first three joints is calculated by computing the joint variables q_1, q_2, and q_3 corresponding to d^3_{0init} and d^3_{0final}. The motion of the final three joint variables is found by computing a set of Euler angles corresponding to R^6_{3init} and R^6_{3final}.

For some purposes, such as obstacle avoidance, the path of the end-effector can be further constrained by the addition of **via points** intermediate to the initial and final configurations as illustrated in Figure 7-27. Additional constraints on the velocity or acceleration between via points, as for example in so-called **guarded motion**, shown in Figure 7-28, can be handled in the joint interpolation schemes treated in the next section.

We discuss here the problem of generating smooth trajectories in joint space. By a trajectory we mean a time history of desired joint positions, velocities, and if needed, joint accelerations.

We suppose that at time t_0 the i-th joint variable satisfies

$$\begin{aligned} q_i(t_0) &= q_0 \\ \dot{q}_i(t_0) &= q'_0 \end{aligned} \tag{7.5.1}$$

and we wish to attain the values at t_f

$$\begin{aligned} q_i(t_f) &= q_1 \\ \dot{q}_i(t_f) &= q'_1 \end{aligned} \tag{7.5.2}$$

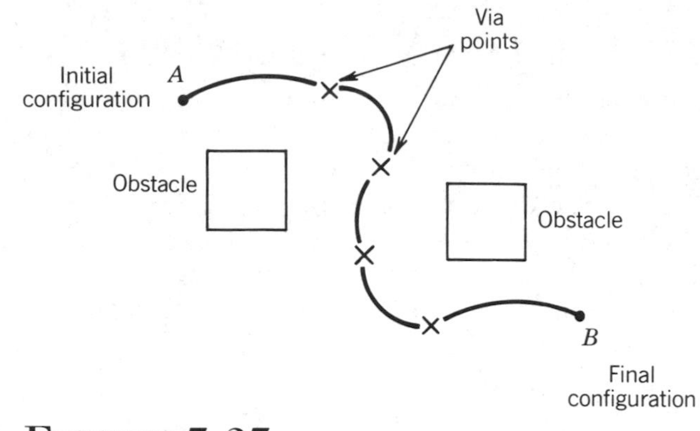

FIGURE 7-27
Via points to plan motion around obstacles.

Figure 7-29 shows a suitable trajectory for this motion.

One way to generate a smooth curve such as that shown is by a polynomial function of t. Since we have four constraints to satisfy,

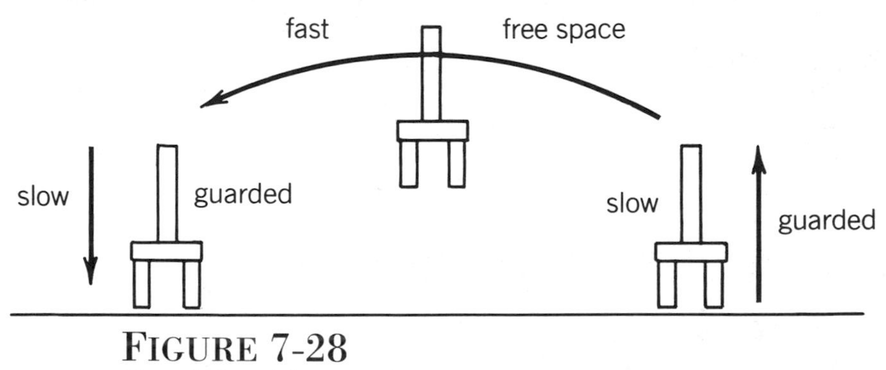

FIGURE 7-28
Guarded and free motions.

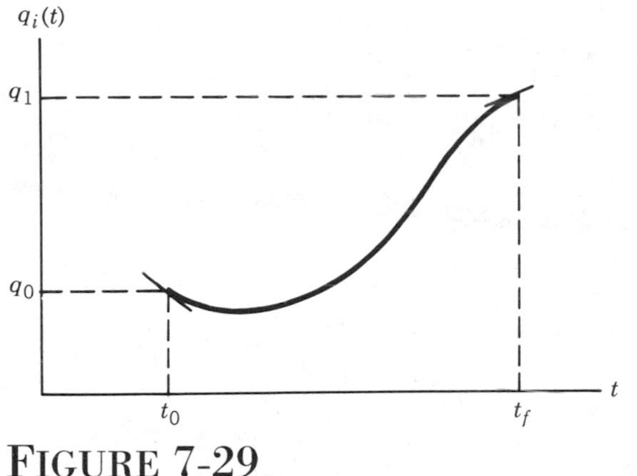

FIGURE 7-29
Typical joint space trajectory.

namely (7.5.1)–(7.5.2) we require a polynomial with four independent coefficients that can be chosen to satisfy these constraints. Thus we consider a cubic trajectory of the form

$$q_i^d = a_0 + a_1 t + a_2 t^2 + a_3 t^3 \tag{7.5.3}$$

Then the desired velocity is automatically given as

$$\dot{q}_i^d = a_1 + 2a_2 t + 3a_3 t^2 \tag{7.5.4}$$

Combining equations (7.5.3) and (7.5.4) with the four constraints yields four equations in four unknowns

$$q_0 = a_0 + a_1 t_0 + a_2 t_0^2 + a_3 t_0^3 \tag{7.5.5}$$

$$q'_0 = a_1 + 2a_2 t_0 + 3a_3 t_0^2 \tag{7.5.6}$$

$$q_1 = a_0 + a_1 t_f + a_2 t_f^2 + a_3 t_f^3 \tag{7.5.7}$$

$$q'_1 = a_1 + 2a_2 t_f + 3a_3 t_f^2 \tag{7.5.8}$$

These four equations can be combined into a single matrix equation

$$\begin{bmatrix} 1 & t_0 & t_0^2 & t_0^3 \\ 0 & 1 & 2t_0 & 3t_0^2 \\ 1 & t_f & t_f^2 & t_f^3 \\ 0 & 1 & 2t_f & 3t_f^2 \end{bmatrix} \begin{bmatrix} a_0 \\ a_1 \\ a_2 \\ a_3 \end{bmatrix} = \begin{bmatrix} q_0 \\ q'_0 \\ q_1 \\ q'_1 \end{bmatrix} \tag{7.5.9}$$

(*iii*) *Example 7.5.1*

Suppose $t_0 = 0$ and $t_f = 1$ sec, with

$$q'_0 = 0 \quad q'_1 = 0 \tag{7.5.10}$$

Thus we want to move from the initial position q_0 to the final position q_1 in 1 second, starting and ending with zero velocity. From the above formula (7.5.9) we obtain

$$\begin{bmatrix} 1 & 0 & 0 & 0 \\ 0 & 1 & 0 & 0 \\ 1 & 1 & 1 & 1 \\ 0 & 1 & 2 & 3 \end{bmatrix} \begin{bmatrix} a_0 \\ a_1 \\ a_2 \\ a_3 \end{bmatrix} = \begin{bmatrix} q_0 \\ 0 \\ q_1 \\ 0 \end{bmatrix} \tag{7.5.11}$$

This is then equivalent to the four equations

$$a_0 = q_0 \tag{7.5.12}$$

$$a_1 = 0 \tag{7.5.13}$$

$$a_2 + a_3 = q_1 - q_0 \tag{7.5.14}$$

$$2a_2 + 3a_3 = 0 \tag{7.5.15}$$

These latter two can be solved to yield

$$a_2 = 3(q_1 - q_0) \tag{7.5.16}$$

$$a_3 = -2(q_1 - q_0) \tag{7.5.17}$$

The required cubic polynomial function is therefore

$$q_i(t) = q_0 + 3(q_1 - q_0)t^2 - 2(q_1 - q_0)t^3 \tag{7.5.18}$$

Figure 7-30 shows this trajectory with $q_0 = 10^o$, $q_1 = -20^o$. The corresponding velocity and acceleration curves are given in Figures 7-31 and 7-32.

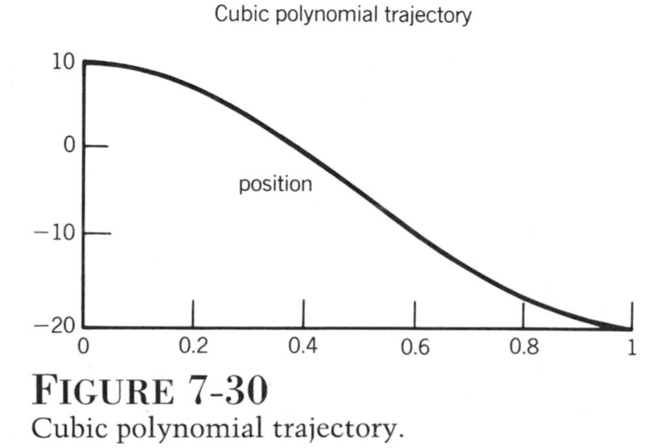

FIGURE 7-30
Cubic polynomial trajectory.

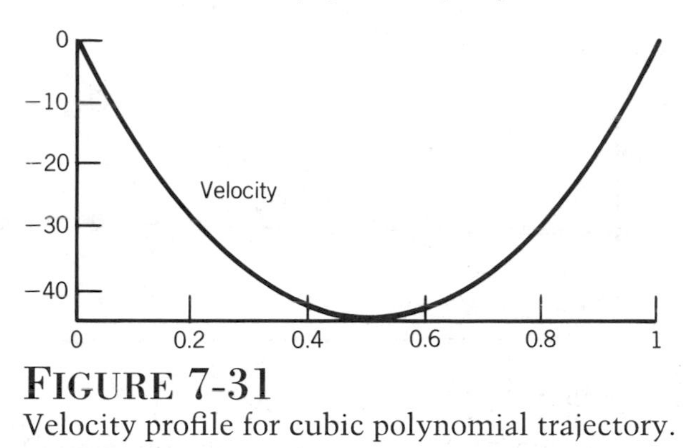

FIGURE 7-31
Velocity profile for cubic polynomial trajectory.

(iv) *General Solution*

It can be shown that the determinant of the coefficient matrix in (7.5.9) is $(t_f - t_0)^4$ and hence as long as $t_0 \neq t_f$ Equation 7.5.9 always has a unique solution. To find the general solution, let us suppose first that $t_0 = 0$. Then (7.5.9) becomes

$$\begin{bmatrix} 1 & 0 & 0 & 0 \\ 0 & 1 & 0 & 0 \\ 1 & t_f & t_f^2 & t_f^3 \\ 0 & 1 & 2t_f & 3t_f^2 \end{bmatrix} \begin{bmatrix} a_0 \\ a_1 \\ a_2 \\ a_3 \end{bmatrix} = \begin{bmatrix} q_0 \\ q'_0 \\ q_1 \\ q'_1 \end{bmatrix} \tag{7.5.19}$$

The first two equations in (7.5.19) yield

$$a_0 = q_0 \; ; \quad a_1 = q'_0 \tag{7.5.20}$$

Similarly, the last two equations can be solved to yield

$$a_2 = \frac{3(q_1 - q_0) - (2q'_0 + q'_1)t_f}{t_f^2} \qquad a_3 = \frac{2(q_0 - q_1) + (q'_0 + q'_1)t_f}{t_f^3} \tag{7.5.21}$$

Figure 7-33 shows a cubic trajectory generated from (7.5.19) and (7.5.20) with $q_0 = 10$, $q_1 = 40$, $q'_0 = q'_1 = -50$. The corresponding velocity and acceleration curves are shown in Figure 7-34 and 7-35, respectively.

If the initial time t_0 is not zero, we may use the above derivation, by simply shifting the time axis to the left by t_0. In other words, we replace t by $t - t_0$ in (7.5.3) and replace t_f by $t_f - t_0$ in (7.5.21).

7.5.1 SUMMARY

Carrying out the above time shift, we have an algorithm for generating cubic trajectories for arbitrary initial and final conditions:

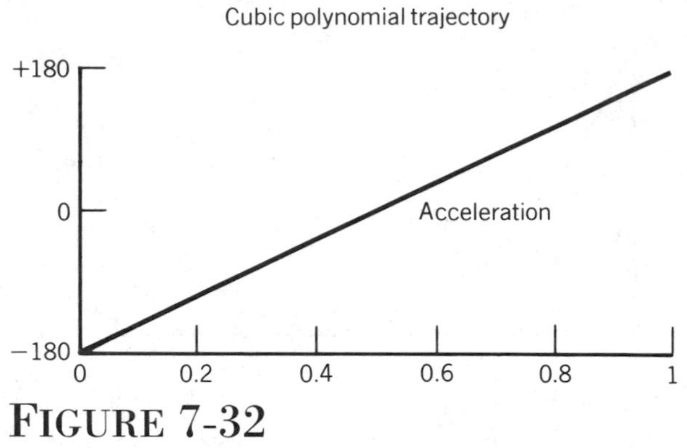

FIGURE 7-32
Acceleration profile for cubic polynomial trajectory.

Given initial and final times, t_0 and t_f, respectively, with

$$q^d(t_0) = q_0 \ ; \ q^d(t_f) = q_1 \tag{7.5.22}$$
$$\dot{q}_d(t_0) = q'_0 \ ; \ \dot{q}^d(t_f) = q'_1$$

the required cubic polynomial $q^d(t)$ can be computed from

$$q^d(t) = a_0 + a_1(t - t_0) + a_2(t - t_0)^2 + a_3(t - t_0)^3 \tag{7.5.23}$$

where

$$a_0 = q_0 \ ; \ a_1 = q'_0$$

$$a_2 = \frac{3(q_1 - q_0) - (2q'_0 + q'_1)(t_f - t_0)}{(t_f - t_0)^2} \qquad a_3 = \frac{2(q_0 - q_1) + (q'_0 + q'_1)(t_f - t_0)}{(t_f - t_0)^3}$$

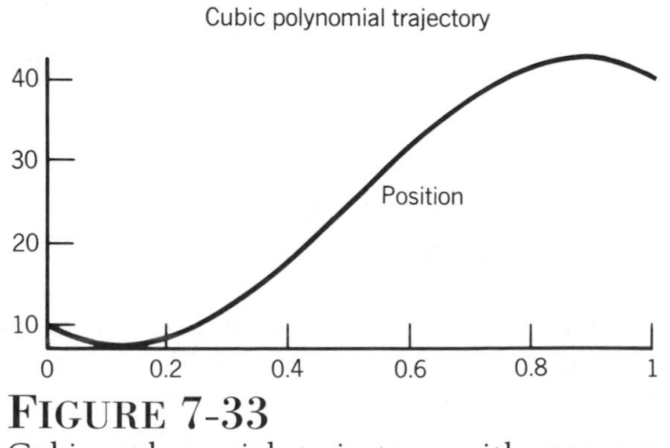

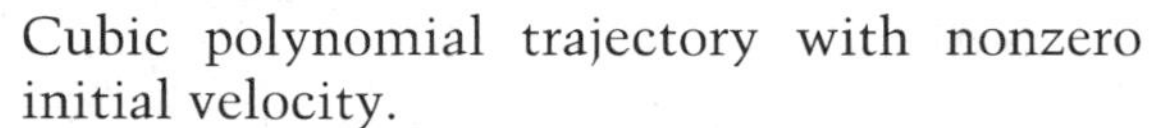

FIGURE 7-33
Cubic polynomial trajectory with nonzero initial velocity.

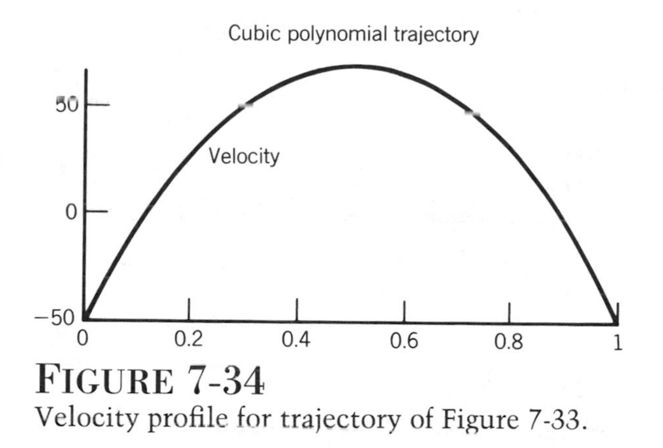

FIGURE 7-34
Velocity profile for trajectory of Figure 7-33.

A sequence of moves can be planned using the above formula by using the end conditions q_1, q'_1 of the i-th move as initial conditions for the i+1-st move. For example, Figure 7-36 shows a 6-second move, computed in three parts using (7.5.23), where the trajectory begins at 10° and is required to reach 40° at 2-seconds, 30° at 4-seconds, and 90° at 6-seconds, with zero velocity at 0,2,4, and 6 seconds.

7.5.2 LINEAR SEGMENTS WITH PARABOLIC BLENDS (LSPB)

Another way to generate suitable joint space trajectories is by so-called **Linear Segments with Parabolic Blends** or (**LSPB**) for short. This type of trajectory is appropriate when a constant velocity is desired along a por-

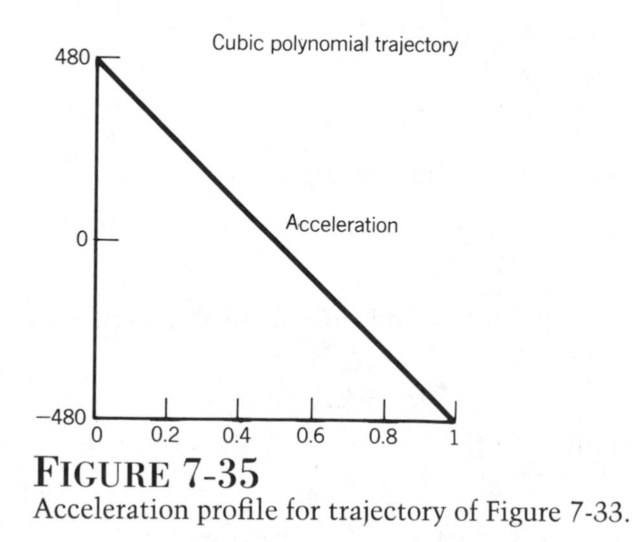

FIGURE 7-35
Acceleration profile for trajectory of Figure 7-33.

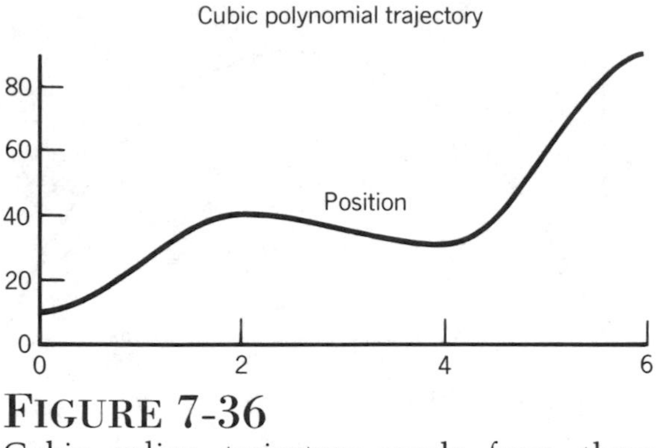

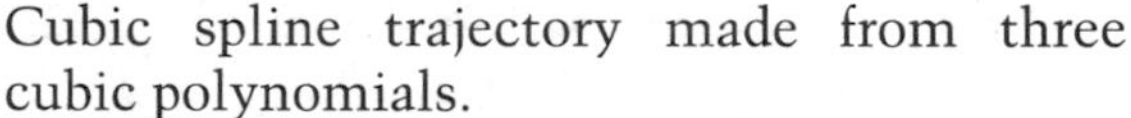

FIGURE 7-36
Cubic spline trajectory made from three cubic polynomials.

tion of the path. The LSPB trajectory is such that the velocity is "ramped up" to its specific value initially and then "ramped down" at the goal position. To achieve this we specify the desired position in three parts. The first part from time t_0 to time t_b is a quadratic polynomial. This results in a linear "ramp" velocity. At time t_b, called the **blend time**, the position trajectory switches to a linear function. This corresponds to a constant velocity. Finally at time $t_f - t_b$ the position trajectory switches once again to a quadratic polynomial so that the velocity is linear.

We choose the blend time t_b so that the position curve is symmetric as shown in Figure 7-37. For convenience suppose that $t_0 = 0$ and $\dot{q}_i(t_f) = 0 = \dot{q}_i(0)$. Then between times 0 and t_b we have

$$q_i^d(t) = a_0 + a_1 t + a_2 t^2 \tag{7.5.24}$$

so that the velocity is

$$\dot{q}_i^d(t) = a_1 + 2a_2 t \tag{7.5.25}$$

The constraints $q_0 = 0$ and $\dot{q}(0) = 0$ imply that

$$a_0 = q_0 \tag{7.5.26}$$

$$a_1 = 0 \tag{7.5.27}$$

Since at time t_b we want the velocity to equal a given constant V we have

$$\dot{q}_i^d(t_b) = 2a_2 t_b = V \tag{7.5.28}$$

which implies that

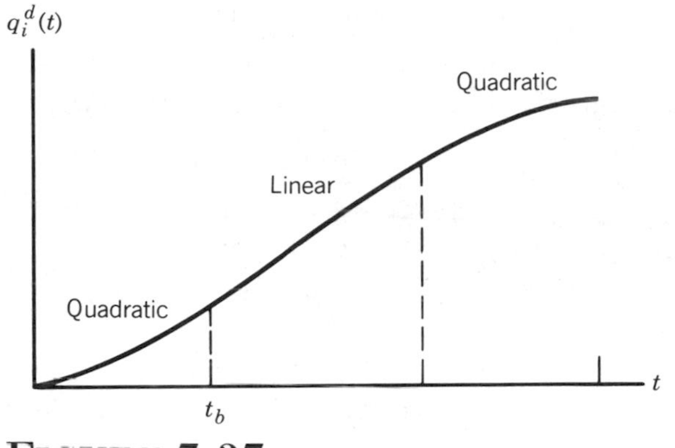

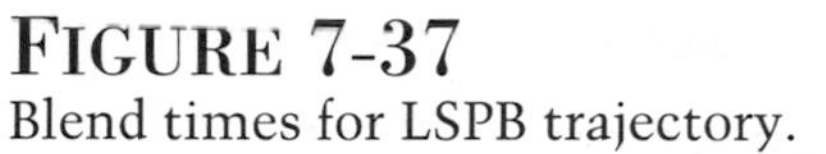
FIGURE 7-37
Blend times for LSPB trajectory.

$$a_2 = \frac{V}{2t_b} \tag{7.5.29}$$

Therefore the required trajectory between 0 and t_b is given as

$$q_i^d(t) = q_0 + \frac{V}{2t_b}t^2 \tag{7.5.30}$$

$$= q_0 + \frac{a}{2}t^2$$

$$\dot{q}_i^d(t) = \frac{V}{t_b}t = at \tag{7.5.31}$$

$$\ddot{q}_i^d = \frac{V}{t_b} = a \tag{7.5.32}$$

where a denotes the acceleration.

Now, between time t_f and $t_f - t_b$, the trajectory is a linear segment (corresponding to a constant velocity V)

$$q_i^d(t) = \alpha_0 + \alpha_1 t = \alpha_0 + Vt \tag{7.5.33}$$

Since, by symmetry,

$$q_i^d\left(\frac{t_f}{2}\right) = \frac{q_0 + q_1}{2} \tag{7.5.34}$$

we have

$$\frac{q_0 + q_1}{2} = \alpha_0 + V\frac{t_f}{2} \tag{7.5.35}$$

which implies that

$$\alpha_0 = \frac{q_0+q_1-Vt_f}{2} \tag{7.5.36}$$

Since the two segments must "blend" at time t_b we require

$$q_0+\frac{V}{2}t_b = \frac{q_0+q_1-Vt_f}{2}+Vt_b \tag{7.5.37}$$

which gives upon solving for the blend time t_b

$$t_b = \frac{q_0-q_1+Vt_f}{V} \tag{7.5.38}$$

Note that we have the constraint $0 < t_b \leq \frac{t_f}{2}$. This leads to the inequality

$$\frac{q_f-q_0}{V} < t_f \leq \frac{2(q_f-q_0)}{V} \tag{7.5.39}$$

To put it another way we have the inequality

$$\frac{q_f-q_0}{t_f} < V \leq \frac{2(q_f-q_0)}{t_f} \tag{7.5.40}$$

Thus the specified velocity must be between these limits or the motion is not possible.

The portion of the trajectory between $t_f - t_b$ and t_f is now found by symmetry considerations (Problem 8-6). The complete LSPB trajectory is given by

$$q_i^d(t) = \begin{cases} q_0+\frac{a}{2}t^2 & 0 \leq t \leq t_b \\ \frac{q_f+q_0-Vt_f}{2}+Vt & t_b < t \leq t_f - t_b \\ q_f - \frac{at_f^2}{2}+at_f t-\frac{a}{2}t^2 & t_f - t_b < t \leq t_f \end{cases} \tag{7.5.41}$$

Figure 7-38 shows such an LSPB trajectory, where the maximum velocity $V = 60$. In this case $t_b = \frac{1}{3}$. The velocity and acceleration curves are given in Figures 7-39 and 7-40, respectively.

7.5.3 MINIMUM TIME TRAJECTORIES

An important variation of this trajectory is obtained by leaving the final time t_f unspecified and seeking the "fastest" trajectory between q_0 and q_1 with a given constant acceleration a, that is, the trajectory with the final time t_f a minimum. This is sometimes called a **Bang-Bang** trajectory since the optimal solution is achieved with the acceleration at its maximum value $+a$ until an appropriate **switching time** t_s at which time it abruptly switches to its minimum value $-a$ (maximum deceleration) from t_s to t_f.

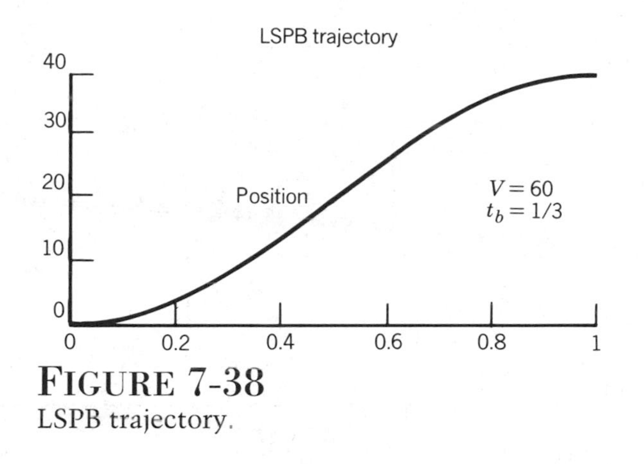

FIGURE 7-38
LSPB trajectory.

Returning to our simple example in which we assume that the trajectory begins and ends at rest, that is, with zero initial and final velocities, symmetry considerations would suggest that the switching time t_s is just $\frac{t_f}{2}$. This is indeed the case. For nonzero initial and/or final velocities, the situation is more complicated and we will not discuss it here.

If we let V_s denote the velocity at time t_s then we have

$$V_s = a\, t_s \tag{7.5.42}$$

and also

$$t_s = \frac{q_0 - q_1 + V_s t_f}{V_s} \tag{7.5.43}$$

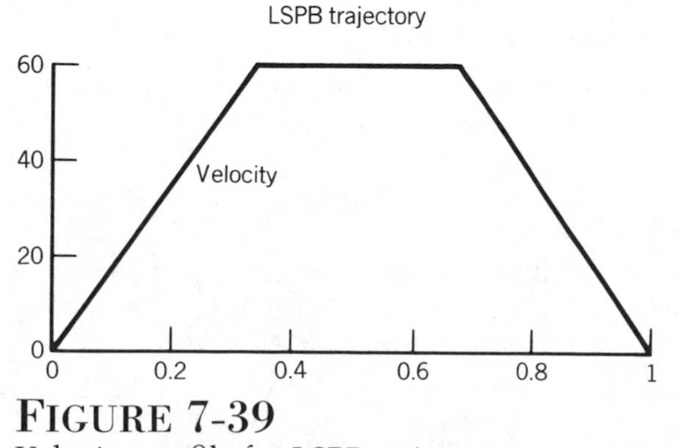

FIGURE 7-39
Velocity profile for LSPB trajectory.

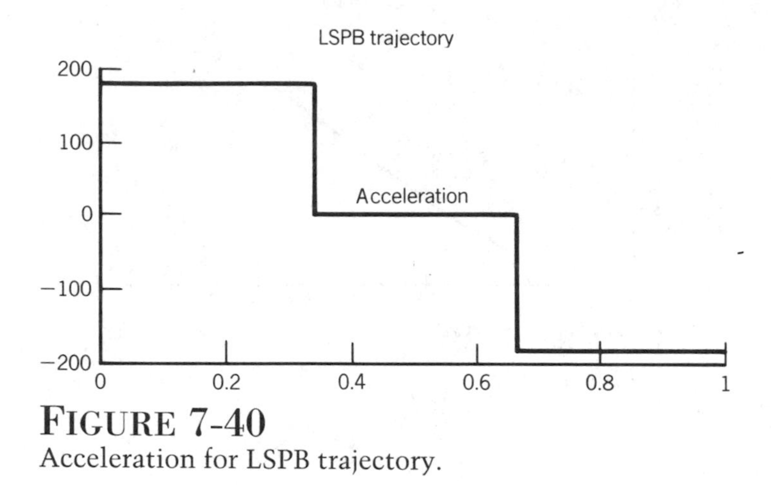

FIGURE 7-40
Acceleration for LSPB trajectory.

The symmetry condition $t_s = \frac{t_f}{2}$ implies that

$$V_s = \frac{q_1 - q_0}{t_s} \tag{7.5.44}$$

Combining these two we have the conditions

$$\frac{q_1 - q_0}{t_s} = a\ t_s \tag{7.5.45}$$

which implies that

$$t_s = \sqrt{q_1 - q_0/2} \tag{7.5.46}$$

7.6 FEEDFORWARD CONTROL AND COMPUTED TORQUE

In this section we introduce the notion of **feedforward control** as a method to track time varying trajectories and reject time varying disturbances.

Suppose that $r(t)$ is an arbitrary reference trajectory and consider the block diagram of Figure 7-41, where $G(s)$ represents the forward transfer function of a given system and $H(s)$ is the compensator transfer function. A feedforward control scheme consists of adding a feedforward path with transfer function $F(s)$ as shown.

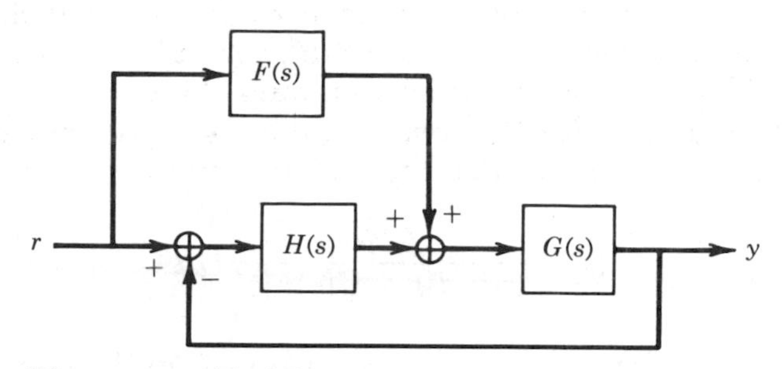

FIGURE 7-41
Feedforward control scheme.

Let each of the three transfer functions be represented as ratios of polynomials

$$G(s) = \frac{q(s)}{p(s)} \quad H(s) = \frac{c(s)}{d(s)} \quad F(s) = \frac{a(s)}{b(s)} \tag{7.6.1}$$

We assume that $G(s)$ is strictly proper and $H(s)$ is proper. Simple block diagram manipulation shows that the closed loop transfer function $T(s) = \frac{Y(s)}{R(s)}$ is given by (Problem 7-9)

$$T(s) = \frac{q(s)(c(s)b(s) + a(s)d(s))}{b(s)(p(s)d(s) + q(s)c(s))} \tag{7.6.2}$$

The closed loop characteristic polynomial of the system is then $b(s)(p(s)d(s) + q(s)c(s))$. For stability of the closed loop system therefore we require that the compensator $H(s)$ and the feedforward transfer function $F(s)$ be chosen so that the polynomials $p(s)d(s) + q(s)c(s)$ and $b(s)$ are Hurwitz. This says that, in addition to stability of the closed loop system the feedforward transfer function $F(s)$ must itself be stable.

If we choose the feedforward transfer function $F(s)$ equal to $\frac{1}{G(s)}$, the inverse of the forward plant, that is, $a(s) = p(s)$ and $b(s) = q(s)$, then the closed loop system becomes

$$q(s)(p(s)d(s) + q(s)c(s))Y(s) = q(s)(p(s)d(s) + q(s)c(s))R(s) \tag{7.6.3}$$

or, in terms of the tracking error $E(s) = R(s) - Y(s)$,

$$q(s)(p(s)d(s) + q(s)c(s))E(s) = 0 \tag{7.6.4}$$

Thus, assuming stability, the output $y(t)$ will track any reference trajectory $r(t)$. Note that we can only choose $F(s)$ in this manner provided that the numerator polynomial $q(s)$ of the forward plant is Hurwitz, that is, as long as all zeros of the forward plant are in the left half plane. Such systems are called **minimum phase**.

If there is a disturbance $D(s)$ entering the system as shown in Figure 7-42, then it is easily shown that the tracking error $E(s)$ is given by

$$E(s) = \frac{q(s)d(s)}{p(s)d(s) + q(s)c(s)} D(s) \tag{7.6.5}$$

We have thus shown that, in the absence of disturbances the closed loop system will track *any* desired trajectory $r(t)$ provided that the closed loop system is stable. The steady state error is thus due only to the disturbance.

Let us apply this idea to the robot model of Section 7.3. Suppose that $\theta^d(t)$ is an arbitrary trajectory that we wish the system to track. In this case we have from (7.3.6) $G(s) = \frac{K}{J_{eff}s^2 + B_{eff}s}$ together with a PD compensator $H(s) = K_P + K_D s$. The resulting system is shown in Figure 7-43. Note that $G(s)$ has no zeros at all and hence is minimum phase. Note also that $G(s)^{-1}$ is not a proper rational function. However, since the derivatives of the reference trajectory θ^d are known and precomputed, the implementation of the above scheme does not require differentiation of an actual signal. It is easy to see from (7.6.5) that the steady state error to a step disturbance is now given by the same expression (7.3.14) independent of the reference trajectory. As before, a PID compensator would result in zero steady state error to a step disturbance. In the time domain the control law of Figure 7-43 can be written as

$$\begin{aligned} V(t) &= \frac{J_{eff}}{K}\ddot{\theta}^d + \frac{B_{eff}}{K}\dot{\theta}^d + K_D(\dot{\theta}^d - \dot{\theta}_m) + K_P(\theta^d - \theta_m) \\ &= f(t) + K_D\dot{e}(t) + K_P e(t) \end{aligned} \tag{7.6.6}$$

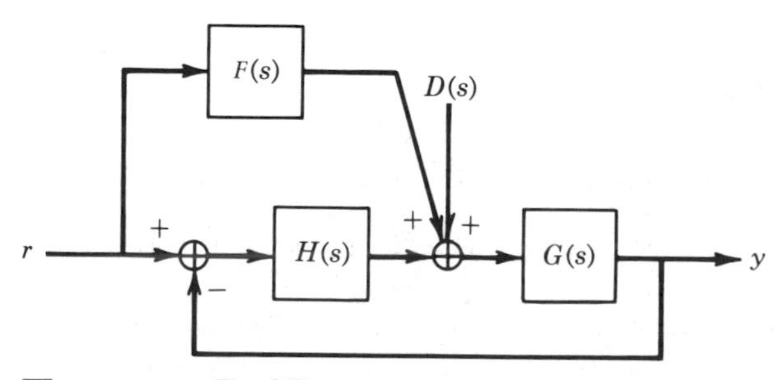

FIGURE 7-42
Feedforward control with disturbance.

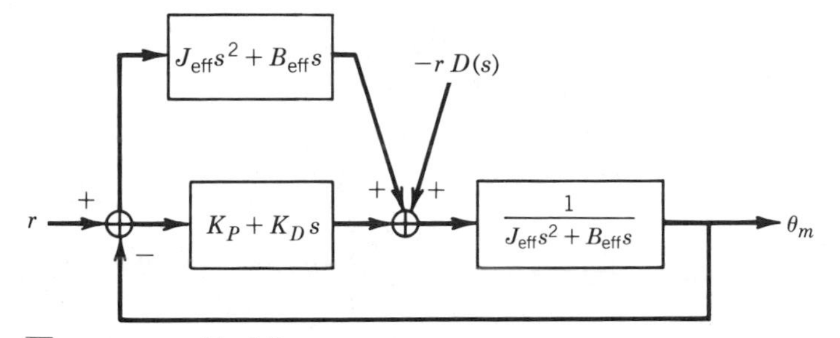

FIGURE 7-43
Feedforward compensator for second order system.

where $f(t)$ is the feedforward signal

$$f(t) = \frac{J_{eff}}{K}\ddot{\theta}^d + \frac{B_{eff}}{K}\dot{\theta}^d \tag{7.6.7}$$

and $e(t)$ is the tracking error $\theta^d(t) - \theta(t)$. Since the forward plant equation is

$$J_{eff}\,\ddot{\theta}_m + B_{eff}\,\dot{\theta}_m = KV(t) - r\,d(t)$$

the closed loop error $e(t) = \theta_m - \theta^d$ satisfies the second order differential equation

$$J_{eff}\,\ddot{e} + (B_{eff} + KK_D)\dot{e} + KK_P e(t) = -r\,d(t) \tag{7.6.8}$$

(i) *Remark 7.6.1*

We note from (7.6.8) that the characteristic polynomial of the closed loop system is identical to (7.3.10). The system now however is written in terms of the tracking error $e(t)$. Therefore, assuming that the closed loop system is stable, the tracking error will approach zero asymptotically for any desired joint space trajectory in the absence of disturbances, that is, if $d=0$.

7.6.1 FEEDFORWARD DISTURBANCE CANCELLATION: COMPUTED TORQUE

We see that the feedforward signal (7.6.7) results in asymptotic tracking of any trajectory in the absence of disturbances but does not otherwise improve the disturbance rejection properties of the system. However, although the term $\mathbf{d}(t)$ in (7.6.8) represents a disturbance, it is not completely unknown since $\mathbf{d}$ satisfies (7.3.7). Thus we may consider adding to the above feedforward signal, a term to anticipate the effects of the disturbance $\mathbf{d}(t)$. Consider the diagram of Figure 7-44. Given a desired trajectory, then we superimpose, as shown, the term

$$d^d := \sum d_{jk}(\mathbf{q}^d)\ddot{q}_j^d + \sum c_{ijk}(\mathbf{q}^d)\dot{q}_i^d\dot{q}_j^d + g_k(\mathbf{q}^d) \tag{7.6.9}$$

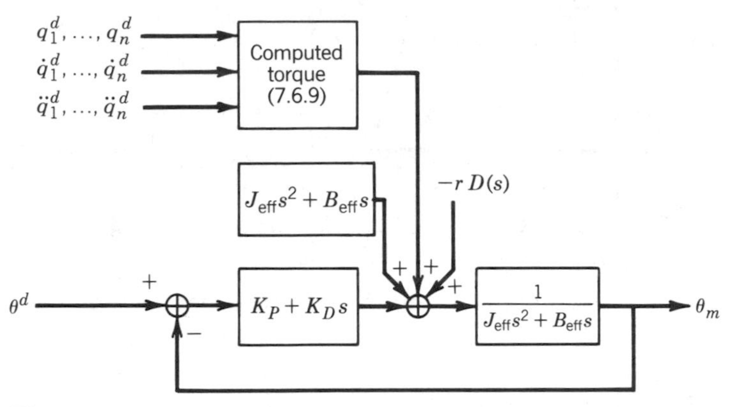

FIGURE 7-44
Feedforward computed torque compensation.

Since d^d has units of torque, the above feedforward disturbance cancellation control is called the **method of computed torque**. The expression (7.6.9) thus compensates in a feedforward manner the nonlinear coupling inertial, coriolis, centripetal, and gravitational forces arising due to the motion of the manipulator. Although the difference $\Delta d := d^d - d$ is zero only in the ideal case of perfect tracking ($\theta = \theta^d$) and perfect computation of (7.6.9), in practice, Δd can be expected to be smaller than d and hence the computed torque has the advantage of reducing the effects of d. Note that the expression (7.6.9) is in general extremely complicated so that the computational burden involved in computing (7.6.9) is of major concern. In fact the problem of real-time implementation of this computed torque control has stimulated a great deal of research. The development of efficient recursive formulations of manipulator dynamics, such as the recursive Newton–Euler equations of Chapter Six, was partly motivated by the need to compute the expression (7.6.9) in real-time.

Since only the values of the desired trajectory need to be known, many of these terms can be precomputed and stored off-line. Thus there is a trade-off between memory requirements and on-line computational requirements. This has led to the development of table look-up schemes to implement (7.6.9) and also to the development of computer programs for the automatic generation and simplification of manipulator dynamic equations.

REFERENCES AND SUGGESTED READING

[1] ALBERT, M., and SPONG, M.W., "Linear and Nonlinear Controller Design for Elastic Joint Manipulators," Coordinated Science Laboratory, University of Illinois, Report UILU-ENG-87-2251, Aug 1987.

[2] ARIMOTO, S., and MIYAZAKI, F., "Stability and Robustness of P.I.D. Feedback Control for Robot Manipulators of Sensory Capability," *First Int. Symp. Robotics Res.*, 1983.

[3] ARIMOTO, S., and MIYAZAKI, F., "Stability and Robustness of PD Feedback Control with Gravity Compensation for Robot Manipulators," *Robotics: Theory and Practice*, DSC-Vol. 3, pp. 67-72, ASME Winter Annual Meeting, Anaheim, Dec. 1986.

[4] AHMAD, S., "Analysis of Robot Drive Train Errors, Their Static Effects, and Their Compensations," preprint, 1987.

[5] BEJCZY, A.K., "Robot Arm Dynamics and Control," *JPL Tech. Memo.* 33-69, Feb. 1974.

[6] FRANKLIN, G.F., POWELL, J.D., and EMAMI-NAEINI, A., *Feedback Control of Dynamic Systems*, Addison–Wesley, Reading, MA, 1986.

[7] GOOD, M.C., SWEET, L.M., and STROBEL, K.L., "Dynamic Models for Control System Design of Integrated Robot and Drive Systems," *ASME J. Dyn. Sys., Meas., and Cont.*, Vol. 107, Mar. 1985.

[8] KAILATH, T., *Linear Systems*, Prentice-Hall, Englewood Cliffs, NJ, 1980.

[9] KARNI, S., and BYATT, W.J., *Mathematical Methods in Continuous and Discrete Systems*, Holt, Rinehart, and Winston, New York, 1982.

[10] KUO, B.C., *Automatic Control Systems*, Fourth Edition, Prentice-Hall, Englewood Cliffs, NJ, 1982.

[11] LUH, J.Y.S., "An Anatomy of Industrial Robots and Their Controls," *IEEE Trans. Aut. Control*, Vol AC-28, No.2, Feb. 1983.

[12] LUH, J.Y.S., "Conventional Controller Design for Industrial Robots—A Tutorial," *IEEE Trans. Sys. Man. Cyber*, SMC-13, NO. 3, May/June 1983.

[13] RIVIN, E. I., *Mechanical Design of Robots*, McGraw-Hill, St. Louis, 1988.

PROBLEMS

7-1 Using block diagram reduction techniques derive the transfer functions (7.2.12) and (7.2.13).

7-2 Derive the transfer functions for the reduced order model (7.2.14)–(7.2.15).

7-3 Derive Equations 7.3.9, 7.3.10 and 7.3.11.

7-4 Derive Equations 7.3.19–7.3.20.

7-5 Derive Equations 7.4.7, 7.4.8, and 7.4.9.

7-6 Given the state space model (7.4.13) show that the transfer function

$$G(s) = \mathbf{c}^T(sI - A)^{-1}\mathbf{b}$$

is identical to (7.4.7).

7-7 Search the control literature (e.g., [8]) and find two or more algorithms for the pole assignment problem for linear systems. Design a state feedback control law for (7.4.13) using the parameter values given in Example 7.4.1 so that the poles are at $s = -10$. Simulate the step response. How does it compare to the response of Example 7.4.1? How do the torque profiles compare?

7-8 Design an observer for the system (7.4.13) using the parameter values of Example 7.4.1. Choose reasonable locations for the observer poles. Simulate the combined observer/state feedback control law using the results of Problem 7-7.

7-9 Derive (7.4.22) and (7.4.30).

7-10 Given a three-link elbow type robot, a three-link SCARA robot and a three-link cartesian robot, discuss the differences in the dynamics of each type of robot as they impact the control problem. Discuss the nature of the coupling nonlinearities, the effect of gravity, and inertial variations as the robots move about. For which manipulator would you expect PD control to work best? worst?

7-11 Consider the two-link cartesian robot of Example 6.4.1. Suppose that each joint is actuated by a permanent magnet DC-motor. Write the complete dynamics of the robot assuming perfect gears.

7-12 Carry out the details of a PID control for the two-link cartesian robot of Problem 7-11. Note that the system is linear and the gravitational forces are configuration independent. What does this say about the validity of this approach to control?

7-13 Simulate the above PID control law. Choose reasonable numbers for the masses, inertias, etc. Also place reasonable limits on the magnitude of the control input. Use various methods, such as root locus, Bode plots, etc. to design the PID gains.

7-14 Search the control literature (e.g., [6]) to find out what is meant by **integrator windup**. Did you experience this problem with the PID control law of Problem 7-13? Find out what is meant by **anti-windup** (or anti-reset windup). Implement the above PID control with anti-reset windup. Is the response better?

7-15 Repeat the above analysis and control design (Problems 7-11—7-14) for the two-link elbow manipulator of Example 6.4.2. Note that you will have to make some assumptions to arrive at a value of the effective inertias J_{eff}.

7-16 Repeat Problem 7-15 for the two-link elbow manipulator with remote drive of Example 6.4.3.

7-17 Include the dynamics of a permanent magnet DC-motor for the system (7.4.1)–(7.4.2). What can you say now about controllability and observability of the system?

7-18 Choose appropriate state variables and write the system (7.2.10)–(7.2.11) in state space. What is the order of the state space?

7-19 Repeat 7-18 for the reduced order system (7.2.16).

7-20 Suppose in the flexible joint system represented by (7.4.1)–(7.4.2) the following parameters are given

$$J_\ell = 10 \quad B_\ell = 1 \quad k = 100$$
$$J_m = 2 \quad B_m = 0.5$$

(a) Sketch the open loop poles of the transfer functions (7.4.7).

(b) Apply a PD compensator to the system (7.4.7). Sketch the root locus for the system. Choose a reasonable location for the compensator zero. Using the Routh criterion find the value of the compensator gain K when the root locus crosses the imaginary axis.

7-21 One of the problems encountered in space applications of robots is the fact that the base of the robot cannot be anchored, that is, cannot be fixed in an inertial coordinate frame. Consider the idealized situation shown in Figure 7-45, consisting of an inertia J_1 connected to the rotor of a motor whose stator is connected to an inertia J_2. For example, J_1 could represent the space shuttle robot arm and J_2 the inertia of the shuttle itself. The simplified equations of motion are thus

$$J_1\ddot{q}_1 = \tau$$
$$J_2\ddot{q}_2 = \tau$$

Write this system in state space form and show that it is uncontrollable. Discuss the implications of this and suggest possible solutions.

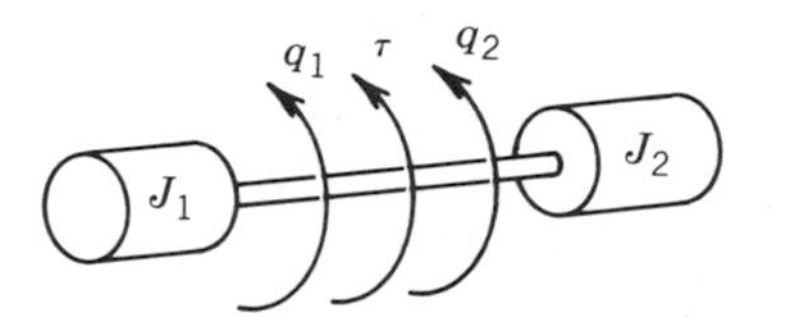

FIGURE 7-45

7-22 Given the linear second order system

$$\begin{bmatrix} \dot{x}_1 \\ \dot{x}_2 \end{bmatrix} = \begin{bmatrix} 1 & -3 \\ 1 & -2 \end{bmatrix} \begin{bmatrix} x_1 \\ x_2 \end{bmatrix} + \begin{bmatrix} 1 \\ -2 \end{bmatrix} u$$

find a linear state feedback control $u = k_1x_1 + k_2x_2$ so that the closed loop system has poles at $s = -2, -2$.

7-23 Repeat the above if possible for the system

$$\begin{bmatrix} \dot{x}_1 \\ \dot{x}_2 \end{bmatrix} = \begin{bmatrix} -1 & 0 \\ 0 & 2 \end{bmatrix} \begin{bmatrix} x_1 \\ x_2 \end{bmatrix} + \begin{bmatrix} 0 \\ 1 \end{bmatrix} u$$

Can the closed loop poles be placed at -2?
Can this system be stabilized? Explain.
[Remark: The system of Problem 7-23 is said to be *stabilizable,* which is a weaker notion than controllability.]

7-24 Repeat the above for the system

$$\begin{bmatrix} \dot{x}_1 \\ \dot{x}_2 \end{bmatrix} = \begin{bmatrix} +1 & 0 \\ 0 & 2 \end{bmatrix} \begin{bmatrix} x_1 \\ x_2 \end{bmatrix} + \begin{bmatrix} 0 \\ 1 \end{bmatrix} u$$

7-25 Show by direct calculation that the determinant of the coefficient matrix in Equation 7.5.9 is $(t_f - t_0)^4$.

7-26 Use Gaussian elimination to reduce the system (7.5.9) to upper triangular form and verify that the solution is indeed given by Equation 7.5.23.

7-27 Suppose we wish a manipulator to start from an initial configuration at time t_0 and track a conveyor. Discuss the steps needed in planning a suitable trajectory for this problem.

7-28 Suppose we desire a joint space trajectory $\dot{q}_i^d(t)$ for the i-th joint (assumed to be revolute) that begins at rest at position q_0 at time t_0 and reaches position q_1 in 2 seconds with a final velocity of 1 radian/sec. Compute a cubic polynomial satisfying these constraints. Sketch the trajectory as a function of time.

7-29 Compute a LSPB trajectory to satisfy the same requirements as in 7-28. Sketch the resulting position, velocity, and acceleration profiles.

7-30 Fill in the details of the computation of the LSPB trajectory. In other words compute the portion of the trajectory between times $t_f - t_b$ and t_f and hence verify Equations 7.5.41.

7-31 Verify Equations 7.6.2 and 7.6.4.

7-32 Consider the block diagram of Figure 7-41. Suppose that $G(s) = \frac{1}{2s^2+s}$, and suppose that it is desired to track a reference signal $r(t) = \sin(t) + \cos(2t)$. If we further specify that the closed loop system should have a natural frequency less than 10 radians with a damping ratio greater than 0.707, compute an appropriate compensator $C(s)$ and feedforward transfer function $F(s)$.

CHAPTER EIGHT

MULTIVARIABLE CONTROL

8.1 INTRODUCTION

In the previous chapter we discussed techniques to derive a control law for each joint of a manipulator based on a single-input/single-output model. Coupling effects among the joints were regarded as disturbances to the individual systems. In reality, the dynamic equations of a robot manipulator form a complex, nonlinear, and multivariable system. In this chapter, therefore, we treat the robot control problem in the context of nonlinear, multivariable control. This approach allows us to provide more rigorous analysis of the performance of control systems, and also allows us to design robust nonlinear control laws that guarantee global stability and tracking of arbitrary trajectories.

8.2 PD CONTROL REVISITED

We first give a more mathematically rigorous discussion of the application of PD control for the set-point control of rigid robots.[1] Let us first reformulate the nonlinear dynamic manipulator equations in a form

[1]The reader should review the discussion on Lyapunov Stability in Appendix C.

more suitable for the discussion to follow. Recall the robot equations of motion (7.3.2) and (7.3.3)

$$\sum_{j=1}^{n} d_{jk}(\mathbf{q})\ddot{q}_j + \sum_{i,j=1}^{n} c_{ijk}(\mathbf{q})\dot{q}_i\dot{q}_j + g_k(\mathbf{q}) = \tau_k \tag{8.2.1}$$

$$J_{m_k}\ddot{\theta}_{m_k} + (B_{m_k} + K_b K_m/R)\dot{\theta}_{m_k} = K_m/Rv_k - r_k\tau_k \tag{8.2.2}$$

Dividing (8.2.2) by r_k and using the fact that

$$\theta_{m_k} = \frac{1}{r_k}q_k \tag{8.2.3}$$

we write Equation 8.2.2 as

$$\frac{1}{r_k^2}J_m\ddot{q}_k + \frac{1}{r_k^2}B\dot{q}_k = \frac{K_m}{r_k R}v_k - \tau_k \tag{8.2.4}$$

where $B = B_{m_k} + K_b L_m/R$. Substituting (8.2.4) into (8.2.1) yields

$$\frac{1}{r_k^2}J_{m_k}\ddot{q}_k + \sum_{j=1}^{n} d_{jk}\ddot{q}_j + \sum_{i,j=1}^{n} c_{ijk}\dot{q}_i\dot{q}_j + \frac{1}{r_k^2}B\dot{q}_k + g_k = \frac{K_m}{r_k R}v_k \tag{8.2.5}$$

In matrix form these equations of motion can be written as

$$(D(\mathbf{q}) + J)\ddot{\mathbf{q}} + C(\mathbf{q},\dot{\mathbf{q}})\dot{\mathbf{q}} + B\dot{\mathbf{q}} + g(\mathbf{q}) = \mathbf{u} \tag{8.2.6}$$

where $D(\mathbf{q})$ is the $n \times n$ inertia matrix of the manipulator, and J is a diagonal matrix with diagonal elements $\frac{1}{r_k^2}J_{m_k}$. The vector $\mathbf{g}(\mathbf{q})$ and the matrix $C(\mathbf{q},\dot{\mathbf{q}})$ are defined by (6.3.12) and (6.3.13), respectively, and the input vector $\mathbf{u}$ has components

$$u_k = \frac{K_m}{r_k R}v_k$$

Note that u_k has units of torque.

An independent joint PD-control scheme can be written in vector form as

$$\mathbf{u} = K_P\tilde{\mathbf{q}} - K_D\dot{\mathbf{q}} \tag{8.2.7}$$

where $\tilde{\mathbf{q}} = \mathbf{q}^d - \mathbf{q}$ is the error between the desired joint displacements $\mathbf{q}^d$ and the actual joint displacements $\mathbf{q}$, and K_P, K_D are diagonal matrices of (positive) proportional and derivative gains, respectively. We first show that, in the absense of gravity, that is, if $\mathbf{g}$ is zero in (8.2.6), the PD control law (8.2.7) achieves asymptotic tracking of the desired joint positions. This, in effect, reproduces the result derived previously, but is more rigorous, in the sense that the nonlinear equations of motion (8.2.1) are not approximated by a constant disturbance but are handled directly.

To show that the above control law achieves zero steady state error consider the Lyapunov function candidate

$$V = 1/2\dot{\mathbf{q}}^T(D(\mathbf{q}) + J)\dot{\mathbf{q}} + 1/2\tilde{\mathbf{q}}^T K_P\tilde{\mathbf{q}} \tag{8.2.8}$$

The first term in (8.2.8) is the kinetic energy of the robot and the second term accounts for the proportional feedback $K_P\tilde{\mathbf{q}}$. Note that V represents the total kinetic energy that would result if the joint actuators were to be replaced by springs with stiffnesses represented by K_P and with equilibrium positions at $\mathbf{q}^d$. Thus V is a positive function except at the "goal" $\mathbf{q} = \mathbf{q}^d$, $\dot{\mathbf{q}} = 0$, at which point V is zero. The idea is to show that along any motion of the robot, the function V is decreasing to zero. This will imply that the robot is moving toward the desired goal configuration.

To show this we note that, since J and $\mathbf{q}^d$ are constant, the time derivative of V is given by

$$\dot{V} = \dot{\mathbf{q}}^T(D(\mathbf{q}) + J)\ddot{\mathbf{q}} + 1/2\,\dot{\mathbf{q}}^T\dot{D}(\mathbf{q})\dot{\mathbf{q}} - \dot{\mathbf{q}}^T K_P\tilde{\mathbf{q}} \tag{8.2.9}$$

Solving for $(D(\mathbf{q}) + J)\ddot{\mathbf{q}}$ in (8.2.6) with $\mathbf{g}(\mathbf{q}) = 0$ and substituting the resulting expression into (8.2.9) yields

$$\begin{aligned}\dot{V} &= \dot{\mathbf{q}}^T(\mathbf{u} - C(\mathbf{q},\dot{\mathbf{q}})\dot{\mathbf{q}} - B\dot{\mathbf{q}}) + 1/2\,\dot{\mathbf{q}}^T\dot{D}(\mathbf{q})\dot{\mathbf{q}} - \dot{\mathbf{q}}^T K_P\tilde{\mathbf{q}} \qquad (8.2.10)\\ &= \dot{\mathbf{q}}^T(\mathbf{u} - B\dot{\mathbf{q}} - K_P\tilde{\mathbf{q}}) + 1/2\,\dot{\mathbf{q}}^T(\dot{D}(\mathbf{q}) - 2C(\mathbf{q},\dot{\mathbf{q}}))\dot{\mathbf{q}}\\ &= \dot{\mathbf{q}}^T(\mathbf{u} - B\dot{\mathbf{q}} - K_P\tilde{\mathbf{q}})\end{aligned}$$

where in the last equality we have used the fact (Theorem 6.3.1) that $\dot{D} - 2C$ is skew symmetric. Substituting the PD control law (8.2.7) for $\mathbf{u}$ into the above yields

$$\dot{V} = -\dot{\mathbf{q}}^T(K_D + B)\dot{\mathbf{q}} \le 0 \tag{8.2.11}$$

The above analysis shows that V is decreasing as long as $\dot{\mathbf{q}}$ is not zero. This, by itself is not enough to prove the desired result since it is conceivable that the manipulator can reach a position where $\dot{\mathbf{q}} = 0$ but $\mathbf{q} \ne \mathbf{q}^d$. To show that this cannot happen we can use LaSalle's Theorem (Appendix C). Suppose $\dot{V} \equiv 0$. Then (8.2.11) implies that $\dot{\mathbf{q}} \equiv 0$ and hence $\ddot{\mathbf{q}} \equiv 0$. From the equations of motion with PD-control

$$(D + J)\ddot{\mathbf{q}} + C(\mathbf{q},\dot{\mathbf{q}})\dot{\mathbf{q}} = -K_P\tilde{\mathbf{q}} - K_D\dot{\mathbf{q}}$$

we must then have

$$0 = -K_P\tilde{\mathbf{q}}$$

which implies that $\tilde{\mathbf{q}} = 0$, $\dot{\mathbf{q}} = 0$. LaSalle's Theorem then implies that the system is asymptotically stable.

In case there are gravitational terms present in (8.2.6) Equation 8.2.10 must be modified to read

$$\dot{V} = \dot{\mathbf{q}}^T(\mathbf{u} - \mathbf{g}(\mathbf{q}) - B\dot{\mathbf{q}} - K_P\tilde{\mathbf{q}}) \tag{8.2.12}$$

The presence of the gravitational term in (8.2.12) means that PD control alone cannot guarantee asymptotic tracking. In practice there will be a steady state error or offset. Assuming that the closed loop system is stable the robot configuration $\mathbf{q}$ that is achieved will satisfy

$$K_P(\mathbf{q}^d - \mathbf{q}) = \mathbf{g}(\mathbf{q}) \tag{8.2.13}$$

The physical interpretation of (8.2.13) is that the configuration $\mathbf{q}$ must be such that the motor generates a steady state "holding torque" $K_P(\mathbf{q}^d - \mathbf{q})$ sufficient to balance the gravitational torque $\mathbf{g}(\mathbf{q})$. Thus we see that the steady state error can be reduced by increasing the position gain K_P.

In order to remove this steady state error we can modify the PD control law as

$$\mathbf{u} = K_P\tilde{\mathbf{q}} - K_D\dot{\mathbf{q}} + \mathbf{g}(\mathbf{q}) \tag{8.2.14}$$

The modified control law (8.2.14), in effect, cancels the effect of the gravitational terms and we achieve the same Equation 8.2.11 as before. The control law (8.2.14) requires microprocessor implementation to compute at each instant the gravitational terms $\mathbf{g}(\mathbf{q})$ from the Lagrangian equations. In the case that these terms are unknown the control law (8.2.14) is unachievable. We will say more about this and related issues later.

8.2.1 THE EFFECT OF JOINT FLEXIBILITY

We next investigate the effect of joint flexibility on the stability of the above PD control scheme. Specifically we show that, in the absence of gravity terms, the same PD control (8.2.7) achieves asymptotic stability for set-point control even when there is joint flexibility present provided that the motor variables are used in the control law.

We first derive a model similar to (8.2.6) to represent the dynamics of an n link robot with joint flexibility. For simplicity, assume that the joints are revolute and are actuated by DC-motors, and model the flexibility of the i-th joint as a linear torsional spring with spring constant k_i for $i = 1, \ldots, n$.

We note that, due to the presence of the joint flexibility, there are now twice as many degrees of freedom as in the rigid joint case. Referring to Figure 8-1, let $\mathbf{q} = (q_1, \cdots q_{2n})^T$ be a set of generalized coordinates for the system where

$$q_{2i-1} = \text{ the angle of link } i \;\; ,i=1, \ldots ,n \tag{8.2.15}$$

$$q_{2i} = -\frac{1}{r_i}\theta_i \;\; ,i=1, \ldots ,n \tag{8.2.16}$$

θ_i being the angular displacement of rotor i and r_i is the gear ratio. In this case then $q_{2i} - q_{2i-1}$ is the elastic displacement of joint i.

Define the n-dimensional vectors $\mathbf{q}_1$ and $\mathbf{q}_2$ as

$$\mathbf{q}_1 = (q_1, q_3, \cdots, q_{2n-1})^T \;, \quad \mathbf{q}_2 = (q_2, q_4, \cdots, q_{2n})^T \tag{8.2.17}$$

Neglecting the effects of the link motion on the kinetic energy of the rotor, it can be shown [24] that the kinetic and potential energies of the system are given as

$$KE = \frac{1}{2}\dot{\mathbf{q}}_1^T D(\mathbf{q}_1)\dot{\mathbf{q}}_1 + \frac{1}{2}\dot{\mathbf{q}}_2^T J\dot{\mathbf{q}}_2 \tag{8.2.18}$$

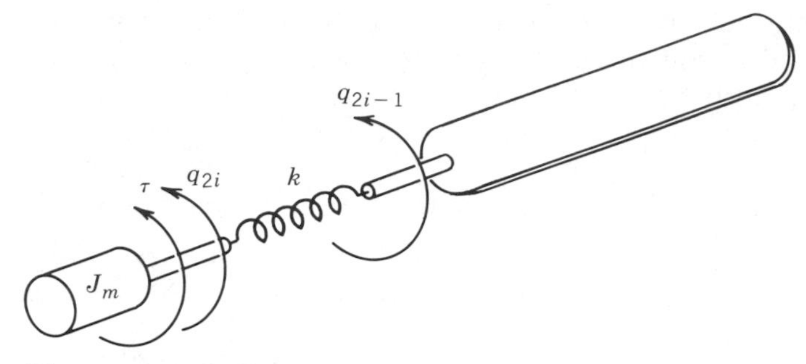

FIGURE 8-1
Model for joint flexibility.

$$PE = V_1(\mathbf{q}_1) + V_2(\mathbf{q}_1 - \mathbf{q}_2) \tag{8.2.19}$$

where $D(\mathbf{q}_1)$ and $V_1(\mathbf{q}_1)$ are, respectively, the inertia matrix and potential energy of the rigid manipulator as derived in Chapter Six, J is a diagonal matrix

$$J = diag(\frac{1}{r_1^2}J_1, \ldots, \frac{1}{r_n^2}J_n) \tag{8.2.20}$$

where J_i is the inertia of the $i-th$ rotor, and V_2 is the elastic potential energy of the joints

$$V_2 = \frac{1}{2}(\mathbf{q}_1 - \mathbf{q}_2)^T K(\mathbf{q}_1 - \mathbf{q}_2) \tag{8.2.21}$$

with K a diagonal matrix whose diagonal entries are equal to the joint spring constants $k_1, \ldots, k_n$.

By forming the corresponding Lagrangian, the equations of motion are found from the Euler–Lagrange equations as (Problem 8-1)

$$D(\mathbf{q}_1)\ddot{\mathbf{q}}_1 + C(\mathbf{q}_1, \dot{\mathbf{q}}_1)\dot{\mathbf{q}}_1 + \mathbf{g}(\mathbf{q}_1) + K(\mathbf{q}_1 - \mathbf{q}_2) = 0 \tag{8.2.22}$$

$$J\ddot{\mathbf{q}}_2 + B\dot{\mathbf{q}}_2 - K(\mathbf{q}_1 - \mathbf{q}_2) = \mathbf{u} \tag{8.2.23}$$

For the problem of set-point tracking let us consider a PD control law of the form

$$\mathbf{u} = K_P\tilde{\mathbf{q}}_2 - K_D\dot{\mathbf{q}}_2 \tag{8.2.24}$$

It is important to note in (8.2.24) that the PD control law is a function of the motor variables $\mathbf{q}_2$. To show asymptotic tracking for the closed loop system consider the Lyapunov function candidate

$$V = 1/2\,\dot{\mathbf{q}}_1^T D(\mathbf{q}_1)\dot{\mathbf{q}}_1 + 1/2\,\dot{\mathbf{q}}_2^T J\dot{\mathbf{q}}_2 + 1/2\,(\mathbf{q}_1 - \mathbf{q}_2)^T K(\mathbf{q}_1 - \mathbf{q}_2) + 1/2\,\tilde{\mathbf{q}}^T K_P\tilde{\mathbf{q}} \tag{8.2.25}$$

where $\tilde{\mathbf{q}} = \mathbf{q}^d - \mathbf{q}_2$.

Assuming that the gravitational term $\mathbf{g}(\mathbf{q}_1) = 0$ in (8.2.22), and again using the fact that $\dot{D} - 2C$ is skew-symmetric, a routine calculation shows that the time derivative of V given by (8.2.25) along trajectories of (8.2.22) (8.2.23) is given by

$$\dot{V} = \dot{\mathbf{q}}_2^T(\mathbf{u} - B\dot{\mathbf{q}}_2 - K_P\tilde{\mathbf{q}}) \tag{8.2.26}$$

Therefore, substituting the PD-control law (8.2.24) into (8.2.26) yields,

$$\dot{V} = -\dot{\mathbf{q}}_2^T(K_D + B)\dot{\mathbf{q}}_2 \leq 0 \tag{8.2.27}$$

The result follows as before by applying Lasalle's Theorem. The details are left as an exercise (Problem 8-2). In case the gravity term $\mathbf{g}$ is present in (8.2.22), we must modify (8.2.26) to read

$$\dot{V} = \dot{\mathbf{q}}_2^T(\mathbf{u} - B\dot{\mathbf{q}}_2 - K_P\tilde{\mathbf{q}}) - \dot{\mathbf{q}}_1^T\mathbf{g}(\mathbf{q}_1) \tag{8.2.28}$$

We see now that gravity compensation cannot be performed in the same manner as was possible in the rigid case. In Chapter Ten we will detail an approach that may be used to compensate the gravitational and other effects.

8.3 INVERSE DYNAMICS

We now consider the application of more complex nonlinear control techniques for trajectory tracking of rigid manipulators. Consider again the dynamic equations of an n-link robot in matrix form from (8.2.6)

$$(D(\mathbf{q}) + J)\ddot{\mathbf{q}} + C(\mathbf{q}, \dot{\mathbf{q}})\dot{\mathbf{q}} + B\dot{\mathbf{q}} + \mathbf{g}(\mathbf{q}) = \mathbf{u} \tag{8.3.1}$$

For simplicity we write the above equation as

$$M(\mathbf{q})\ddot{\mathbf{q}} + \mathbf{h}(\mathbf{q}, \dot{\mathbf{q}}) = \mathbf{u} \tag{8.3.2}$$

where $M = D + J$, $\mathbf{h} = C\dot{\mathbf{q}} + B\dot{\mathbf{q}} + \mathbf{g}$. The idea of inverse dynamics is to seek a nonlinear feedback control law

$$u = f(\mathbf{q}, \dot{\mathbf{q}}) \tag{8.3.3}$$

which, when substituted into (8.3.2), results in a linear closed loop system. For general nonlinear systems such a control law may be quite difficult or impossible to find. In the case of the manipulator dynamic equations (8.3.2), however, the problem is actually easy. By inspecting (8.3.2) we see that if we choose the control $\mathbf{u}$ according to the equation

$$\mathbf{u} = M(\mathbf{q})\mathbf{v} + \mathbf{h}(\mathbf{q}, \dot{\mathbf{q}}) \tag{8.3.4}$$

then, since the inertia matrix M is invertible, the combined system (8.3.2)–(8.3.4) reduces to

$$\ddot{\mathbf{q}} = \mathbf{v} \tag{8.3.5}$$

The term $\mathbf{v}$ represents a new input to the system which is yet to be chosen. Equation 8.3.5 is known as the **double integrator system** as it represents n uncoupled double integrators. The nonlinear control law (8.3.4) is called the **inverse dynamics control**[2] and achieves a rather remarkable result, namely that the "new" system (8.3.5) is linear, and decoupled. This means that each input v_k can be designed to control a scalar linear system. Moreover, assuming that v_k is a function only of q_k and its derivatives, then v_k will affect the q_k *independently* of the motion of the other links.

Since v_k can now be designed to control a simple linear second order system, the obvious choice is to set

$$\mathbf{v} = -K_0\mathbf{q} - K_1\dot{\mathbf{q}} + \mathbf{r} \tag{8.3.6}$$

where K_0 and K_1 are diagonal matrices with diagonal elements consisting of position and velocity gains, respectively. The closed loop system is then the linear system

$$\ddot{\mathbf{q}} + K_1\dot{\mathbf{q}} + K_0\mathbf{q} = \mathbf{r} \tag{8.3.7}$$

Now, given a desired trajectory

$$t \rightarrow (\mathbf{q}^d(t), \dot{\mathbf{q}}^d(t)) \tag{8.3.8}$$

if one chooses the reference input $r(t)$ as[3]

$$\mathbf{r}(t) = \ddot{\mathbf{q}}^d(t) + K_0\mathbf{q}^d(t) + K_1\dot{\mathbf{q}}^d(t) \tag{8.3.9}$$

then the tracking error $\mathbf{e}(t) = \mathbf{q}^d - \mathbf{q}$ satisfies

$$\ddot{\mathbf{e}}(t) + K_1\mathbf{e}(t) + K_0\mathbf{e}(t) = 0 \tag{8.3.10}$$

An obvious choice for the gain matrices K_0 and K_1 is

$$\begin{aligned} K_0 &= diag\{\omega_1^2, \cdots, \omega_n^2\} \\ K_1 &= diag\{2\omega_1, \cdots, 2\omega_n\} \end{aligned} \tag{8.3.11}$$

which results in a closed loop system which is globally decoupled, with each joint response equal to the response of a critically damped linear second order system with natural frequency ω_i. As before, the natural frequency ω_i determines the speed of response of the joint, or equivalently, the rate of decay of the tracking error.

The inverse dynamics approach is extremely important as a basis for control of robot manipulators and it is worthwhile trying to see it from alternative viewpoints. We can give a second interpretation of the control law (8.3.4) as follows. Consider again the manipulator dynamic equations (8.3.2). Since $M(\mathbf{q})$ is invertible for $\mathbf{q} \in \mathbb{R}^n$ we may solve for the acceleration $\ddot{\mathbf{q}}$ of the manipulator as

$$\ddot{\mathbf{q}} = M^{-1}\{\mathbf{u} - \mathbf{h}(\mathbf{q}, \dot{\mathbf{q}})\} \tag{8.3.12}$$

---

[2]We should point out that in the research literature the control law (8.3.4) is frequently called **computed torque** as well.

[3]Compare this with the feedforward expression (7.6.6).

Suppose we were able to specify the acceleration as the input to the system. That is, suppose we had actuators capable of producing directly a commanded acceleration (rather than indirectly by producing a force or torque). Then the dynamics of the manipulator, which is after all a position control device, would be given as

$$\ddot{\mathbf{q}}(t) = \mathbf{v}(t) \tag{8.3.13}$$

where $\mathbf{v}(t)$ is the input acceleration vector. This is again the familiar double integrator system. Note that (8.3.13) is not an approximation in any sense; rather it represents the actual open loop dynamics of the system provided that the acceleration is chosen as the input. The control problem for the system (8.3.13) is now easy and the acceleration input $\mathbf{v}$ can be chosen as before according to (8.3.6).

In reality, however, our physical universe prohibits such "acceleration actuators" and we must be content with the ability to produce a generalized force (torque) u_i at each joint i. Comparing equations (8.3.12) and (8.3.13) we see that the torque $\mathbf{u}$ and the acceleration $\mathbf{v}$ of the manipulator are related by

$$M^{-1}\{\mathbf{u}(t) - \mathbf{h}(\mathbf{q}(t), \dot{\mathbf{q}}(t))\} = \mathbf{v}(t) \tag{8.3.14}$$

By the invertibility of the inertia matrix we may solve for the input torque $\mathbf{u}(t)$ as

$$\mathbf{u}(t) = M(\mathbf{q}(t))\mathbf{v}(t) + \mathbf{h}(\mathbf{q}(t), \dot{\mathbf{q}}(t)) \tag{8.3.15}$$

which is the same as the previously derived expression (8.3.4). Thus the inverse dynamics can be viewed as an input transformation which transforms the problem from one of choosing torque input commands, which is difficult, to one of choosing acceleration input commands, which is easy.

Note that the implementation of this control scheme requires the computation at each sample instant of the inertia matrix $M(\mathbf{q})$ and the vector of coriolis, centripetal, gravitational and damping terms $\mathbf{h}(\mathbf{q}, \dot{\mathbf{q}})$. Unlike the computed torque scheme (7.6.8), however, the inverse dynamics *must* be computed on-line. In other words, as a feedback control law, it cannot be precomputed off-line and stored as can the computed torque (7.6.9). An important issue therefore in the control system implementation is the design of the computer architecture for the above computations. An attractive method to implement this scheme, and one which will undoubtedly become common in the future, is to use a dedicated hardwire interface, that is, a special purpose VLSI chip to perform the required computations in real time. Such a scheme is shown in Figure 8-2.

Figure 8-2 illustrates the notion of **inner-loop/outer-loop** control. By this we mean that the computation of the nonlinear control (8.3.4) is performed in an inner loop, perhaps with a dedicated hardwire interface, with the vectors $\mathbf{q}$, $\dot{\mathbf{q}}$, and $\mathbf{v}$ as its inputs and $\mathbf{u}$ as output. The outer loop in the system is then the computation of the additional

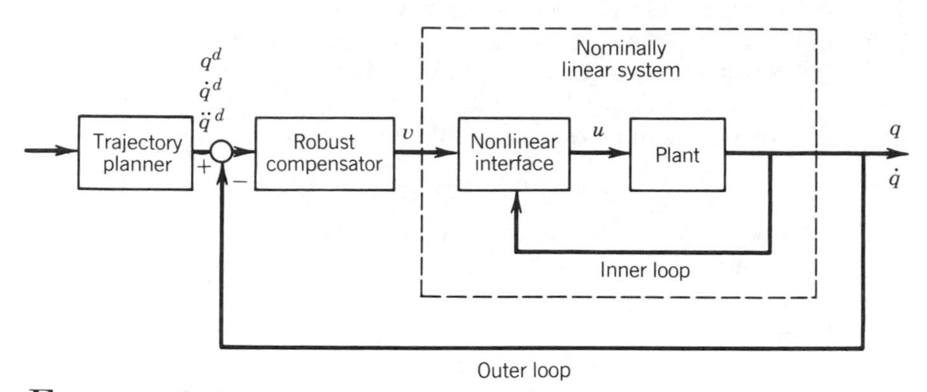

FIGURE 8-2

Inner loop/outer control architecture.

input term $\mathbf{v}$. Note that the outer loop control $\mathbf{v}$ is more in line with the notion of a feedback control in the usual sense of being error driven. The design of the outer loop feedback control is in theory greatly simplified since it is designed for the plant represented by the dotted lines in Figure 8-2, which is now a linear or nearly linear system.

The method of inverse dynamics is thus very attractive from a control standpoint since the highly nonlinear coupled dynamics of the manipulator are canceled and replaced by a simple decoupled linear second order system. However, such exact cancellation schemes leave open many issues of sensitivity and robustness that must be addressed. We shall address some of these issues in the next section.

8.4 IMPLEMENTATION AND ROBUSTNESS ISSUES

Practical implementation of the inverse dynamics control law (8.3.4) requires both that the parameters in the dynamic model of the system be known precisely and also that the complete equations of motion be computable in real-time, typically 60-100Hz. The above requirements are difficult to satisfy in practice. In any physical system there is a degree of uncertainty regarding the values of various parameters. In the case of a system as complicated as a robot, this is particularly true, especially if the robot is carrying unknown loads. Practically speaking there will always be inexact cancellation of the nonlinearities in the system due to this uncertainty and also due to computational round-off, etc. In addition, the burden of computing the complete model of the manipulator may be prohibitively expensive or impossible within the bounds imposed by the available computer architecture. In such cases it is desirable to simplify the equations of motion as much as possible by ignoring certain of the terms in the equations in order to speed the computation of the control law.

Therefore it is much more reasonable to suppose that, instead of (8.3.4), the nonlinear control law is actually of the form

$$\mathbf{u}(t) = \hat{M}(\mathbf{q})\mathbf{v} + \hat{\mathbf{h}}(\mathbf{q}, \dot{\mathbf{q}}) \tag{8.4.1}$$

where $\hat{M}, \hat{\mathbf{h}}$ represent nominal or computed versions of $M, \mathbf{h}$, respectively. The uncertainty or modeling error, represented by

$$\Delta M := \hat{M}(\mathbf{q}) - M(\mathbf{q}) \tag{8.4.2}$$

$$\Delta \mathbf{h} := \hat{\mathbf{h}}(\mathbf{q}, \dot{\mathbf{q}}) - \mathbf{h}(\mathbf{q}, \dot{\mathbf{q}}) \tag{8.4.3}$$

is then due to (and is a measure of) the problems of parameter uncertainty, etc. Note also that the above modeling error may also arise due to intentional model simplification. In fact, in the extreme case, one may take $\hat{M} = I$ and $\hat{\mathbf{h}} = 0$ in which the entire control law consists of the outer loop control $\mathbf{v}$, which for example could be simply an independent PID control for each joint.

With the nonlinear control law (8.4.1), and dropping arguments for simplicity, the system becomes

$$M\ddot{\mathbf{q}} + \mathbf{h} = \hat{M}\mathbf{v} + \hat{\mathbf{h}} \tag{8.4.4}$$

Thus $\ddot{\mathbf{q}}$ can be expressed as

$$\begin{aligned} \ddot{\mathbf{q}} &= M^{-1}\hat{M}\mathbf{v} + M^{-1}\Delta\mathbf{h} \\ &= \mathbf{v} + (M^{-1}\hat{M} - I)\mathbf{v} + M^{-1}\Delta\mathbf{h} \end{aligned} \tag{8.4.5}$$

where $\Delta\mathbf{h} = \hat{\mathbf{h}} - \mathbf{h}$. Defining

$$E = M^{-1}\hat{M} - I \tag{8.4.6}$$

and setting

$$\mathbf{v} = \ddot{\mathbf{q}}^d - K_1(\dot{\mathbf{q}} - \dot{\mathbf{q}}^d) - K_0(\mathbf{q} - \mathbf{q}^d) \tag{8.4.7}$$

we can write the above equation for the error $\mathbf{e} = \mathbf{q} - \mathbf{q}^d$ as

$$\ddot{\mathbf{e}} + K_1\dot{\mathbf{e}} + K_0\mathbf{e} = \eta \tag{8.4.8}$$

where η, hereafter called the **uncertainty**, is given by the expression

$$\begin{aligned} \eta &= E\,\mathbf{v} + M^{-1}\Delta h \\ &= E(\ddot{\mathbf{q}}^d - K_1\dot{\mathbf{e}} - K_0\mathbf{e}) + M^{-1}\Delta h \end{aligned} \tag{8.4.9}$$

Notice that the system (8.4.8) is still a coupled nonlinear system since η is a nonlinear function of $\mathbf{e}$. Therefore it is not obvious that system (8.4.8) is stable nor can one simply raise the gains sufficiently high in (8.4.8) and claim stability since the nonlinear function η also depends on $\mathbf{v}$, given by (8.4.7), and hence may increase with larger gains.

(*i*) *Example 8.4.1*

Consider the single-link robot of Figure 8-3, modeled for simplicity by the equation

$$I\ddot{\theta} + MgL\, sin(\theta) = u \tag{8.4.10}$$

where I is the total moment of inertia about the joint axis, M is the total mass, and L is the distance from the joint axis to the center of mass of the system. The inverse dynamics control law is then

$$u = I\, v + MgL\, sin(\theta) \tag{8.4.11}$$

$$v = \ddot{\theta}^d - k_1(\dot{\theta} - \dot{\theta}^d) - k_0(\theta - \theta^d) \tag{8.4.12}$$

Suppose that the coefficients I and MgL are unknown but that

$$5 \le I \le 10 \quad ; \quad 5 \le MgL \le 10 \qquad (8.4.12)$$

If we choose the control law

$$u = \hat{I} v + \widehat{MgL}\, \sin(\theta) \tag{8.4.13}$$

with $\hat{I} = 5$ and $\widehat{MgL} = 5$, then the response for a given desired trajectory is shown in Figure 8-4. The tracking error is due to the inexact cancellation in the inner loop control.

The situation at this point is not very satisfactory, since the response of the system (8.4.8) is difficult to determine. As long as we restrict the outer loop control to be of the form (8.4.7) this is the best that we can do. In order to overcome the effects of the uncertainty represented by η we must introduce additional feedback into the system. This is known as the **robust control problem** and can be done in several ways. In the next section we discuss one method for modifying the outer loop control $\mathbf{v}$ to overcome the effects of inexact cancellation of nonlinearities in the inner loop.

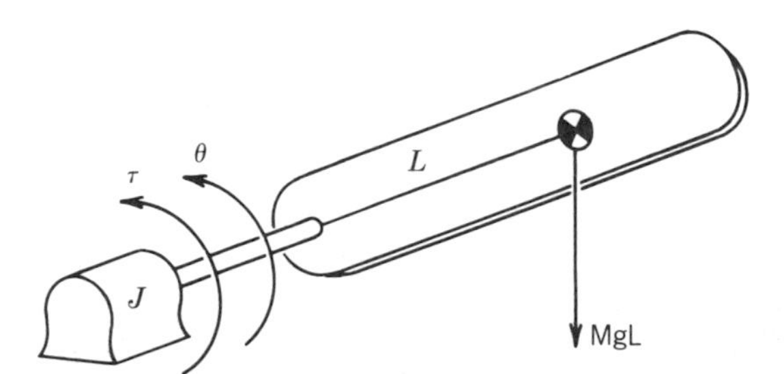

FIGURE 8-3
Single-link robot example.

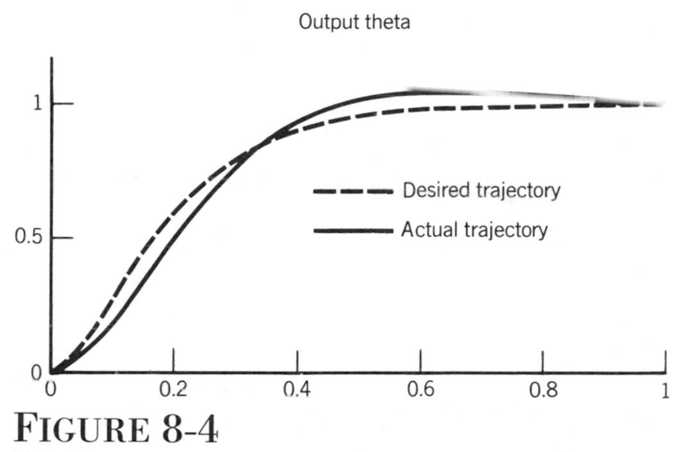

FIGURE 8-4
Tracking response—approximate inverse dynamics control.

8.5 ROBUST OUTER LOOP DESIGN

In this section we discuss the redesign of the outer loop control law $\mathbf{v}$ in (8.4.7) to achieve robust tracking. There are many approaches that can be used at this point. In fact, the robust control of manipulators is currently a very active area of research. For illustration purposes we shall illustrate only one technique for robust control in this chapter, using the so-called **Second Method of Lyapunov**. The technique that we discuss is based on the theory of guaranteed stability of uncertain systems as developed in [12] and other references. The derivations and proofs in this section are taken mostly from these references to which the interested reader is referred for further details.

Recall from above that, given the nonlinear dynamic equations for an n-link robot

$$M(\mathbf{q})\ddot{\mathbf{q}} + \mathbf{h}(\mathbf{q}, \dot{\mathbf{q}}) = \mathbf{u} \tag{8.5.1}$$

we choose an inner loop control law of the form

$$\mathbf{u} = \hat{M}(\mathbf{q})\,\mathbf{v} + \hat{\mathbf{h}}(\mathbf{q}, \dot{\mathbf{q}}) \tag{8.5.2}$$

which results in the system

$$\ddot{\mathbf{q}} = \mathbf{v} + \eta\,(\mathbf{v}, \mathbf{q}, \dot{\mathbf{q}}) \tag{8.5.3}$$

where η is given by

$$\eta = E(\mathbf{q})\,\mathbf{v} + M^{-1}\Delta\mathbf{h} \tag{8.5.4}$$

with $E = M^{-1}\hat{M} - I$ and $\Delta\mathbf{h} = \hat{\mathbf{h}} - \mathbf{h}$.

In state space then the system (8.5.3) becomes

$$\dot{\mathbf{x}} = A\mathbf{x} + B\{\mathbf{v} + \eta\} \tag{8.5.5}$$

where

$$A = \begin{bmatrix} 0 & I \\ 0 & 0 \end{bmatrix} \quad B = \begin{bmatrix} 0 \\ I \end{bmatrix} \quad \mathbf{x} = \begin{pmatrix} \mathbf{x}_1 \\ \mathbf{x}_2 \end{pmatrix} = \begin{pmatrix} \mathbf{q} \\ \dot{\mathbf{q}} \end{pmatrix} \tag{8.5.6}$$

Suppose that a desired trajectory $t \to \mathbf{q}^d(t)$ is given in joint space. We form the error vectors

$$\begin{aligned} \mathbf{e}_1 &= \mathbf{x}_1 - \mathbf{x}_1^d = \mathbf{q} - \mathbf{q}^d \\ \mathbf{e}_2 &= \mathbf{x}_2 - \mathbf{x}_2^d = \dot{\mathbf{q}} - \dot{\mathbf{q}}^d \end{aligned} \tag{8.5.7}$$

Then the tracking error satisfies the equations

$$\begin{aligned} \dot{\mathbf{e}}_1 &= \mathbf{e}_2 \\ \dot{\mathbf{e}}_2 &= \ddot{\mathbf{q}} - \ddot{\mathbf{q}}^d \\ &= \mathbf{v} + \eta - \ddot{\mathbf{q}}^d \end{aligned} \tag{8.5.8}$$

which we write in vector form as

$$\dot{\mathbf{e}} = A\mathbf{e} + B\{\mathbf{v} + \eta - \ddot{\mathbf{q}}^d\} \tag{8.5.9}$$

Therefore the problem of tracking the desired trajectory $t \to \mathbf{q}^d(t)$ becomes one of stabilizing the (time-varying, nonlinear) system (8.5.9). The control design to follow is based on the premise that although the uncertainty η is unknown, it may be possible to estimate "worst case" bounds on its effects on the tracking performance of the system. The control law $\mathbf{v}$ is then designed to guarantee stability of (8.5.9) provided that these bounds on η are satisfied.

In order to estimate a worst case bound on the function η we make the following assumptions.

Assumption 1. $\sup_{t \geq 0} \| \ddot{\mathbf{q}}^d \| < Q_1 < \infty$.

Assumption 2. $\| E \| = \| M^{-1}\hat{M} - I \| \leq \alpha < 1$ for some α, for all $\mathbf{q} \in \mathbb{R}^n$.

Assumption 3. $\| \Delta\mathbf{h} \| \leq \boldsymbol{\phi}(\mathbf{e}, t)$ for a known function $\boldsymbol{\phi}$, bounded in t.

Assumption 2 is the most restrictive and shows how accurately the inertia of the manipulator must be estimated in order to use this approach. It turns out, however, that there is always a simple choice for $\hat{M}$ satisfying Assumption 2. Since the inertia matrix $M(\mathbf{q})$ is uniformly positive definite for all $\mathbf{q}$ there exist positive constants $\underline{M}$ and $\overline{M}$ such that

$$\underline{M} \leq \| M^{-1}(\mathbf{q}) \| \leq \overline{M} \quad \text{for all } \mathbf{q} \in \mathbb{R}^n \tag{8.5.10}$$

If we therefore choose

$$\hat{M} = \frac{1}{c} I \quad \text{where } c = \frac{\overline{M} + \underline{M}}{2} \tag{8.5.11}$$

it can be shown that

$$\| M^{-1}\hat{M} - I \| \leq \frac{\overline{M} - \underline{M}}{\overline{M} + \underline{M}} =: \alpha < 1 \tag{8.5.12}$$

The point is that there is always at least one choice of $\hat{M}$ satisfying Assumption 2. In practice one would like to choose $\hat{M}$ based on knowledge of the inertial parameters and range of expected loads so that α is as small as possible.

The following algorithm may now be used to generate a stabilizing control $\mathbf{v}$:

Step 1: Since the matrix A in (8.5.9) is unstable we first set

$$\mathbf{v} = \ddot{\mathbf{q}}^d - K_1\mathbf{e}_1 - K_2\mathbf{e}_2 + \Delta\mathbf{v} \tag{8.5.13}$$

where

$$K_1 = diag\{\omega_1^2, \cdots, \omega_n^2\} \tag{8.5.14}$$

$$K_2 = diag\{2\zeta_1\omega_1, \cdots, 2\zeta_n\omega_n\}$$

Then we have, with $K = [K_1, K_2]$,

$$\dot{\mathbf{e}} = \bar{A}\,\mathbf{e} + B\{\Delta\mathbf{v} + \bar{\eta}\} \tag{8.5.15}$$

where $\bar{A} = A - BK$ is Hurwitz and

$$\bar{\eta} = E\Delta\mathbf{v} + E(\ddot{\mathbf{q}}^d - K\mathbf{e}) + M^{-1}\Delta\mathbf{h} \tag{8.5.16}$$

Note that the outer loop term $\mathbf{v}$ is equal to the outer loop term $\mathbf{v}$ in the inverse dynamics control law (8.3.4)–(8.3.6) with the addition of an extra term $\Delta\mathbf{v}$. This additional feedback will be used to overcome the effects of the uncertainty. In the ideal case that $\bar{\eta} = 0$ we may take $\Delta\mathbf{v} = 0$ and the control law reduces to the previous inverse dynamics control.

Step 2: Given the system (8.5.15) with $\bar{A}$ Hurwitz, suppose we can find a continuous function $\rho(\mathbf{e}, t)$, which is bounded in t, satisfying the inequalities

$$\| \Delta\mathbf{v} \| < \rho(\mathbf{e}, t) \tag{8.5.17}$$

$$\| \bar{\eta} \| < \rho(\mathbf{e}, t) \tag{8.5.18}$$

The function ρ can be defined implictly as follows. Using Assumptions 1-3 and (8.5.17) we have the estimate

$$\begin{aligned} \| \bar{\eta} \| &\leq \| E\Delta\mathbf{v} + E\ddot{\mathbf{q}}^d - Ke + M^{-1}\Delta\mathbf{h} \| \\ &\leq \alpha\rho(\mathbf{e}, t) + \alpha Q_1 + \| K \| \cdot \| \mathbf{e} \| + \overline{M}\phi(\mathbf{e}, t) \\ &=: \rho(\mathbf{e}, t) \end{aligned} \tag{8.5.19}$$

This definition of ρ makes sense since $0 < \alpha < 1$ and we may solve for ρ as

$$\rho(\mathbf{e}, t) = \frac{1}{1-\alpha}\{\alpha Q_1 + \|K\| \cdot \|\mathbf{e}\| + \overline{M}\phi(\mathbf{e}, t)\} \tag{8.5.20}$$

Note that whatever $\Delta\mathbf{v}$ is now chosen must satisfy (8.5.17).

Step 3: Since $\bar{A}$ is Hurwitz, choose a $n \times n$ symmetric, positive definite matrix Q and let P be the unique positive definite symmetric solution to the Lyapunov equation

$$\bar{A}^T P + P\bar{A} + Q = 0 \tag{8.5.21}$$

Step 4: Choose the outer loop control $\Delta\mathbf{v}$ according to

$$\Delta\mathbf{v} = \begin{cases} -\rho(\mathbf{e}, t)\dfrac{B^T P\mathbf{e}}{\|B^T P\mathbf{e}\|} & \text{if } \|B^T P\mathbf{e}\| \neq 0 \\ 0 & \text{if } \|B^T P\mathbf{e}\| = 0 \end{cases} \tag{8.5.22}$$

Note that $\Delta\mathbf{v}$ does indeed satisfy (8.5.17). We can now prove the following:

(ii) *Theorem 8.5.1*

Set $V(\mathbf{e}) = \mathbf{e}^T P\mathbf{e}$. Then with the outer loop control law (8.5.22) $\dot{V}(\mathbf{e}) < 0$ along solution trajectories of the system (8.5.15).

Proof: A calculation shows that along trajectories of (8.5.15), $\dot{V}$ satisfies

$$\begin{aligned} \dot{V}(\mathbf{e}) &= \dot{\mathbf{e}}^T P\mathbf{e} + \mathbf{e}^T P\dot{\mathbf{e}} \\ &= \mathbf{e}^T(\bar{A}^T P + P\bar{A})\mathbf{e} + 2\mathbf{e}^T PB(\Delta\mathbf{v} + \bar{\eta}) \\ &= -\mathbf{e}^T Q\mathbf{e} + 2\mathbf{e}^T PB(\Delta\mathbf{v} + \bar{\eta}) \end{aligned} \tag{8.5.23}$$

For simplicity set $\mathbf{w} = B^T P\mathbf{e}$. Then the second term above can be written (ignoring the factor 2) as

$$\mathbf{w}^T(\Delta\mathbf{v} + \bar{\eta}) \tag{8.5.24}$$

If $\mathbf{w} = 0$ then this term is zero and

$$\dot{V} = -\mathbf{e}^T Q\mathbf{e} < 0 \tag{8.5.25}$$

If $\mathbf{w} \neq 0$ then

$$\Delta\mathbf{v} = -\rho\frac{\mathbf{w}}{\|\mathbf{w}\|} \tag{8.5.26}$$

and (8.5.24) becomes

$$\begin{aligned} \mathbf{w}^T\left(-\rho\frac{\mathbf{w}}{\|\mathbf{w}\|} + \bar{\eta}\right) &= \frac{-\rho\mathbf{w}^T\mathbf{w}}{\|\mathbf{w}\|} + \mathbf{w}^T\bar{\eta} \\ &\leq -\rho\|\mathbf{w}\| + \|\mathbf{w}\| \cdot \|\bar{\eta}\| \\ &= \|\mathbf{w}\|(-\rho + \|\bar{\eta}\|) \leq 0 \end{aligned} \tag{8.5.27}$$

since $\|\bar{\eta}\| \leq \rho$. Combining this with the first term in (8.5.23), which is strictly negative, concludes the proof of the theorem.

We would like to conclude from this that $V(\mathbf{e})$ is a Lyapunov function for the system and that the origin $\mathbf{e} = 0$ is globally asymptotically stable. However we have shown only that $\dot{V}$ is negative along solution trajectories of the system. The argument for stability breaks down since the control law $\Delta \mathbf{v}$ is discontinuous. This means that we cannot guarantee the existence of a solution in the traditional sense. Using the results of Gutman[9] one can show the existence of a solution to (8.5.15) in a generalized (set-theoretic) sense. We will not go into this here.

(iii) *Example 8.5.2*

We will illustrate the above approach to robust design by providing a detailed treatment of the system of Example 8.4.1. With the equation of motion

$$I\ddot{\theta} + MgL \sin(\theta) = u \tag{8.5.28}$$

we choose the control law

$$u = \hat{I}(v + \Delta v) + \widehat{MgL}\, sin(\theta) \tag{8.5.29}$$

where v is given by (8.4.12) and Δv is to be designed according to (8.5.22). In terms of the tracking error $e_1 = \theta - \theta^d$, $e_2 = \dot{\theta} - \dot{\theta}^d$, we can write (8.5.28)–(8.5.29) together as

$$\dot{e}_1 = e_2 \tag{8.5.30}$$

$$\begin{aligned}\dot{e}_2 &= \frac{1}{I}u - \frac{MgL}{I}\sin(\theta) - \ddot{\theta}^d \\ &= \frac{1}{I}(\hat{I}(v + \Delta v) + \widehat{MgL}\sin(\theta)) - \frac{MgL}{I}\sin(\theta) - \ddot{\theta}^d \\ &= v + \Delta v + (\frac{\hat{I}}{I} - 1)(v + \Delta v) + \frac{\Delta MgL}{I}\sin(\theta) - \ddot{\theta}^d\end{aligned}$$

With $v = \ddot{\theta}^d - 20e_2 - 100e_1$ we have

$$\begin{aligned}\dot{\mathbf{e}} &= \begin{bmatrix} 0 & 1 \\ -100 & -20 \end{bmatrix}\mathbf{e} + \begin{bmatrix} 0 \\ 1 \end{bmatrix}(\Delta v + \bar{\eta}) \\ &= \bar{A}\,\mathbf{e} + \mathbf{b}(\Delta v + \bar{\eta})\end{aligned} \tag{8.5.31}$$

where

$$\bar{\eta} = (\frac{\hat{I}}{I} - 1)(v + \Delta v) + \frac{\Delta MgL}{I}\sin(\theta) \tag{8.5.32}$$

With worst case bounds

$$5 \leq I \leq 10 \quad 5 \leq MgL \leq 10 \tag{8.5.33}$$

and choosing

$$\hat{I} = 5 \quad \widehat{MgL} = 5 \tag{8.5.34}$$

we have the estimate

$$|\frac{\hat{I}}{I} - 1| \leq \frac{1}{2} = \alpha \tag{8.5.35}$$

$$|\frac{\Delta MgL}{I}\sin(\theta)| \leq \frac{2}{5} \tag{8.5.36}$$

Therefore,

$$\|\bar{\eta}\| \leq \alpha(\|v\| + \rho) + \frac{2}{5} := \rho \tag{8.5.37}$$

and so, ρ is given as

$$\rho = \frac{\alpha}{1-\alpha}\|v\| + \frac{1}{1-\alpha}\frac{2}{5} = \|v\| + \frac{4}{5} \tag{8.5.38}$$

Next, set $Q = I$ and solve the Lyapunov equation

$$\bar{A}^T P + P\bar{A} + I = 0 \tag{8.5.39}$$

where

$$\bar{A} = \begin{bmatrix} 0 & 1 \\ -100 & -20 \end{bmatrix} \tag{8.5.40}$$

The unique positive definite solution is

$$P = \begin{bmatrix} 2.625 & 0.005 \\ 0.005 & 0.02525 \end{bmatrix} \tag{8.5.41}$$

Set

$$w = \mathbf{b}^T P \mathbf{e} = 0.005e_1 + 0.02525e_2 \tag{8.5.42}$$

and chose Δv according to

$$\Delta v = \begin{cases} -\rho\frac{w}{\|w\|} & \text{if } \|w\| \neq 0 \\ 0 & \text{if } \|w\| = 0 \end{cases} \tag{8.5.43}$$

The response is shown in Figure 8-5. Comparing this with Figure 8-4 we see that the addition of the term Δv reduces the tracking error caused by the inexact cancellation in the inner loop inverse dynamics control law.

Figure 8-6 shows the tracking errors as a function of time with and without the additional compensation Δv. Figure 8-7 shows the control signal Δv which is added to the inverse dynamics control. This control signal illustrates a phenomenon known as **chattering**, which is characteristic of discontinuous control laws. Chattering is often undesirable since the high frequency component in the control can excite unmodeled dynamic effects such as the joint flexibility.

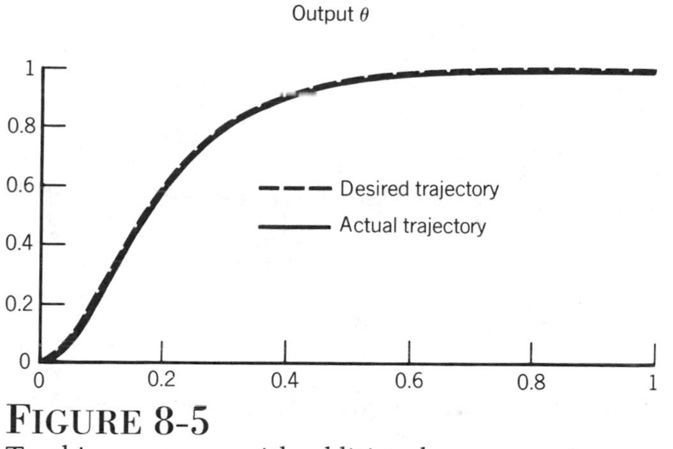

FIGURE 8-5

Tracking response with additional compensation.

An alternative procedure, which avoids having to treat generalized solutions of (8.5.1), and which eliminates the control chattering, is to approximate the discontinuous control (8.5.8) by a continuous control as follows. Choose $\varepsilon > 0$ and set

$$\Delta \mathbf{v} = \begin{cases} -\rho(\mathbf{e}, t)\dfrac{B^T P \mathbf{e}}{\| B^T P \mathbf{e} \|} & \text{if } \| B^T P \mathbf{e} \| \geq \varepsilon \\ -\dfrac{\rho(\mathbf{e}, t)}{\varepsilon} B^T P \mathbf{e} & \text{if } \| B^T P \mathbf{e} \| < \varepsilon \end{cases} \tag{8.5.44}$$

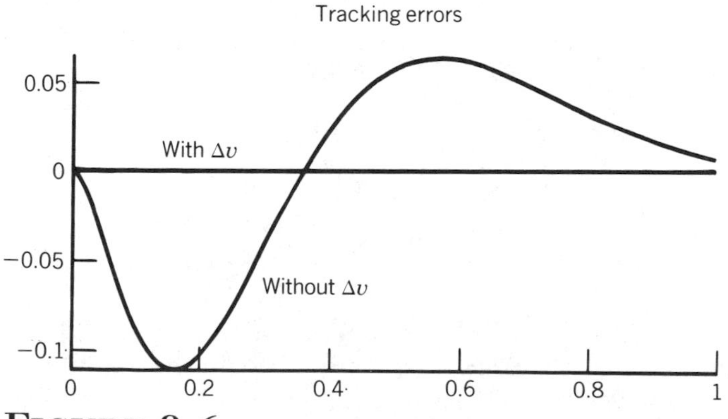

FIGURE 8-6

Tracking errors with and without additional compensation.

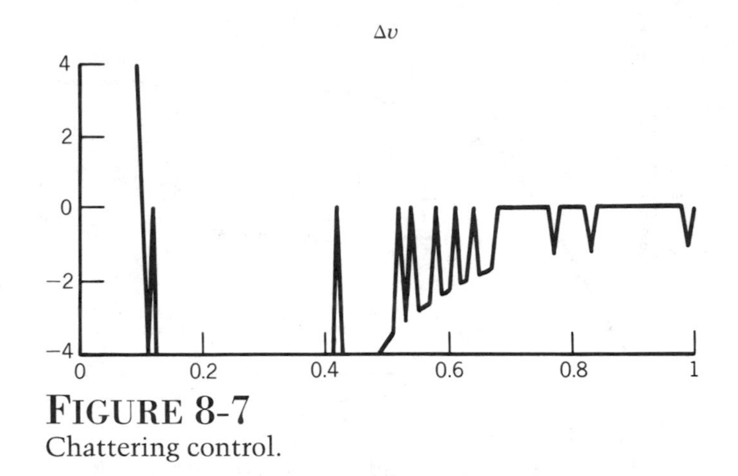

FIGURE 8-7
Chattering control.

In this case since the control signal (8.5.44) is continuous, a solution to the system (8.5.15) exists and can be shown to be **uniformly ultimately bounded** (u.u.b.)(Appendix C).

(*iv*) *Theorem 8.5.3*

(Leitmann [12]) The solution $\mathbf{e}(t)$ of the system (8.5.15) with initial condition $\mathbf{e}(t_0) = \mathbf{e}_0$ is u.u.b. with respect to the set S defined below using the control law (8.5.44).

Proof: For $\| B^T P \mathbf{e} \| \geq \varepsilon$

$$\| \Delta \mathbf{v} \| = \rho \tag{8.5.45}$$

and for $\| B^T P \mathbf{e} \| < \varepsilon$

$$\| \Delta \mathbf{v} \| = \frac{\rho}{\varepsilon} \| B^T P \mathbf{e} \| < \rho \tag{8.5.46}$$

As before choose $V(\mathbf{e}) = \mathbf{e}^T P \mathbf{e}$ and compute

$$\begin{aligned} \dot{V} &= -\mathbf{e}^T Q \mathbf{e} + 2\mathbf{e}^T PB(\Delta \mathbf{v} + \bar{\eta}) \\ &= \mathbf{e}^T Q \mathbf{e} + 2(B^T P \mathbf{e})^T (\Delta \mathbf{v} + \bar{\eta}) \\ &\leq -\mathbf{e}^T Q \mathbf{e} + 2(B^T P \mathbf{e})^T (\Delta \mathbf{v} + \rho \frac{B^T P \mathbf{e}}{\| B^T P \mathbf{e} \|}) \end{aligned} \tag{8.5.47}$$

This last inequality is a clever argument due to Leitmann[12] and is the key to the proof. The reader should try to verify this for himself.

Now for $\| B^T P \mathbf{e} \| < \varepsilon$ the second term becomes

$$2(B^T P \mathbf{e})^T (-\frac{\rho}{\varepsilon} B^T P \mathbf{e} + \rho \frac{B^T P \mathbf{e}}{\| B^T P \mathbf{e} \|}) \tag{8.5.48}$$

which attains a maximum value of $\varepsilon \frac{\rho}{2}$ when $\| B^T P \mathbf{e} \| = \frac{\varepsilon}{2}$. Thus we have that

$$\dot{V} \leq -\mathbf{e}^T Q \mathbf{e} + \frac{\varepsilon}{2}\rho < 0 \tag{8.5.49}$$

provided

$$\mathbf{e}^T Q \mathbf{e} > \frac{\varepsilon}{2}\rho \tag{8.5.50}$$

Using the relationship

$$\lambda_{min}(Q) \| \mathbf{e} \|^2 \leq \mathbf{e}^T Q \mathbf{e} \leq \lambda_{max}(Q) \| \mathbf{e} \|^2 \tag{8.5.51}$$

where $\lambda_{min}(Q)$, $\lambda_{max}(Q)$ denote the minimum and maximum eigenvalues, respectively, of a matrix, we have that $\dot{V} < 0$ if

$$\lambda_{min}(Q) \| \mathbf{e} \|^2 > \frac{\varepsilon}{2}\rho \tag{8.5.52}$$

or, equivalently

$$\| \mathbf{e} \| \geq \left(\frac{\varepsilon\rho}{2\lambda_{min}(Q)}\right)^{\frac{1}{2}} := \omega \tag{8.5.53}$$

Let S denote the smallest level surface of V containing the ball $B(\omega)$ of radius ω centered at $\mathbf{e}=0$. If $\mathbf{e}(t_0) \in S$ then the solution remains in S. If $\mathbf{e}(t_0) \notin S$ then V is decreasing along solutions of (8.5.15) as long as $\mathbf{e}(t) \notin S$ and the solution reaches the boundary ∂S in finite time.

Let $S(k_0)$ be the level surface of V defined by

$$k_0 = \mathbf{e}_0^T P \mathbf{e}_0 \tag{8.5.54}$$

and define the constant c_0 by

$$c_0 = \min\{\mathbf{e}^T Q \mathbf{e} - \frac{\varepsilon}{2}\rho(\mathbf{e}, t) \mid \mathbf{e} \in S(k_0) - int\, S\} \tag{8.5.55}$$

where $int\, S$ denotes the interior of S. Let k be such that $k = \mathbf{e}^T P \mathbf{e}$ defines S. Then the solution reaches the boundary of S in time

$$\bar{t} = t_0 + \frac{k_0 - k}{c_0} := t_0 + T(\mathbf{e}_0, S(k)) \tag{8.5.56}$$

(v) *Example 8.5.4*

In the previous example, we choose $\varepsilon = 0.001$ and implement the continuous control law (8.5.44). The response is shown in Figure 8-8. The tracking error is shown in Figure 8-9 and the control signal Δv is shown in Figure 8-10. Note that we obtain nearly the same performance from the system while removing the control chattering.

The previous analysis has been given to illustrate what can be done to overcome the effects of uncertainty in control laws based on inverse dynamics control schemes. There are many other approaches that can be used to achieve the same result. The list of references at the end of this chapter contains several alternative approaches for robust design.

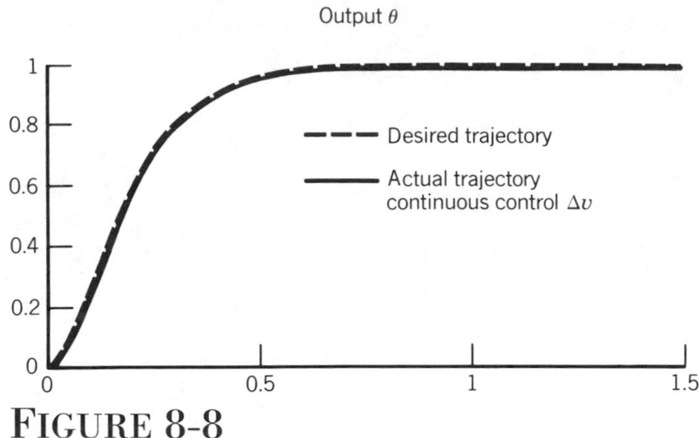

FIGURE 8-8
Tracking response with continuous approximation to discontinuous control.

REFERENCES AND SUGGESTED READING

[1] BALESTRINO, A., DEMARIA, G., and ZINOBER, A.S.I., "Nonlinear Adaptive Model-following Control," *Automatica*, Vol. 20, No. 5, pp. 559–568, 1984.

[2] BECKER, N., and GRIMM, W.M., "On L_2– and L_∞– Stability Approaches for the Robust Control of Robot Manipulators," *IEEE Trans. on Automatic Control*, Vol. 33, No. 1, pp. 118-121, Jan. 1988.

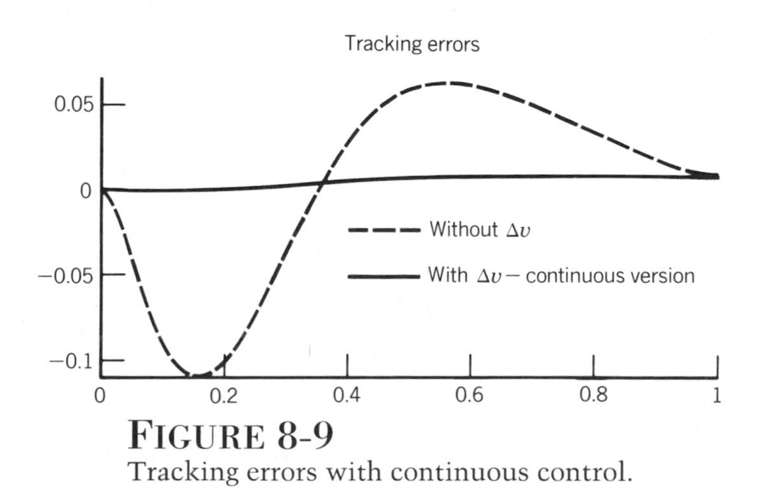

FIGURE 8-9
Tracking errors with continuous control.

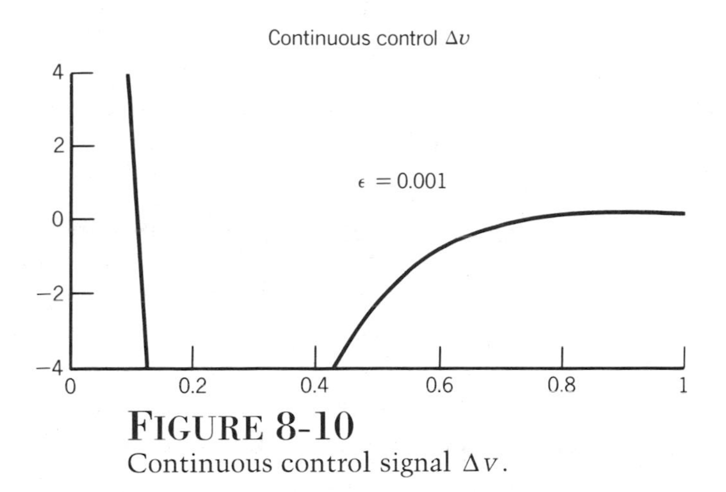

FIGURE 8-10
Continuous control signal Δv.

[3] CORLISS, M., and LEITMANN, G., "Continuous State Feedback Guaranteeing Uniform Ultimated Boundedness for Uncertain Dynamic Systems," *IEEE Trans. on Automatic Control*, AC-26, No. 5, pp. 1139–1144, Oct. 1981.

[4] CRAIG, J., *Adaptive Control of Mechanical Manipulators*, Addison–Wesley, Reading, MA, 1988.

[5] CRAIG, J.J., HSU, P., and SASTRY, S., "Adaptive Control of Mechanical Manipulators," *Proc. IEEE Int. Conf. on Robotics and Automation*, San Francisco, Mar. 1986.

[6] CVETKOVIC, V., and VUKOBRATOVIC, M., "One Robust, Dynamic Control Algorithm for Manipulation Systems," *The Int. J. of Robotics Research*, Vol. 1, No. 4, pp. 15–28, Winter 1982.

[7] FREUND, E.,"Fast Nonlinear Control with Arbitrary Pole-Placement for Industrial Robots and Manipulators," *Int. J. Robotics Res.*, Vol.1, No.1, pp. 65–78, 1982.

[8] GOOD, M., STROBEL, K., and SWEET, L.M., "Dynamics and Control of Robot Drive Systems," General Electric Company, Schenectady, NY, 1983.

[9] GUTMAN, S., "Uncertain Dynamical Systems — A Lyapunov Min–Max Approach," *IEEE Trans. Aut. Control*, AC-24, pp. 437–443, 1979.

[10] HA, I. J., GILBERT, E.G., "Robust Tracking in Nonlinear Systems and Its Application to Robotics," The University of Michigan, preprint, 1985.

[11] KODITCSHEK, D., "Natural Motion of Robot Arms," *Proc. IEEE Conf. on Decision* and *Control* , Las Vegas, 1984.

[12] LEITMANN, G., "On the Efficacy of Nonlinear Control in Uncertain Linear Systems," *J. Dyn Sys, Meas. and Cont,* Vol. 103, pp. 95–102, 1981.

[13] LUH, J. Y. S., WALKER, M., and PAUL, R. P., "Resolved Acceleration Control of Mechanical Manipulators," *IEEE Trans. Automatic Control,* Vol. AC-25, pp. 468–474, 1980.

[14] LUH, J. Y. S., WALKER, M., and PAUL, R. P., "On-line Computational Schemes for Mechanical Manipulators," *J. Dyn. Sys. Meas. and Control,* Vol. 102, pp. 69–76, 1980.

[15] LUO, G-L., and SARIDIS, G.N., "Robust Compensation of Optimal Control for Manipulators," *Proc. 21st IEEE CDC,* Orlando, FL, Dec. 1982.

[16] MIDDLETON, R.H., and GOODWIN, G., "Adaptive Computed Torque Control of Robot Manipulators," *Systems and Control Letters,* 1987.

[17] RYAN, E.P., LEITMANN, G., and CORLESS, M. J., "Practical Stabilizability of Uncertain Dynamical Systems: Application to Robotic Tracking," preprint, 1986.

[18] SAMSON, C., "Robust Non Linear Control of Robotic Manipulators," *Proc. 22nd IEEE CDC, San Antonio, Dec. 1983.*

[19] SHOURESHI, R., ROESLER, M.D., and CORLESS, M.J., "Control of Industrial Manipulators with Bounded Uncertainties," *Japan-U.S.A. Symposium on Flexible Automation,* Osaka, 1986.

[20] SLOTINE, J.J.E., and LI, W., "On the Adaptaive Control of Robot Manipulators," *Int. J. Robotics Research,* Vol. 6, No. 3, 1987.

[21] SLOTINE, J.J.E., and LI, W., "Theoretical Issues in Adaptive Manipulator Control," *Fifth Yale Workshop on Applications of Adaptive Systems Theory,* Yale Univ., 1987.

[22] SLOTINE, J.J., "Robust Control of Robot Manipulators," *Int. J. Robotics Res.,* Vol.4, No.2, 1985.

[23] SLOTINE, J.J., and SASTRY, S.S., "Tracking Control of Nonlinear Systems Using Sliding Surfaces with Application to Robot Manipulators," *Int. J. Control,* Vol. 38, pp. 465-492, 1983.

[24] SPONG, M.W., "Modeling and Control of Elastic Joint Robots," *Proc. 1986 ASME Winter Annual Meeting,* Anaheim, CA, Dec. 1986.

[25] SPONG, M.W., THORP, J.S., and KLEINWAKS, J.W., "Robust Microprocessor Control of Robot Manipulators," *Automatica*, Vol. 23, No. 3, pp. 373–379, 1987.

[26] SPONG, M.W., and VIDYASAGAR, M., "Robust Linear Compensator Design for Nonlinear Robotic Control," *IEEE Journal of Robotics and Automation*, Vol. RA-3, No. 4, pp. 345–351, Au 1987.

[27] SWEET, L.M., and GOOD, M.C., "Re-Definition of the Robot Motion Control Problem: Effects of Plant Dynamics, Drive System Constraints, and User Requirements," *Proc. 23rd IEEE Conf. on Decision and Control*, pp.724–731, Las Vegas, NV, Dec. 1984.

[28] TARN, T.J., BEJCZY, A.K., ISIDORI, A., CHEN, Y., "Nonlinear Feedback in Robot Arm Control,", *Proc. 23rd IEEE CDC*, Las Vegas, NV, Dec. 1984.

[29] WOLOVICH, W.A., *Robotics: Basic Analysis and Design*, Holt–Rinehart, & Winston, 1985.

[30] WOLOVICH, W.A., *Linear Multivariable Systems*, Springer –Verlag, New York, 1974.

PROBLEMS

8-1 Form the Lagrangian for an n-link manipulator with joint flexibility using (8.2.18)–(8.2.19). From this derive the dynamic equations of motion (8.2.22)–(8.2.23).

8-2 Complete the proof of stability of PD-control for the flexible joint robot without gravity terms using (8.2.27) and LaSalle's Theorem.

8-3 Suppose that the PD control law (8.2.24) is implemented using the link variables, that is,

$$\mathbf{u} = K_P\tilde{\mathbf{q}}_1 - K_D\dot{\mathbf{q}}_1$$

What can you say now about stability of the system? Note: This is a difficult problem. Try to prove the following conjecture: Suppose that $B_2 = 0$ in (8.2.23). Then with the above PD control law, the system is unstable. [Hint: Use Lyapunov's First Method, that is, show that the equilibrium is unstable for the linearized system.]

8-4 Using the control law (8.2.24) for the system (8.2.22)–(8.2.23), what is the steady state error if the gravity terms are present?

8-5 Simulate an inverse dynamics control law for a two-link elbow manipulator whose equations of motion were derived in Chapter Six. Investigate what happens if there are bounds on the maximum available input torque.

8-6 For the system of Problem 8-5 what happens to the response of the system if the coriolis and centrifugal terms are dropped from the inverse dynamics control law in order to facilitate computation? What happens if incorrect values are used for the link masses? Investigate via computer simulation.

8-7 Add an outer loop correction term $\Delta \mathbf{v}$ to the control law of Problem 8-6 to overcome the effects of uncertainty. Base your design on the Second Method of Lyapunov as in Section 8.5.

8-8 Consider the coupled nonlinear system

$$\ddot{y}_1 + 3y_1 y_2 + y_2^2 = u_1 + y_2 u_2$$

$$\ddot{y}_2 + \cos y_1 \dot{y}_2 + 3(y_1 - y_2) = u_2 - 3(\cos y_1)^2 y_2 u_1$$

where u_1, u_2 are the inputs and y_1, y_2 are the outputs.

a) What is the dimension of the state space?

b) Choose state variables and write the system as a system of first order differential equations in state space.

c) Find an inverse dynamics control so that the closed loop system is linear and decoupled, with each subsystem having natural frequency 10 radians and damping ratio 1/2.

CHAPTER NINE

FORCE CONTROL

9.1 INTRODUCTION

Position control strategies are adequate for tasks such as materials transfer and spot welding where the manipulator is not interacting significantly with objects in the workplace (hereafter referred to as the **environment**). However, tasks such as assembly, grinding, and deburring, which involve extensive contact with the environment, are better handled by controlling the **forces**[1] of interaction between the manipulator and the environment directly. For example, consider an application where the manipulator is required to wash a window, or to write with a felt tip marker. In both cases a pure position control scheme is unlikely to work. Slight deviations of the end-effector from the planned trajectory will cause the manipulator either to lose contact with the surface or to press too strongly on the surface. For a highly rigid structure such as a robot, a slight position error could lead to extremely large forces of interaction with disastrous consequences (broken window, smashed pen, damaged end-effector, etc.). The above applications are typical in that they involve both force control and tra-

[1]Hereafter we use **force** to mean force and/or torque, **position** to mean position and/or orientation, and **motion** to mean translation and/or rotation.

jectory control. In the writing application we clearly need to control the forces normal to the plane of the paper and position in the plane of the paper in order to generate the appropriate text. It is clear that we should modify the position commands based on externally sensed force information. Therefore a force feedback control algorithm should accept force and motion commands, measure forces and positions, and produce motion commands to the manipulator.

There are three main types of sensors for force feedback, **wrist force** sensors, **joint torque** sensors, and **tactile** or hand sensors. A wrist force sensor such as that shown in Figure 9-1 usually consists of an array of strain gauges and can delineate the three components of the vector force along the three axes of the sensor coordinate frame, and the three components of the torque about these axes. A joint torque sensor likewise consists of strain gauges located on the actuator shaft. In the case of electric drives one can also use the motor current as an indication of the torque at the motor shaft. Tactile sensors are usually located on the fingers of the gripper and are useful for sensing gripping force and for shape detection. For the purposes of controlling the end-effector/environment interactions, the six-axis wrist sensor usually gives the best results and we shall henceforth assume that the manipulator is equipped with such a device.

9.1.1 STATIC FORCE/TORQUE RELATIONSHIPS

Interaction of the manipulator with the environment will produce forces and moments at the end-effector or tool. Let $\mathbf{F} = (F_x, F_y, F_z, n_x, n_y, n_z)^T$ represent the vector of forces and torques at the end-effector, expressed in the tool frame. Thus F_x, F_y, F_z are the

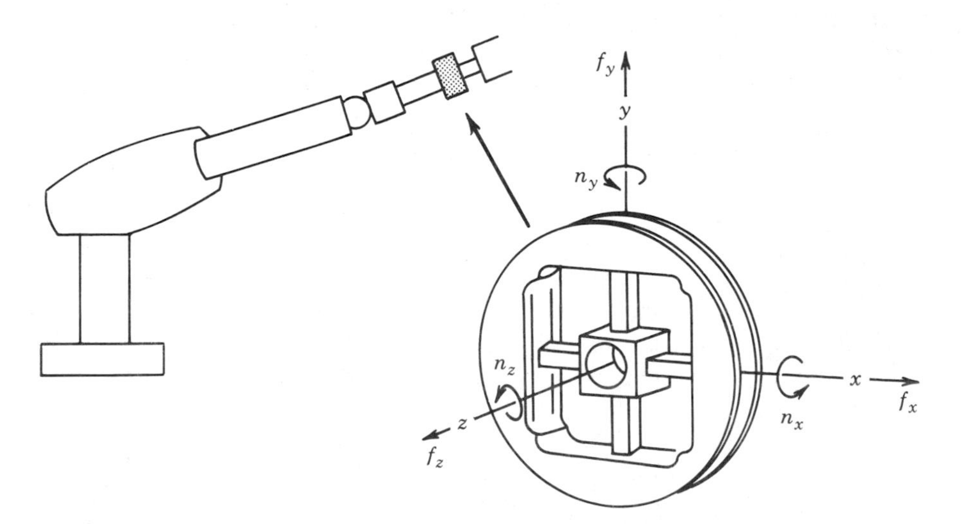

FIGURE 9-1
A typical wrist force sensor.

components of the force at the end-effector, and n_x, n_y, n_z are the components of the torque at the end-effector as shown in Figure 9-2. Let $\boldsymbol{\tau}$ denote the vector of joint torques, and let $\delta\mathbf{X}$ represent a virtual end-effector displacement caused by the force $\mathbf{F}$. Finally, let $\delta\mathbf{q}$ represent the corresponding virtual joint displacement. These virtual displacements are related through the manipulator Jacobian $J(\mathbf{q})$ according to

$$\delta\mathbf{X} = J(\mathbf{q})\delta\mathbf{q} \tag{9.1.1}$$

The virtual work δw of the system is

$$\delta w = \mathbf{F}^T\delta\mathbf{X} - \boldsymbol{\tau}^T\delta\mathbf{q} \tag{9.1.2}$$

Substituting (9.1.1) into (9.1.2) yields

$$\delta w = (\mathbf{F}^T J - \boldsymbol{\tau}^T)\delta\mathbf{q} \tag{9.1.3}$$

which is equal to zero if the manipulator is in equilibrium. Since the generalized coordinates $\mathbf{q}$ are independent we have the equality

$$\boldsymbol{\tau} = J(\mathbf{q})^T\mathbf{F} \tag{9.1.4}$$

In other words the end-effector forces are related to the joint torques by the **transpose** of the manipulator Jacobian according to (9.1.4).

(*i*) *Example 9.1.1*

Consider the two-link planar manipulator of Figure 9-3, with a force $\mathbf{F} = (F_x, F_y)^T$ applied at the end of link two as shown. The Jacobian of this manipulator is given by Equation 5.2.4. The resulting joint torques $\boldsymbol{\tau} = (\tau_1, \tau_2)^T$ are then given as

$$\begin{bmatrix}\tau_1\\ \tau_2\end{bmatrix} = \begin{bmatrix} -a_1s_1 - a_2s_{12} & a_1c_1 + a_2c_{12} & 0 & 0 & 0 & 1 \\ -a_2s_{12} & a_2c_{12} & 0 & 0 & 0 & 1\end{bmatrix}\begin{bmatrix}F_x\\ F_y\\ F_z\\ n_x\\ n_y\\ n_z\end{bmatrix} \tag{9.1.5}$$

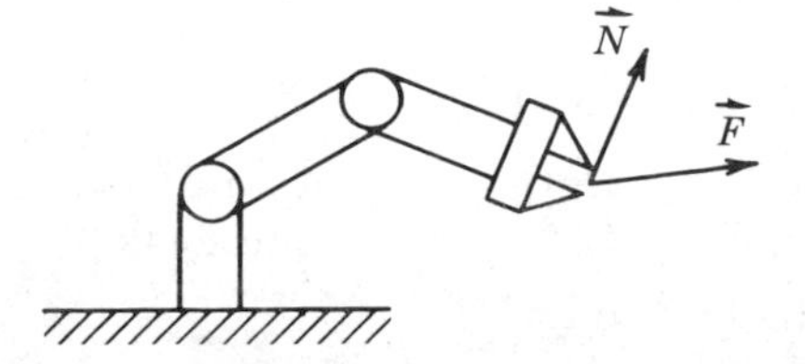

FIGURE 9-2
End-effector force and torque vectors.

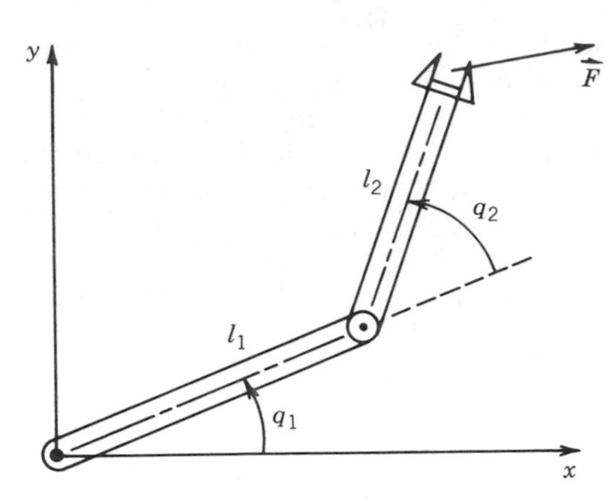

FIGURE 9-3
Two-link planar robot.

9.1.2 SINGULARITIES

We see that the Jacobian plays an important role in the forces and torques acting on a manipulator. We should expect, therefore, that singular configurations give rise to special behavior with regard to these forces and torques. At a singular configuration, i.e. at a configuration **q** where the null space of J^T is non-empty, a vector **F** in this nullspace does not produce any torque at the joints. Likewise at a singular configuration, there are directions in cartesian space in which the manipulator cannot exert forces.

(i) *Example 9.1.2*

The planar manipulator of Figure 9-4 is in a singular configuration. No set of joint torques will produce a force along the direction parallel to the links as shown. Conversely, a force acting in this direction will not produce any torque (and therefore no rotation) about the joint axes.

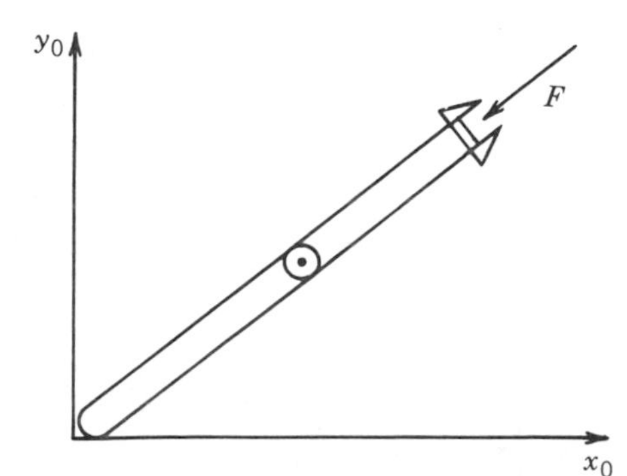

FIGURE 9-4
Singular configuration and force that produces zero torque about the joints.

9.2 NATURAL AND ARTIFICIAL CONSTRAINTS

Force control tasks can be thought of in terms of constraints imposed on the manipulator by its interaction with the environment. A manipulator moving through free space within its workspace is unconstrained in position and can exert *no* forces since there is no source of reaction force from the environment. A wrist force sensor in such a case would record only the inertial forces due to any acceleration of the end-effector. As soon as the manipulator comes in contact with, say a rigid surface as shown in Figure 9-5, a degree of freedom in position is lost since the manipulator cannot move through the surface. At the same time the manipulator can exert forces normal to the surface.

In order to describe force control tasks it is customary to introduce a **compliance frame** $o_c x_c y_c z_c$ (also called a **constraint frame**) in which the task to be performed is easily described. For example in the window washing application we can define a frame at the tool with the z_c-axis along the surface normal direction. The task specification would then be expressed in terms of maintaining a constant force in the z_c direction while following a prescribed trajectory in the $x_c - y_c$ plane. More specifically, a compliance frame is a (time-varying) coordinate frame whose coordinate axes decompose the task into directions along which either a pure position command or a pure force command can be specified.

With respect to the compliance frame we may associate certain sets of constraints known as **natural constraints** and **artificial constraints**, respectively, that define the task. For example, a position constraint in the z_c direction arising from the presence of a rigid surface is a natural constraint, while the $x_c - y_c$ trajectory necessary to wash the window is an artificial constraint. Figure 9-6 shows a typical task, that of turning a crank. We may define a compliance frame either at the end-effector or at the axis of rotation of the crank shaft. These compliance frames are shown in Figures 9-7 and 9-8 together with their respective sets of natural and artificial constraints. The artificial constraints must themselves be compatible with the natural constraints. The number of natural constraints and the number of artificial constraints are both equal

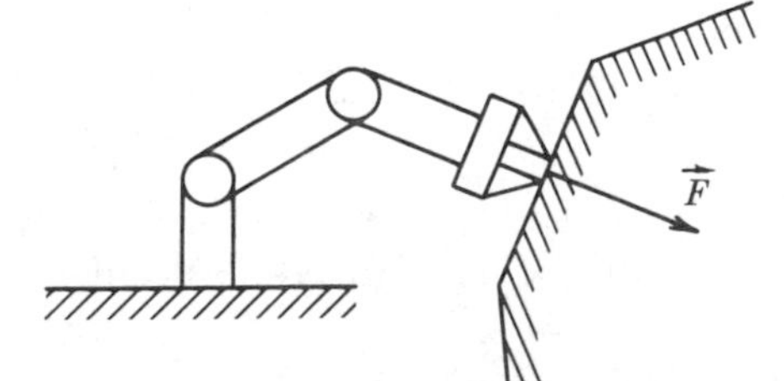

FIGURE 9-5
Manipulator in contact with the environment.

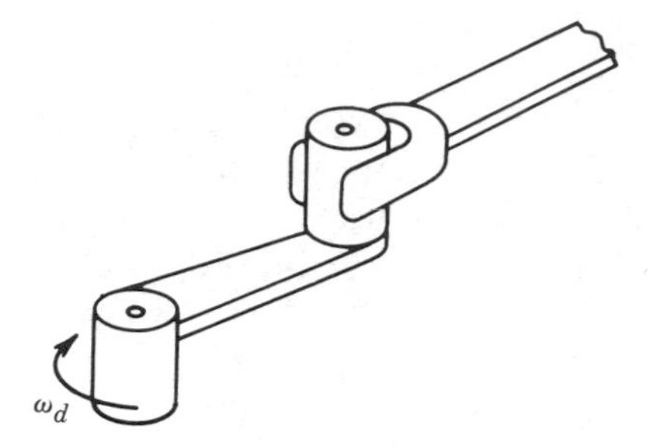

FIGURE 9-6
Turning a crank. (Source: *Robot Analysis and Control*, by H. Asada and J.J.E. Slotine. Copyright 1986, John Wiley & Sons, Inc.)

to the total number of degrees-of-freedom of the task space (in general, six). Figures 9-9 and 9-10 show natural and artificial constraints for two additional tasks, that of turning a screw and inserting a peg into a hole, respectively.

9.3 STIFFNESS AND COMPLIANCE

Robot manipulators are mechanically very rigid by design. This rigidity is necessary for high positioning accuracy and stability unless very advanced control techniques combined with direct end-point sensing are employed. However, force control applications are extremely difficult to accomplish with such rigid structures. One way to alleviate these problems is with the use of **passive compliance**. By this is meant a mechanical device composed of springs and dampers as shown in Figure 9-11 for the purpose of reducing the end-point stiffness. With such a device certain applications such as inserting a peg in a hole or writing can be achieved with pure position control. This provides a simple,

Natural Constraints	Artificial Constraints
$v_x = 0$	$f_x = 0$
$f_y = 0$	$v_y = 0$
$v_z = 0$	$f_z = 0$
$\omega_x = 0$	$\tau_x = 0$
$\omega_y = 0$	$\tau_y = 0$
$\tau_z = 0$	$\omega_z = \omega_d$

FIGURE 9-7
Compliance frame at handle of the crank. (Source: *Robot Analysis and Control*, by H. Asada and J.J.E. Slotine. Copyright 1986, John Wiley & Sons, Inc.)

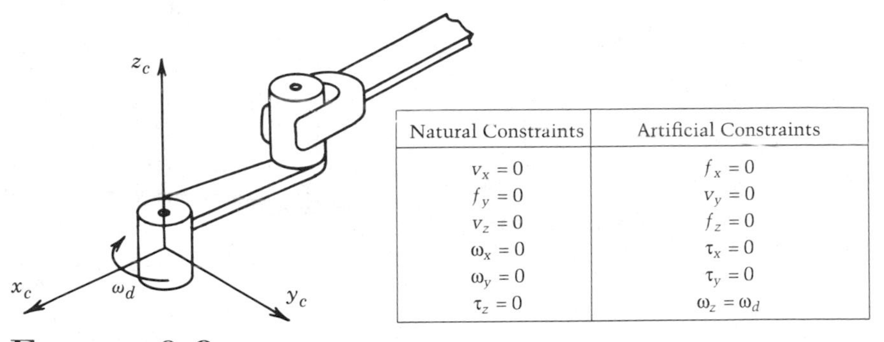

Natural Constraints	Artificial Constraints
$v_x = 0$	$f_x = 0$
$f_y = 0$	$v_y = 0$
$v_z = 0$	$f_z = 0$
$\omega_x = 0$	$\tau_x = 0$
$\omega_y = 0$	$\tau_y = 0$
$\tau_z = 0$	$\omega_z = \omega_d$

FIGURE 9-8
Compliance frame at axis of rotation of the crank. (Source: *Robot Analysis and Control*, by H. Asada and J.J.E. Slotine. Copyright 1986, John Wiley & Sons, Inc.)

inexpensive solution for certain applications that otherwise could not be accomplished with position control alone. The disadvantage of passively compliant devices is that they are limited in their range of applicability. The Remote Center Compliance (RCC) of Figure 9-11, for example, can only insert pegs of a certain length and orientation with respect to the hand. To achieve a wider applicability active control of end-point compliance is necessary.

9.3.1 STIFFNESS CONTROL OF A SINGLE DEGREE-OF-FREEDOM

As a simple illustration of an active force control strategy to regulate the end-point stiffness, consider the single degree-of-freedom system shown in Figure 9-12. We assume that the manipulator is in contact with the environment whose position is indicated as x_e in Figure 9-12.

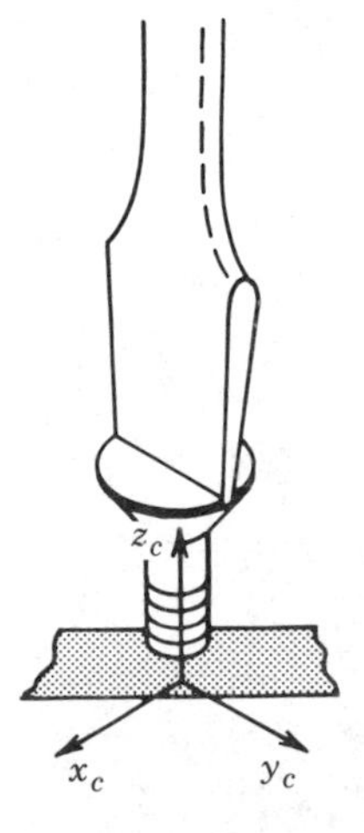

Natural Constraints	Artificial Constraints
$v_x = 0$	$f_x = 0$
$v_y = 0$	$v_y = 0$
$f_z = 0$	$v_z = p\,\omega_d$
$\omega_x = 0$	$\tau_x = 0$
$\omega_y = 0$	$\tau_y = 0$
$\tau_z = 0$	$\omega_z = \omega_d$

FIGURE 9-9
Turning a screw. (Source: *Robot Analysis and Control*, by H. Asada and J.J.E. Slotine. Copyright 1986, John Wiley & Sons, Inc.)

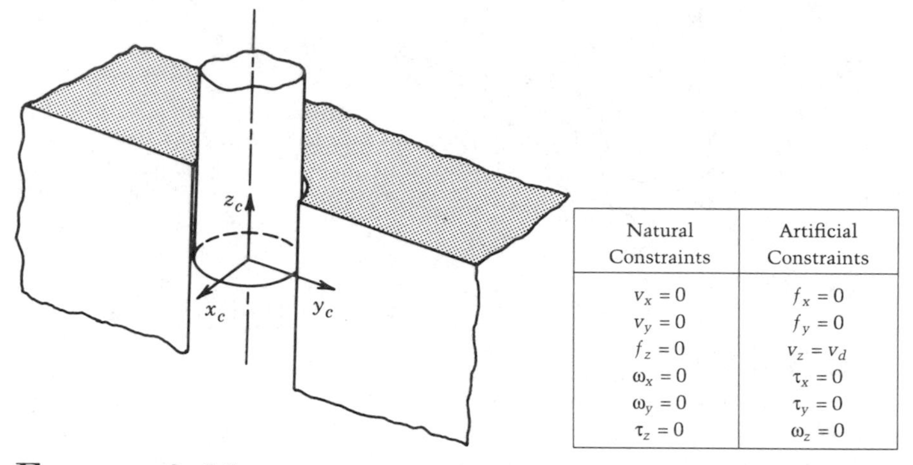

Natural Constraints	Artificial Constraints
$v_x = 0$	$f_x = 0$
$v_y = 0$	$f_y = 0$
$f_z = 0$	$v_z = v_d$
$\omega_x = 0$	$\tau_x = 0$
$\omega_y = 0$	$\tau_y = 0$
$\tau_z = 0$	$\omega_z = 0$

FIGURE 9-10

Inserting a peg into a hole. (Source: *Robot Analysis and Control*, by H. Asada and J.J.E. Slotine. Copyright 1986, John Wiley & Sons, Inc.)

If the position of the manipulator is $x > x_e$ then the force exerted on the environment is given by

$$f_e = k_e(x - x_e) \tag{9.3.1}$$

where k_e is the environmental stiffness. Note that in reality the above deformation $x - x_e$ includes the deformation in the manipulator, its supporting structure, the tool, etc., as well as the deformation of the

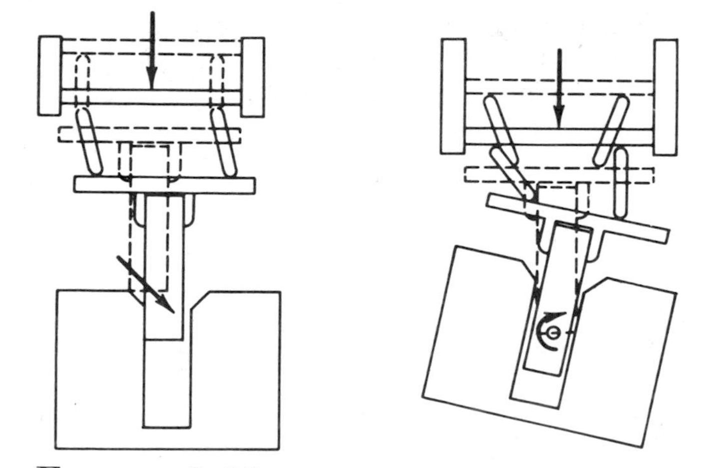

FIGURE 9-11

Remote center compliance (RCC) device. (Source: J.L. Nevins and D.E. Whitney, "Computer Controled Assembly," *Scientific American Magazine*, February 1978, page 67, Copyright 1978, Scientific American, Inc. All rights reserved.)

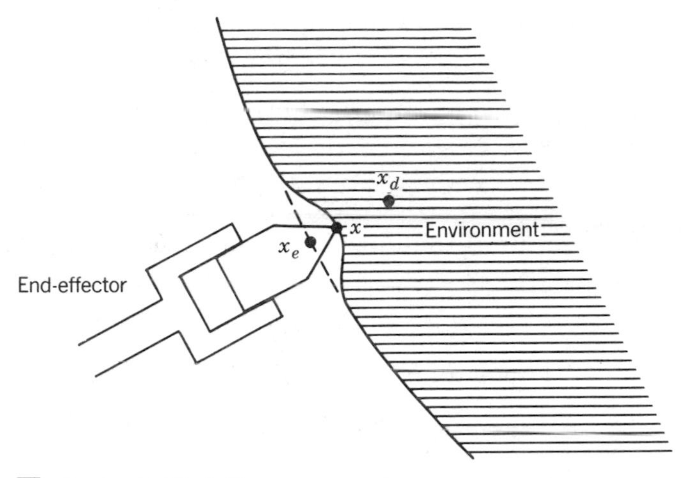

FIGURE 9-12
Stiffness control.

contacted surface. The overall system is governed by the equation

$$m\ddot{x} + k_e(x - x_e) = f \tag{9.3.2}$$

where f is the input force.

Now with x^d as shown it is easy to show (Problem 9-1) that the PD-control law

$$f = k_p(x^d - x) - k_v\dot{x} \tag{9.3.3}$$

results in a stable system if the gains are positive and that the steady state force exerted on the environment is

$$f_e = \frac{k_p k_e}{k_p + k_e}(x^d - x_e) \tag{9.3.4}$$

If the environmental stiffness k_e is large, then the force given by (9.3.4) will be approximately

$$f_e = k_p(x^d - x_e) \tag{9.3.5}$$

This suggests that, in order to exert a given force f_e on an object we can command a desired position slightly inside the object and use a position control scheme. The control law (9.3.3) in trying to eliminate the position error will then cause the force f_e in (9.3.5) to be exerted on the object in the steady state. The position gain k_p now has the interpretation of the desired "stiffness" of the manipulator.

In the case of an n-degree-of-freedom manipulator, an analogous **stiffness control** scheme can be implemented as follows. With a compliance frame $o_c x_c y_c z_c$ specified, suppose a desired stiffness is chosen along each degree of freedom. This can be represented by a 6×6 diagonal matrix K_x whose diagonal elements are stiffness constants representing linear and torsional stiffness relative to $o_c x_c y_c z_c$. For example, in the writing application we would desire a low stiffness in the z_c direction and high stiffness along the x_c and y_c directions as well as high torsional stiffness about each coordinate axis. Given K_x a desired restoring force F in response to a virtual displacement $\delta\mathbf{X}$ is given by

$$\mathbf{F} = K_x \delta\mathbf{X} \tag{9.3.6}$$

If $\delta\mathbf{q}$ denotes the resulting joint displacement, then for small displacements we have

$$\delta\mathbf{X} = J(\mathbf{q})\delta\mathbf{q} \tag{9.3.7}$$

where $J(\mathbf{q})$ is the manipulator Jacobian. Similarly

$$\boldsymbol{\tau} = J(\mathbf{q})^T \mathbf{F} \tag{9.3.8}$$

Combining the above equations yields

$$\boldsymbol{\tau} = J^T K_x J \delta\mathbf{q} =: K_{\mathbf{q}}(\mathbf{q})\delta\mathbf{q} \tag{9.3.9}$$

The configuration dependent matrix $K_{\mathbf{q}}(\mathbf{q})$ is called the **joint stiffness matrix**. The joint stiffness matrix expresses the end-point stiffnesses, which are expressed relative to the compliance frame, in terms of the joint torques. These stiffness can then be achieved by appropriately controlling the joint torques. Even though K_x is typically diagonal, $K_{\mathbf{q}}$ will not be diagonal in general. Thus in order to achieve the appropriate stiffness in the task space directions, the joint torques must be coordinated as specified by (9.3.9). When the arm is in a singular configuration then the stiffness is infinite in at least one direction (assuming, of course that the links themselves are rigid).

Figure 9-13 illustrates a joint control architecture which attempts to achieve a desired stiffness. The control law to be implemented is given as

$$\boldsymbol{\tau} = J^T K_x J(\mathbf{q}^d - \mathbf{q}) + \boldsymbol{\tau}_b \tag{9.3.10}$$

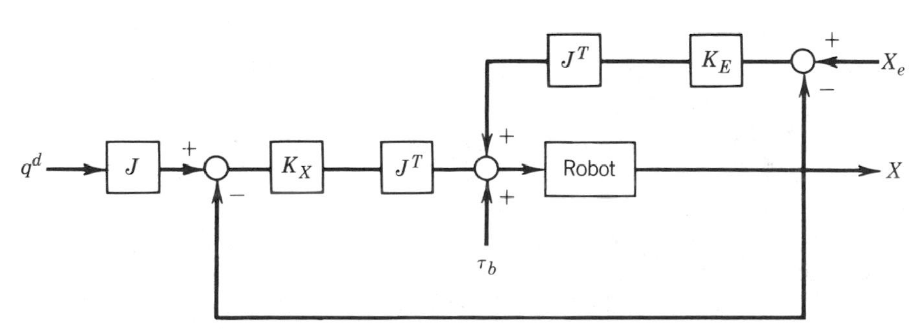

FIGURE 9-13
Stiffness control block diagram.

The additional term $\boldsymbol{\tau}_b$ can be used to provide additional damping for stability, or to compensate for say, gravitational and frictional torques. In the steady state, assuming stability, then the force on the environment is equal to (Problem 9-2)

$$\mathbf{F} = K_x \delta\mathbf{X} \tag{9.3.11}$$

as desired.

The above analysis, while describing the steady state response of stable systems, does not adequately describe nor control the transient response, that is, the transient forces resulting from the dynamic behavior of the system. We shall detail other schemes in the next few sections that take the transient behavior of the system into account.

9.4 INVERSE DYNAMICS IN TASK SPACE

Since a manipulator task specification, such as grasping an object, or inserting a peg into a hole, is typically given relative to the end-effector, it is natural to attempt to derive the control algorithm directly in the task space, rather than joint space or actuator space. This is even more important in force control than it is in position control due to the nature of the interaction of the manipulator with the environment.

If the manipulator is in contact with the environment then the dynamic equations of Chapter Six must be modified to include the reaction torque $J^T\mathbf{F}$ corresponding to the end-effector force $\mathbf{F}_e$. Thus the equations of motion of the manipulator in joint space are given by

$$M(\mathbf{q})\ddot{\mathbf{q}} + \mathbf{h}(\mathbf{q}, \dot{\mathbf{q}}) + J^T(\mathbf{q})\mathbf{F}_e = \mathbf{u} \tag{9.4.1}$$

We can rewrite these equations directly in terms of the end-effector coordinates as follows. The relationship between the vector of end-effector coordinates $\mathbf{X}$ and the vector of joint coordinates $\mathbf{q}$ is given as

$$\mathbf{X} = f(\mathbf{q}) \tag{9.4.2}$$

where the function f represents the forward kinematic equations. In general we can only determine such a function f uniquely in a region free of kinematic singularities. Taking the first and second derivatives of (9.4.1) yields

$$\dot{\mathbf{X}} = J(\mathbf{q})\dot{\mathbf{q}} \tag{9.4.3}$$

$$\ddot{\mathbf{X}} = J(\mathbf{q})\ddot{\mathbf{q}} + \dot{J}(\mathbf{q})\dot{\mathbf{q}} \tag{9.4.4}$$

where $J(\mathbf{q})$ is the manipulator Jacobian. Substituting into (9.4.4) the expression

$$\ddot{\mathbf{q}} = M^{-1}(\mathbf{q})(\mathbf{u} - \mathbf{h}(\mathbf{q}, \dot{\mathbf{q}}) - J^T\mathbf{F}_e) \tag{9.4.5}$$

from (9.4.1) and suppressing arguments for brevity, we obtain

$$\ddot{\mathbf{X}} = JM^{-1}(\mathbf{u} - \mathbf{h} - J^T\mathbf{F}_e) + \dot{J}\dot{\mathbf{q}} \tag{9.4.6}$$

Equation 9.4.6 can now be rewritten concisely as

$$\overline{M}(\mathbf{X})\ddot{\mathbf{X}} + \bar{\mathbf{h}}(\mathbf{X}, \dot{\mathbf{X}}) = \mathbf{F} - \mathbf{F}_e \tag{9.4.7}$$

where $\mathbf{F} = (J^T)^{-1}\mathbf{u}$ is the input expressed in task space and

$$\overline{M}(\mathbf{X}) := (J^T)^{-1}MJ^{-1} \;\; ; \;\; \bar{\mathbf{h}}(\mathbf{X}, \dot{\mathbf{X}}) = (J^T)^{-1}\mathbf{h} - (J^T)^{-1}\dot{J}J^{-1}\dot{\mathbf{X}} \tag{9.4.8}$$

Now, to obtain the double integrator system in task space

$$\ddot{\mathbf{X}} = \mathbf{f} \tag{9.4.9}$$

it is easy to see that the control $\mathbf{F}$ in (9.4.7) should be chosen according to

$$\mathbf{F} = \mathbf{F}_e + \overline{M}\mathbf{f} + \bar{\mathbf{h}} \tag{9.4.10}$$

Equation 9.4.10 corresponds to the joint torque control $\mathbf{u}$ given by

$$\mathbf{u} = J^T\mathbf{F} \tag{9.4.11}$$

We see that the control law (9.4.10) is more complicated than the inverse dynamics control in joint space due to the presence of the Jacobian, its inverse and its time derivative. The control law (9.4.10) can also be expected to be poorly behaved near manipulator singularities.

9.5 IMPEDANCE CONTROL

In this section we discuss an approach to force control known as **impedance control**. Impedance control does not attempt to track motion or force trajectories but rather attempts to regulate the relationship **between** the velocity and force, in other words, the **mechanical impedance**. Using the common mechanical/electrical analog that equates force with voltage[2] and velocity with current[3] the ratio of force to velocity (torque to angular velocity) is referred to as the **mechanical impedance** of the system. In the frequency domain this is represented by

$$\frac{F_e(s)}{V(s)} = Z(s) \tag{9.5.1}$$

where $F_e(s)$, $V(s)$, and $Z(s)$ are, respectively, the force, velocity and impedance. In terms of the position $X(s)$ we may write

$$\frac{F_e(s)}{X(s)} = sZ(s) \tag{9.5.2}$$

In the linear case a desired impedance might be specified as

$$sZ(s) = -(Ms^2 + Bs + K) \tag{9.5.3}$$

---

[2]Such variables are known as **through variables** in network theory.

[3]Such variables are likewise called **across variables**.

The transfer function relation (9.5.2) then corresponds to the differential equation

$$M\ddot{x} + B\dot{x} + Kx = -\mathbf{F}_e \quad (9.5.4)$$

Equation 9.5.4 then represents the desired response of the manipulator. The constants M, B, and K represent desired inertia, damping, and stiffness, respectively, and it is the task of the impedance control system to produce the actual response of the system represented by (9.5.4).

Impedance control represents a unification of sorts of the various control strategies discussed. The stiffness control strategy discussed in Section 9.3 can be thought of as a special case of impedance control that considers only the steady state force/displacement relationship. Note that an end-effector moving in free space exerts zero force on the environment for a given velocity and hence has zero impedance, while conversely a manipulator rigidly attached to a wall is motionless for any applied force and hence has infinite impedance. Thus, pure position control and pure force control can also be considered as special cases of impedance control, corresponding to infinite impedance and zero impedance, respectively.

The impedance control strategy may be implemented in conjunction with the inverse dynamics in task space of Section 9.4. Consider the equations of motion in task space

$$\overline{M}(\mathbf{X})\ddot{\mathbf{X}} + \bar{h}(\mathbf{X},\dot{\mathbf{X}}) = F - F_e \quad (9.5.5)$$

The desired impedance $Z(s)$ represented by the system (9.5.4) may be obtained by the inner loop/outer loop control law

$$F = F_e + \overline{M}(\mathbf{X})\mathbf{f} + \bar{h}(\mathbf{X},\dot{\mathbf{X}}) \quad (9.5.6)$$

$$\mathbf{f} = -M^{-1}[K(x - x^d) + B\dot{x} + \mathbf{F}_e] \quad (9.5.7)$$

The required control law to be implemented in joint space is then

$$\mathbf{u} = J^T(\mathbf{q})F \quad (9.5.8)$$

9.5.1 Implementation and Robustness Issues

The impedance control law (9.5.6)–(9.5.7) can be implemented in practice using measurements of the joint variables together with the end-effector forces and torques. The control law first cancels the force F_e which we assume is measured directly by a force/torque sensor. The remaining feedback terms are functions of $\mathbf{X}$ and $\dot{\mathbf{X}}$, which can be calculated from the measured joint variables using the forward kinematic and Jacobian expressions previously derived.

The robustness issues involved here are even more difficult than the robustness of the inverse dynamics control law in joint space for pure position control. First, the measured force $\mathbf{F}_e$ in (9.5.5) is typically noisier than joint angle measurements. Second, the inverse dynamics calculations in task space are more involved and hence require more computational power than the inverse dynamics in joint space. Third, there are ever-present problems associated with singularities of the Jacobian, and finally, if the terms $\mathbf{X}$ and $\dot{\mathbf{X}}$ are computed from joint angle measurements rather than measured directly, then the performance of the control law will depend heavily on the accuracy with which these terms are computed. The implementation and performance of impedance control strategies is currently rather poorly understood. Further progress and understanding requires additional research.

9.6 HYBRID POSITION/FORCE CONTROL

In this section we discuss the so-called **Hybrid Position/Force Control** strategy of Raibert and Craig [3]. Given our previous discussions, this approach can easily be understood. Given a compliance frame we wish to design a position control law along force constrained directions and a force control law along position constrained directions.

9.6.1 FORCE CONTROL ALONG A SINGLE DEGREE-OF-FREEDOM

We have so far discussed regulating forces through the position gains in a pure position control system, and by regulating the manipulator impedance. In this section we discuss pure force control, which in theory should be the best way to control both the transient and steady state forces exerted by the manipulator on the environment.

We discuss here the control along a single degree-of-freedom. Given a compliance frame together with a set of natural constraints this approach gives a method of controlling the end-effector force along directions in which a natural position constraint exists.

Consider the simplified system shown in Figure 9-14 consisting of an end-effector contacting the environment along the direction labeled x. The environment is modeled as a second order system consisting of inertia M_e, stiffness k_e and damping B_e. The environment inertia M_e consists of everything beyond the wrist force sensor, for example the inertia of the end-effector itself plus the inertia of any tool in the gripper, etc. The stiffness k_e includes the compliance of the surface being contacted as well as any passive compliance of the wrist and the stiffness of the force sensor itself. The equation governing the behavior of this system is then

$$M_e\ddot{x} + B_e\dot{x} + k_e x = F - F_{dist} \tag{9.6.1}$$

where F is the (input) force exerted by the end-effector and F_{dist} is a disturbance force.

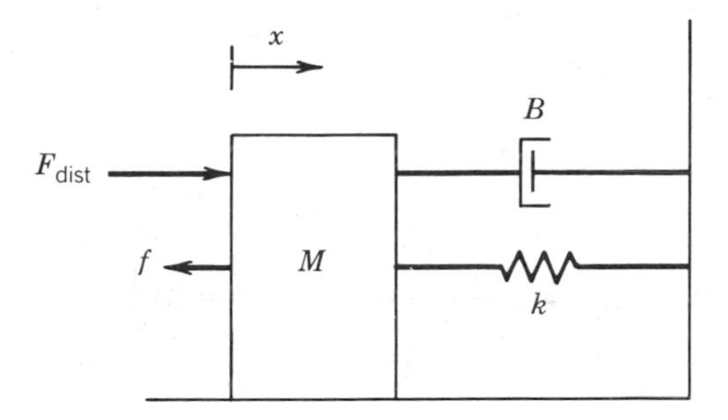

FIGURE 9-14
Mass-spring damper system.

Suppose we desire to regulate the force exerted on the environment. In the absence of disturbances then this force is given by $F_e = k_e x$. In terms of the variable F_e equation (9.6.1) can be written

$$\ddot{F}_e + \frac{B_e}{M_e}\dot{F}_e + \frac{k_e}{M_e}F_e = \frac{k_e}{M_e}(F - F_{dist}) \tag{9.6.2}$$

If a desired force trajectory is now given as $F_e^d(t)$ then a feedforward/feedforward control scheme analogous to the position control law of Chapter Seven can now be employed. In other words by choosing the input force

$$F = -K_1 F_e - K_2 \dot{F}_e + \phi(t) \tag{9.6.3}$$

where $\phi(t)$ is as yet an unspecified feedforward signal, we may achieve tracking and disturbance rejection in the same way as in the position control of Chapter Seven. The design details are left to the reader (Problem 9-3).

In the case of the force control law (9.6.3), however, the robustness issues are again quite difficult. Since the force measurements F_e are noisy in practice, it is difficult to obtain the term $\dot{F}_e$ in (9.6.3). Using the linear relationship $F_e = k_e x$, we may use $k_e \dot{x}$ in place of $\dot{F}_e$ but this requires first that the environment force be a linear function of the position x and second that the stiffness k_e be accurately known. As in the case of impedance control, pure force control remains an important area of future research.

9.6.2 HYBRID CONTROL

To implement the hybrid position/force scheme we now design both a position and a force control law for *each* degree of freedom and implement the overall control through the use of so-called **selection matrices**. Such an overall control scheme is shown in Figure 9-15. The inner loop control law consists of an inverse dynamics control computed in task space.

The selection matrix S shown is a diagonal matrix with a 1 on the diagonal entries corresponding to degrees of freedom in the compliance frame that are to be position controlled. Note then that $I - S$ automati-

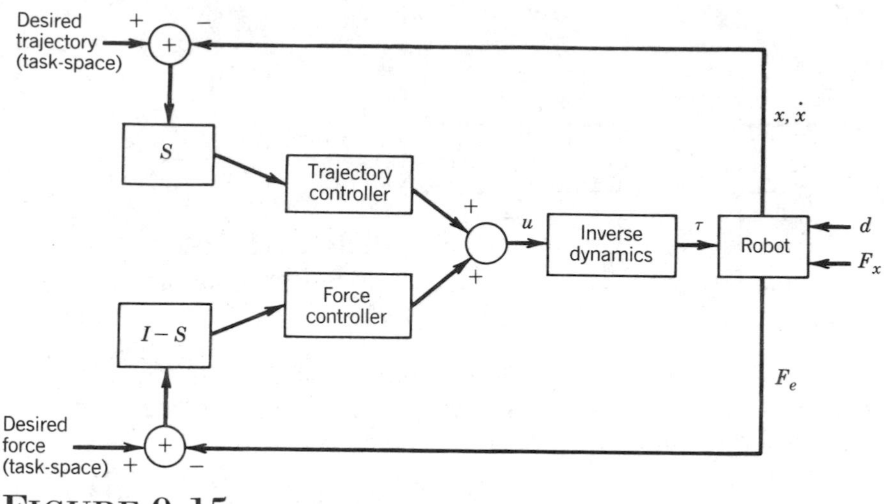

FIGURE 9-15
Hybrid control architecture.

cally is a diagonal matrix with a 1 on the diagonal entries corresponding to degrees-of-freedom that are to be force controlled. In this way each degree-of-freedom is uniquely specified as being either position controlled or force controlled. The previously derived position and force control strategies can now be inserted into this control architecture.

REFERENCES AND SUGGESTED READING

[1] ANDERSON, R.J., "Hybrid Admittance/Impedance Force Control of Robotic Manipulators," M.S. Thesis, University of Illinois at Urbana–Champaign, Urbana, IL, Oct. 1986.

[2] HOGAN, N., "Impedance Control: An Approach to Manipulation, Parts I, II, and III," J. Dyn Sys, Meas and Cont, Vol. 107, Mar. 1985.

[3] RAIBERT, M., and CRAIG, J., "Hybrid Position/Force Control of Manipulators," *J of Dyn Sys, Meas, and Cont*, June, 1981.

[4] SALISBURY, J.K., and CRAIG, J., "Active Stiffness Control of a Manipulator in Cartesian Coordinates," *Proc. 19th IEEE Conf. on Decision and Control*, Dec., 1980.

[5] WHITNEY, D., "Force Feedback Control of Manipulator Fine Motions," *Proc. JACC, San Francisco, CA, 1976.*

PROBLEMS

9-1 Verify the expression (9.3.4).

9-2 Verify the expression (9.3.11).

9-3 Carry out the details of the force control law (9.6.3). Assume it is desired to track a force trajectory $f^d(t)$. Place the poles of the error equations at –2,–2. What is the steady state error due to the disturbance f_d? Can you suggest ways to reduce the effects of the disturbance?

9-4 Given the two-link planar manipulator of Figure 9-3, find the joint torques τ_1 and τ_2 corresponding to the end-effector force vector $(-1,-1)^T$.

9-5 Consider the two-link planar manipulator with remotely driven links shown in Figure 9-16. Find an expression for the motor torques needed to balance a force **F** at the end-effector. Assume that the motor gear ratios are r_1, r_2, respectively.

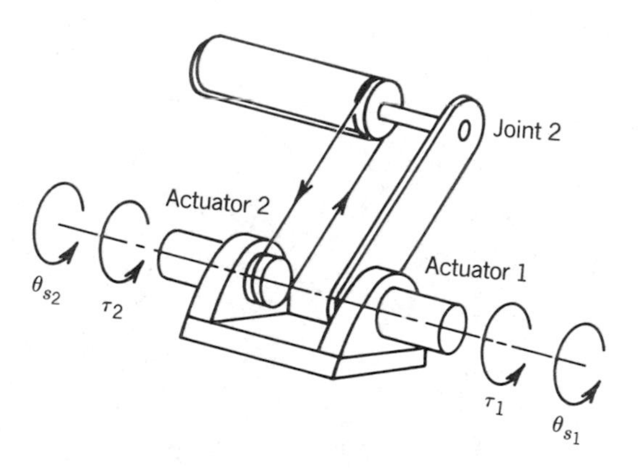

FIGURE 9-16
Two-link manipulator with remotely driven link.

9-6 What are the natural and artificial constraints for the task of inserting a square peg into a square hole? Sketch the compliance frame for this task.

9-7 Describe the natural and artificial constraints associated with the task of opening a box with a hinged lid. Sketch the compliance frame.

9-8 Discuss the task of opening a long two-handled drawer. How would you go about performing this task with two manipulators? Discuss the problem of coordinating the motion of the two arms. Define compliance frames for the two arms and describe the natural and artificial constraints.

9-9 For the two-link planar manipulator of Section 6.4.2 write the dynamic equations in task space coordinates.

CHAPTER TEN

FEEDBACK LINEARIZATION

10.1 INTRODUCTION

In this chapter we discuss the notion of **feedback linearization of nonlinear systems**. This approach generalizes the concept of inverse dynamics of rigid manipulators discussed in Chapter Eight. The basic idea of feedback linearization is to construct a nonlinear control law as a so-called **inner loop control** which, in the ideal case, exactly linearizes the nonlinear system after a suitable state space change of coordinates. The designer can then design a second stage or **outer loop control** in the new coordinates to satisfy the traditional control design specifications such as tracking, disturbance rejection, and so forth.

In the case of rigid manipulators the inverse dynamics control of Chapter Eight and the feedback linearizing control are the same. However, as we shall see, the full power of the feedback linearization technique for manipulator control becomes apparent if one includes in the dynamic description of the manipulator the transmission dynamics, such as elasticity resulting from shaft windup, gear elasticity, etc.

To introduce the idea of feedback linearization consider the following simple system,

$$\begin{aligned} \dot{x}_1 &= a \sin(x_2) \\ \dot{x}_2 &= -x_1^2 + u \end{aligned} \tag{10.1.1}$$

Note that we cannot simply choose u in the above system to cancel the nonlinear term $a\sin(x_2)$. However, if we first change variables by setting

$$y_1 = x_1 \tag{10.1.2}$$

$$y_2 = a\sin(x_2) = \dot{x}_1$$

then, by the chain rule, y_1 and y_2 satisfy

$$\dot{y}_1 = y_2 \tag{10.1.3}$$

$$\dot{y}_2 = a\cos(x_2)(-x_1^2 + u)$$

We see that the nonlinearities can now be cancelled by the input

$$u = \frac{1}{a\cos(x_2)}v + x_1^2 \tag{10.1.4}$$

which result in the linear system in the (y_1, y_2) coordinates

$$\dot{y}_1 = y_2 \tag{10.1.5}$$

$$\dot{y}_2 = v$$

The term v has the interpretation of an outer loop control and can be designed to place the poles of the second order linear system (10.1.5) in the coordinates (y_1, y_2). For example the outer loop control

$$v = -k_1 y_1 - k_2 y_2 \tag{10.1.6}$$

applied to (10.1.5) results in the closed loop system

$$\dot{y}_1 = y_2 \tag{10.1.7}$$

$$\dot{y}_2 = -k_1 y_1 - k_2 y_2$$

which has characteristic polynomial

$$p(s) = s^2 + k_2 s + k_1 \tag{10.1.8}$$

and hence the closed loop poles of the system with respect to the coordinates (y_1, y_2) are completely specified by the choice of k_1 and k_2. Figure 10-1 illustrates the inner loop/outer loop implementation of the above control strategy. The response in the y variables is easy to determine. The corresponding response of the system in the original coordinates (x_1, x_2) can be found by inverting the transformation (10.1.2), in this case

$$x_1 = y_1 \tag{10.1.9}$$

$$x_2 = \sin^{-1}(y_2/a) \qquad -a < y_2 < +a$$

This example illustrates several important features of feedback linearization. The first thing to note is the local nature of the result. We see from (10.1.3) and (10.1.4) that the transformation and the control make sense only in the region $-\infty < x_1 < \infty$, $-\frac{\pi}{2} < x_2 < \frac{\pi}{2}$. Second,

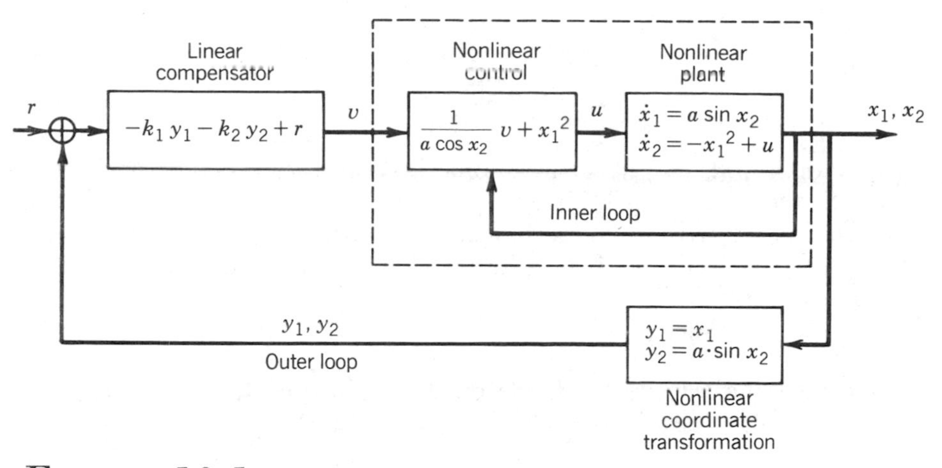

FIGURE 10-1
Architecture of feedback linearization controller.

in order to control the linear system (10.1.5), the coordinates (y_1, y_2) must be available for feedback. This can be accomplished by measuring them directly if they are physically meaningful variables, or by computing them from the measured (x_1, x_2) coordinates using the transformation (10.1.2). In the latter case the parameter a must be known precisely.

10.2 BACKGROUND: THE FROBENIUS THEOREM

In this section we give some background from differential geometry that is necessary to understand the feedback linearization results to follow. In recent years an impressive volume of literature has emerged in the area of differential geometric methods for nonlinear systems, treating not only feedback linearization but also other problems such as disturbance decoupling, estimation, etc. The reader is referred to [3] for a comprehensive treatment of the subject. Most of the results in this area are intended to give abstract, coordinate-free descriptions of various geometric properties of nonlinear systems and as such are difficult for the non-mathematician to follow. It is our intent here to give only that portion of the theory that finds an immediate application to the manipulator control problem, and even then to give only the simplest versions of the results.

We restrict our attention to the class of nonlinear systems of the form

$$\begin{aligned}\dot{\mathbf{x}} &= \mathbf{f}(\mathbf{x}) + \sum_{i=1}^{m} u_i \mathbf{g}_i(\mathbf{x}) \\ &= \mathbf{f}(\mathbf{x}) + G(\mathbf{x})u\end{aligned} \qquad (10.2.1)$$

where $\mathbf{f}(\mathbf{x}), \mathbf{g}_1(\mathbf{x}), \cdots, \mathbf{g}_m(\mathbf{x})$ are smooth vector fields on $\mathbb{R}^n$. By a **smooth vector field on** $\mathbb{R}^n$ we will mean a function $\mathbf{f}: \mathbb{R}^n \to \mathbb{R}^n$ which is infinitely differentiable. Henceforth, whenever we use the term function or vector field, it is assumed that the given function or vector field is smooth.

(i) *Definition 10.2.1*

Let $\mathbf{f}$ and $\mathbf{g}$ be two vector fields on $\mathbb{R}^n$. The **Lie Bracket** of $\mathbf{f}$ and $\mathbf{g}$, denoted by $[\mathbf{f}, \mathbf{g}]$, is a third vector field defined by

$$[\mathbf{f}, \mathbf{g}] = \frac{\partial \mathbf{g}}{\partial \mathbf{x}} \mathbf{f} - \frac{\partial \mathbf{f}}{\partial \mathbf{x}} \mathbf{g} \qquad (10.2.2)$$

where $\frac{\partial \mathbf{g}}{\partial \mathbf{x}}$ (respectively, $\frac{\partial \mathbf{f}}{\partial \mathbf{x}}$) denotes the $n \times n$ Jacobian matrix whose ij-th entry is $\frac{\partial g_i}{\partial x_j}$ (respectively $\frac{\partial f_i}{\partial x_j}$).

(ii) *Example 10.2.2*

Suppose that vector fields $\mathbf{f}(\mathbf{x})$ and $\mathbf{g}(\mathbf{x})$ on $\mathbb{R}^3$ are given as

$$\mathbf{f}(\mathbf{x}) = \begin{bmatrix} x_2 \\ \sin x_1 \\ x_3^2 + x_1 \end{bmatrix} \qquad \mathbf{g}(\mathbf{x}) = \begin{bmatrix} 0 \\ x_2^2 \\ 1 \end{bmatrix}$$

Then the vector field $[\mathbf{f}, \mathbf{g}]$ is computed according to (10.2.2) as

$$\begin{aligned}[\mathbf{f}, \mathbf{g}] &= \begin{bmatrix} 0 & 0 & 0 \\ 0 & 2x_2 & 0 \\ 0 & 0 & 0 \end{bmatrix} \begin{bmatrix} x_2 \\ \sin x_1 \\ x_1 + x_3^2 \end{bmatrix} - \begin{bmatrix} 0 & 1 & 0 \\ \cos x_1 & 0 & 0 \\ 1 & 0 & 2x_3 \end{bmatrix} \begin{bmatrix} 0 \\ x_2^2 \\ 1 \end{bmatrix} \\ &= \begin{bmatrix} x_2^2 \\ 2x_2 \sin x_1 \\ 2x_3 \end{bmatrix}\end{aligned} \qquad (10.2.3)$$

We also denote $[\mathbf{f}, \mathbf{g}]$ as $\mathbf{ad}_{\mathbf{f}}(\mathbf{g})$ and define $\mathbf{ad}_{\mathbf{f}}^k(\mathbf{g})$ inductively by

$$\mathbf{ad}_{\mathbf{f}}^k(\mathbf{g}) = [\mathbf{f}, \mathbf{ad}_{\mathbf{f}}^{k-1}(\mathbf{g})] \qquad (10.2.4)$$

with $\mathbf{ad}_{\mathbf{f}}^0(\mathbf{g}) = \mathbf{g}$.

(iii) *Definition 10.2.3*

Let $h : \mathbb{R}^n \rightarrow \mathbb{R}$ be a scalar function. The **gradient** of h, denoted **dh**, is the row vector

$$\mathbf{dh} = \left[\frac{\partial h}{\partial x_1}, \ldots, \frac{\partial h}{\partial x_n} \right] \tag{10.2.5}$$

For a scalar function h and a vector field $\mathbf{f} = (f_1, \ldots, f_n)$ the **dual product** of **dh** and **f** is defined as

$$<\mathbf{dh}, \mathbf{f}> = \frac{\partial h}{\partial x_1} f_1 + \cdots + \frac{\partial h}{\partial x_n} f_n \tag{10.2.6}$$

The following lemma gives a relationship among the Lie Bracket, gradient, and dual product and is crucial to the subsequent development.

(iv) *Lemma 10.2.4*

Let $h : \mathbb{R}^n \rightarrow \mathbb{R}$ be a scalar function and **f** and **g** be vector fields on $\mathbb{R}^n$. Then we have the following identity

$$< \mathbf{dh}\, , [\mathbf{f}, \mathbf{g}] > = < \mathbf{d} < \mathbf{dh}, \mathbf{g} > , \mathbf{f} > - < \mathbf{d} < \mathbf{dh}, \mathbf{f} > , \mathbf{g} > \tag{10.2.7}$$

Proof: Expand equation (10.2.7) in terms of the coordinates $x_1, \ldots, x_n$ and equate both sides. The $i\text{-}th$ component $[\mathbf{f}, \mathbf{g}]_i$ of the vector field $[\mathbf{f}, \mathbf{g}]$ is given as

$$[\mathbf{f}, \mathbf{g}]_i = \sum_{j=1}^{n} \frac{\partial g_i}{\partial x_j} f_j - \sum_{j=1}^{n} \frac{\partial f_i}{\partial x_j} g_j$$

Therefore, the left-hand side of (10.2.7) is

$$\begin{aligned}
< \mathbf{dh}, [\mathbf{f}, \mathbf{g}] > &= \sum_{i=1}^{n} \frac{\partial h}{\partial x_i} [\mathbf{f}, \mathbf{g}]_i \\
&= \sum_{i=1}^{n} \frac{\partial h}{\partial x_i} \left(\sum_{j=1}^{n} \frac{\partial g_i}{\partial x_j} f_j - \sum_{j=1}^{n} \frac{\partial f_i}{\partial x_j} g_j \right) \\
&= \sum_{i=1}^{n} \sum_{j=1}^{n} \frac{\partial h}{\partial x_i} \left(\frac{\partial g_i}{\partial x_j} f_j - \frac{\partial f_i}{\partial x_j} g_j \right)
\end{aligned}$$

If the right hand side of (10.2.7) is expanded similarly it can be shown, with a little algebraic manipulation, that the two sides are equal. The details are left as an exercise (Problem 10-1).

In order to discuss the idea of feedback linearization we first present a basic result in differential geometry known as the **Frobenius Theorem**. The Frobenius Theorem can be thought of as an existence theorem for solutions to certain systems of first order partial differential equations. Although a rigorous proof of this theorem is beyond the scope of this text, we can gain an intuitive understanding of it by considering the following simple system of equations

$$\frac{\partial z}{\partial x} = f(x, y, z) \qquad (10.2.8)$$
$$\frac{\partial z}{\partial y} = g(x, y, z)$$

In this example there are two partial differential equations in a single unknown z. A solution to (10.2.8) is a function $z = \phi(x, y)$ satisfying

$$\frac{\partial \phi}{\partial x} = f(x, y, \phi(x, y)) \qquad (10.2.9)$$
$$\frac{\partial \phi}{\partial y} = g(x, y, \phi(x, y))$$

We can think of the function $z = \phi(x, y)$ as defining a surface in $\mathbb{R}^3$ as in Figure 10-2. The function $\Phi : \mathbb{R}^2 \rightarrow \mathbb{R}^3$ defined by

$$\Phi(x, y) = (x, y, \phi(x, y)) \qquad (10.2.10)$$

then characterizes both the surface and the solution to the system of equations (10.2.8). At each point (x , y) the tangent plane to the surface is spanned by two vectors found by taking partial derivatives of Φ in the x and y directions, respectively, that is, by

$$\mathbf{X}_1 = (1 , 0 , f(x, y, \phi(x, y))) \qquad (10.2.11)$$
$$\mathbf{X}_2 = (0 , 1 , g(x, y, \phi(x, y))).$$

The vector fields $\mathbf{X}_1$ and $\mathbf{X}_2$ are linearly independent and span a two dimensional subspace at each point. Notice that $\mathbf{X}_1$ and $\mathbf{X}_2$ are com-

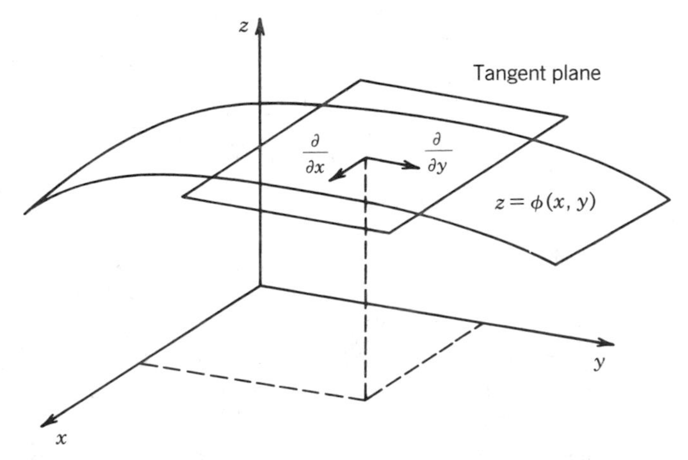

FIGURE 10-2
Integral manifold in R^3.

pletely specified by the system of equations (10.2.8). Geometrically, one can now think of the problem of solving the system of first order partial differential equations (10.2.8) as the problem of finding a surface in $\mathbb{R}^3$ whose tangent space at each point is spanned by the vector fields $\mathbf{X}_1$ and $\mathbf{X}_2$. Such a surface, if it can be found, is called an **integral manifold** for the system (10.2.8). If such an integral manifold exists then the set of vector fields, equivalently, the system of partial differential equations, is called **completely integrable**.

Let us reformulate this problem in yet another way. Suppose that $z = \phi(x,y)$ is a solution of (10.2.8). Then it is a simple computation (Problem 10-2) to check that the function

$$h(x,y,z) = z - \phi(x,y) \tag{10.2.12}$$

satisfies the system of partial differential equations

$$\begin{aligned} <\mathbf{dh}\,,\mathbf{X}_1> &= 0 \\ <\mathbf{dh}\,,\mathbf{X}_2> &= 0 \end{aligned} \tag{10.2.13}$$

Conversely, suppose a scalar function h can be found satisfying (10.2.13), and suppose that we can solve the equation

$$h(x,y,z) = 0 \tag{10.2.14}$$

for z, as $z = \phi(x,y)$.[1] Then it can be shown that ϕ satisfies (10.2.8). (Problem 10-3) Hence, complete integrability of the set of vector fields $\{\mathbf{X}_1, \mathbf{X}_2\}$ is equivalent to the existence of h satisfying (10.2.13).

With the preceding discussion as background we state the following

(v) *Definition 10.2.5*

A linearly independent set of vector fields $\{\mathbf{X}_1, \ldots, \mathbf{X}_m\}$ on $\mathbb{R}^n$ is said to be **completely integrable** if and only if there are $n-m$ linearly independent functions $h_1, \cdots, h_{n-m}$ satisfying the system of partial differential equations

$$<\mathbf{dh}_i, \mathbf{X}_j> = 0 \qquad \text{for } 1 \le i \le j\,;\ 1 \le j \le m \tag{10.2.15}$$

(vi) *Definition 10.2.6*

A linearly independent set of vector fields $\{\mathbf{X}_1, \ldots, \mathbf{X}_m\}$ is said to be **involutive** if and only if there are scalar functions $\alpha_{ijk} : \mathbb{R}^n \to \mathbb{R}$ such that

$$[\mathbf{X}_i, \mathbf{X}_j] = \sum_{k=1}^{m} \alpha_{ijk} \mathbf{X}_k \ \text{ for all } i,j,k \tag{10.2.16}$$

Involutivity simply means that if one forms the Lie Bracket of any pair of vector fields from the set $\{\mathbf{X}_1, \ldots, \mathbf{X}_m\}$ then the resulting vector

---

[1]The so-called **Implicit Function Theorem** states that (10.2.14) can be solved for z as long as $\dfrac{\partial h}{\partial z} \neq 0$.

field can be expressed as a linear combination of the original vector fields $\mathbf{X}_1, \ldots, \mathbf{X}_m$. Note that the coefficients in this linear combination are allowed to be smooth functions on $\mathbb{R}^n$. In the simple case of (10.2.8) one can show that if there is a solution $z = \phi(x,y)$ of (10.2.8) then involutivity of the set $\{\mathbf{X}_1, \mathbf{X}_2\}$ defined by (10.2.15) is equivalent to interchangeability of the order of partial derivatives of ϕ, that is, $\dfrac{\partial^2\phi}{\partial x \partial y} = \dfrac{\partial^2\phi}{\partial y \partial x}$.

The Frobenius Theorem, stated next, gives the conditions for the existence of a solution to the system of partial differential equations (10.2.15).

(vii) Theorem 10.2.7 Frobenius

Let $\{\mathbf{X}_1, \ldots, \mathbf{X}_m\}$ be a set of vector fields that are linearly independent at each point. Then the set of vector fields is completely integrable if and only if it is involutive.

Proof: See, for example, Boothby [1].

10.3 SINGLE-INPUT SYSTEMS

The idea of feedback linearization is easiest to understand in the context of single-input systems. In this section we derive the feedback linearization result of Su[8] for single-input nonlinear systems. As an illustration we apply this result to the control of a single-link manipulator with joint elasticity.

(viii) Definition 10.3.1

A single-input nonlinear system

$$\dot{\mathbf{x}} = \mathbf{f}(\mathbf{x}) + \mathbf{g}(\mathbf{x})u \tag{10.3.1}$$

where $\mathbf{f}(\mathbf{x})$ and $\mathbf{g}(\mathbf{x})$ are smooth vector fields on $\mathbb{R}^n$, $\mathbf{f}(0) = 0$, and $u \in \mathbb{R}$, is said to be **feedback linearizable** if there exists a region U in $\mathbb{R}^n$ containing the origin, a diffeomorphism[2] $T : U \to \mathbb{R}^n$, and nonlinear feedback

$$u = \alpha(\mathbf{x}) + \beta(\mathbf{x})v \tag{10.3.2}$$

with $\beta(\mathbf{x}) \neq 0$ on U such that the transformed variables

$$\mathbf{y} = \mathbf{T}(\mathbf{x}) \tag{10.3.3}$$

satisfy the system of equations

$$\dot{\mathbf{y}} = A\mathbf{y} + \mathbf{b}v \tag{10.3.4}$$

[2] A **diffeomorphism** is simply a differentiable function whose inverse exists and is also differentiable. We shall assume both the function and its inverse to be infinitely differentiable. Such functions are customarily referred to as $\mathbf{C}^\infty$ **diffeomorphisms**.

where

$$A = \begin{bmatrix} 0 & 1 & 0 & & 0 \\ 0 & 0 & 1 & & . \\ . & . & . & . & . \\ . & . & . & . & . \\ . & . & . & & 1 \\ 0 & 0 & . & . & 0 & 0 \end{bmatrix} \qquad \mathbf{b} = \begin{bmatrix} 0 \\ 0 \\ . \\ . \\ . \\ 1 \end{bmatrix} \tag{10.3.5}$$

(ix) *Remark 10.3.2*

The nonlinear transformation (10.3.3) and the nonlinear control law (10.3.2), when applied to the nonlinear system (10.3.1), result in a linear controllable system (10.3.4). The diffeomorphism $\mathbf{T}(\mathbf{x})$ can be thought of as a nonlinear change of coordinates in the state space. The idea of feedback linearization is then that if one first changes to the coordinate system $\mathbf{y} = \mathbf{T}(\mathbf{x})$, then there exists a nonlinear control law to cancel the nonlinearities in the system. The feedback linearization is said to be **global** if the region U is all of $\mathbb{R}^n$.

We next derive necessary and sufficient conditions on the vector fields $\mathbf{f}$ and $\mathbf{g}$ in (10.3.1) for the existence of such a transformation. Let us set

$$\mathbf{y} = \mathbf{T}(\mathbf{x}) \tag{10.3.6}$$

and see what conditions the transformation $\mathbf{T}(\mathbf{x})$ must satisfy. Differentiating both sides of (10.3.6) with respect to time yields

$$\dot{\mathbf{y}} = \frac{\partial \mathbf{T}}{\partial \mathbf{x}}\dot{\mathbf{x}} \tag{10.3.7}$$

where $\frac{\partial \mathbf{T}}{\partial \mathbf{x}}$ is the Jacobian of the transformation $\mathbf{T}(\mathbf{x})$. Using (10.3.1) and (10.3.4), Equation 10.3.7 can be written as

$$\frac{\partial \mathbf{T}}{\partial \mathbf{x}}(\mathbf{f}(\mathbf{x}) + \mathbf{g}(\mathbf{x})u) = A\,\mathbf{y} + \mathbf{b}v \tag{10.3.8}$$

In component form with

$$\mathbf{T} = \begin{pmatrix} T_1 \\ . \\ . \\ . \\ T_n \end{pmatrix} \qquad A = \begin{bmatrix} 0 & 1 & 0 & & 0 \\ 0 & 0 & 1 & & . \\ . & . & . & . & . \\ . & . & . & . & . \\ . & . & . & & 1 \\ 0 & 0 & . & . & 0 & 0 \end{bmatrix} \qquad \mathbf{b} = \begin{bmatrix} 0 \\ 0 \\ . \\ . \\ . \\ 1 \end{bmatrix} \tag{10.3.9}$$

we see that the first equation in (10.3.8) is

$$\frac{\partial T_1}{\partial x_1}\dot{x}_1 + \cdots + \frac{\partial T_1}{\partial x_n}\dot{x}_n = T_2 \tag{10.3.10}$$

which can be written compactly as

$$< \mathbf{d}T_1, \dot{\mathbf{x}} > = < \mathbf{d}T_1, \mathbf{f}(\mathbf{x}) + \mathbf{g}(\mathbf{x})u > = T_2 \tag{10.3.11}$$

Similarly, the other components of $\mathbf{T}$ satisfy

$$< \mathbf{d}T_2, \mathbf{f}(\mathbf{x}) + \mathbf{g}(\mathbf{x})u > = T_3 \tag{10.3.12}$$

$$\vdots$$

$$< \mathbf{d}T_n, \mathbf{f}(\mathbf{x}) + \mathbf{g}(\mathbf{x})u > = v$$

This leads to the system of partial differential equations

$$< \mathbf{d}T_1, \mathbf{f} > + < \mathbf{d}T_1, \mathbf{g} > u = T_2 \tag{10.3.13}$$

$$< \mathbf{d}T_2, \mathbf{f} > + < \mathbf{d}T_2, \mathbf{g} > u = T_3$$

$$\vdots$$

$$< \mathbf{d}T_n, \mathbf{f} > + < \mathbf{d}T_n, \mathbf{g} > u = v$$

Since we assume that $T_1, \cdots, T_n$ are independent of u while v is not independent of u we conclude from (10.3.13) that

$$< \mathbf{d}T_1, \mathbf{g} > = 0 \,, \ldots, < \mathbf{d}T_{n-1}, \mathbf{g} > = 0 \,, < dT_n, \mathbf{g} > \neq 0 \tag{10.3.14}$$

$$< \mathbf{d}T_i, \mathbf{f} > = T_{i+1} \ ; \ i=1, \ldots, n-1 \tag{10.3.15}$$

Using Lemma 10.2.4 and the conditions (10.3.14) and (10.3.15) we can derive a system of partial differential equations in terms of T_1 alone as follows. Using $h = T_1$ in Lemma 10.2.4 we have

$$< \mathbf{d}T_1, [\mathbf{f}, \mathbf{g}] > = < \mathbf{d} < \mathbf{d}T_1, \mathbf{g} >, \mathbf{f} > - < \mathbf{d} < \mathbf{d}T_1, \mathbf{f} >, \mathbf{g} > \tag{10.3.16}$$

$$= 0 - < dT_2, \mathbf{g} > = 0$$

Thus we have shown

$$< dT_1, [\mathbf{f}, \mathbf{g}] > = 0 \tag{10.3.17}$$

By proceeding inductively it can be shown (Problem 10-4) that

$$< dT_1, \mathbf{ad}_{\mathbf{f}}^k(\mathbf{g}) > = 0 \ ; \ k = 0, 1, \ldots, n-2 \tag{10.3.18}$$

$$< \mathbf{d}T_1, \mathbf{ad}_{\mathbf{f}}^{n-1}(\mathbf{g}) > \neq 0 \tag{10.3.19}$$

If we can find T_1 satisfying the system of partial differential equations (10.3.18), then $T_2, \ldots, T_n$ are found inductively from (10.3.15) and the control input u is found from

$$< \mathbf{d}T_n, \mathbf{f} > + < \mathbf{d}T_n, \mathbf{g} > u = v \tag{10.3.20}$$

as

$$u = \frac{1}{< \mathbf{d}T_n, \mathbf{g} >}(v - < \mathbf{d}T_n, \mathbf{f} >) \tag{10.3.21}$$

We have thus reduced the problem to solving the system (10.3.18) for T_1. When does such a solution exist? First note that the vector fields $\mathbf{g}, \mathbf{ad}_\mathbf{f}(\mathbf{g}), \cdots, \mathbf{ad}_\mathbf{f}^{n-1}(\mathbf{g})$ must be linearly independent. If not, that is, if for some index i

$$\mathbf{ad}_\mathbf{f}^{i}(\mathbf{g}) = \sum_{k=0}^{i-1} \alpha_k \, \mathbf{ad}_\mathbf{f}^{k}(\mathbf{g}) \tag{10.3.22}$$

then $\mathbf{ad}_\mathbf{f}^{n-1}(\mathbf{g})$ would be a linear combination of $\mathbf{g}, \mathbf{ad}_\mathbf{f}(\mathbf{g}), \cdots, \mathbf{ad}_\mathbf{f}^{n-2}(\mathbf{g})$ and (10.3.19) could not hold. Now by the Frobenius Theorem (10.3.18) has a solution if and only if the set of vector fields $\{\mathbf{g}, \mathbf{ad}_\mathbf{f}(\mathbf{g}), \ldots, \mathbf{ad}_\mathbf{f}^{n-2}(\mathbf{g})\}$ is involutive. Putting this together we have shown the following.

(x) *Theorem 10.3.3 (Su [8])*

The nonlinear system

$$\dot{\mathbf{x}} = \mathbf{f}(\mathbf{x}) + \mathbf{g}(\mathbf{x})u \tag{10.3.23}$$

with $\mathbf{f}(\mathbf{x})$, $\mathbf{g}(\mathbf{x})$ smooth vector fields, and $\mathbf{f}(0) = 0$ is feedback linearizable if and only if there exists a region U containing the origin in $\mathbb{R}^n$ in which the following conditions hold:

1. The vector fields $\{\mathbf{g}, \mathbf{ad}_\mathbf{f}(\mathbf{g}), \ldots, \mathbf{ad}_\mathbf{f}^{n-1}(\mathbf{g})\}$ are linearly independent in U
2. The set $\{\mathbf{g}, \mathbf{ad}_\mathbf{f}(\mathbf{g}), \ldots, \mathbf{ad}_\mathbf{f}^{n-2}(\mathbf{g})\}$ is involutive in U.

(xi) *Example 10.3.4 (Marino and Spong [6])*

Consider the single link manipulator with flexible joint shown in Figure 10-3. Ignoring damping for simplicity the equations of motion are

$$\begin{aligned} I\ddot{q}_1 + MgL\sin(q_1) + k(q_1 - q_2) &= 0 \\ J\ddot{q}_2 - k(q_1 - q_2) &= u \end{aligned} \tag{10.3.24}$$

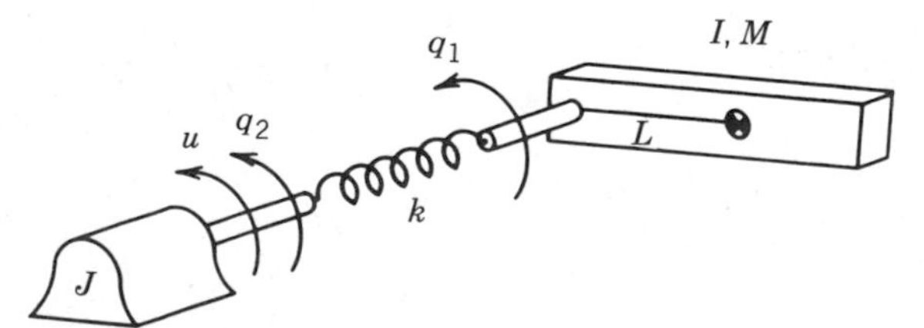

FIGURE 10-3
Single-link robot with joint flexibility.

Note that since the nonlinearity enters into the first equation the control u cannot simply be chosen to cancel it as in the case of the rigid manipulator equations. In other words, there is no obvious analogue of the inverse dynamics control for the system in this form.

In state space we set

$$x_1 = q_1 \qquad x_2 = \dot{q}_1 \qquad (10.3.25)$$
$$x_3 = q_2 \qquad x_4 = \dot{q}_2$$

and write the system (10.3.24) as

$$\begin{aligned}
\dot{x}_1 &= x_2 \\
\dot{x}_2 &= -\frac{MgL}{I}\sin(x_1) - \frac{k}{I}(x_1 - x_3) \\
\dot{x}_3 &= x_4 \\
\dot{x}_4 &= \frac{k}{J}(x_1 - x_3) + \frac{1}{J}u
\end{aligned} \qquad (10.3.26)$$

The system is thus of the form (10.3.1) with

$$\mathbf{f}(\mathbf{x}) = \begin{bmatrix} x_2 \\ -\frac{MgL}{I}\sin(x_1) - \frac{k}{I}(x_1 - x_3) \\ x_4 \\ \frac{k}{J}(x_1 - x_3) \end{bmatrix} \qquad \mathbf{g}(\mathbf{x}) = \begin{bmatrix} 0 \\ 0 \\ 0 \\ \frac{1}{J} \end{bmatrix} \qquad (10.3.27)$$

Therefore $n = 4$ and the necessary and sufficient conditions for feedback linearization of this system are that

$$rank\,\{\mathbf{g}, \mathbf{ad_f}(\mathbf{g}), \mathbf{ad}^2_\mathbf{f}(\mathbf{g}), \mathbf{ad}^3_\mathbf{f}(\mathbf{g})\} = 4 \qquad (10.3.28)$$

and that the set

$$\{\mathbf{g}, \mathbf{ad_f}(\mathbf{g}), \mathbf{ad}^2_\mathbf{f}(\mathbf{g})\} \qquad (10.3.29)$$

be involutive. Performing the indicated calculations it is easy to check that (Problem 10-7)

$$[\mathbf{g}, \mathbf{ad_f}(\mathbf{g}), \mathbf{ad}^2_\mathbf{f}(\mathbf{g}), \mathbf{ad}^3_\mathbf{f}(\mathbf{g})] = \begin{bmatrix} 0 & 0 & 0 & \frac{k}{IJ} \\ 0 & 0 & \frac{k}{IJ} & 0 \\ 0 & \frac{1}{J} & 0 & -\frac{k}{J^2} \\ \frac{1}{J} & 0 & -\frac{k}{J^2} & 0 \end{bmatrix} \qquad (10.3.30)$$

which has rank 4 for $k > 0$, $I, J < \infty$. Also, since the vector fields $\{\mathbf{g}, \mathbf{ad_f}(\mathbf{g}), \mathbf{ad_f^2}(\mathbf{g})\}$ are constant, they form an involutive set. To see this it suffices to note that the Lie Bracket of two constant vector fields is zero. Hence the Lie Bracket of any two members of the set of vector fields in (10.3.29) is zero which is trivially a linear combination of the vector fields themselves. It follows that the system (10.3.24) is feedback linearizable. The new coordinates

$$y_i = T_i \qquad i = 1, \dots, 4 \tag{10.3.31}$$

are found from the conditions (10.3.18), (10.3.19) with $n = 4$, that is,

$$< \mathbf{d}T_1, \mathbf{g} > = 0 \tag{10.3.32}$$

$$< \mathbf{d}T_1, [\mathbf{f}, \mathbf{g}] > = 0 \tag{10.3.33}$$

$$< \mathbf{d}T_1, \mathbf{ad_f^2}(\mathbf{g}) > = 0 \tag{10.3.34}$$

$$< \mathbf{d}T_1, \mathbf{ad_f^3}(\mathbf{g}) > \neq 0 \tag{10.3.35}$$

Carrying out the above calculations leads to the system of equations (Problem 10-8)

$$\frac{\partial T_1}{\partial x_2} = 0, \quad \frac{\partial T_1}{\partial x_3} = 0, \quad \frac{\partial T_1}{\partial x_4} = 0 \tag{10.3.36}$$

and

$$\frac{\partial T_1}{\partial x_1} \neq 0 \tag{10.3.37}$$

From this we see that the function T_1 should be a function of x_1 alone. Therefore, we take the simplest solution

$$y_1 = T_1 = x_1 \tag{10.3.38}$$

and compute from (10.3.15) (Problem 10-9)

$$y_2 = T_2 = < \mathbf{d}T_1, \mathbf{f} > = x_2 \tag{10.3.39}$$

$$y_3 = T_3 = < \mathbf{d}T_2, \mathbf{f} > = -\frac{MgL}{I}\sin(x_1) - \frac{k}{I}(x_1 - x_3) \tag{10.3.40}$$

$$y_4 = T_4 = < \mathbf{d}T_3, \mathbf{f} > = -\frac{MgL}{I}\cos(x_1)\cdot x_2 - \frac{k}{I}(x_2 - x_4) \tag{10.3.41}$$

The feedback linearizing control input u is found from the condition

$$u = \frac{1}{< \mathbf{d}T_4, \mathbf{g} >}(v - < \mathbf{d}T_4, \mathbf{f} >) \tag{10.3.42}$$

as (Problem 10-10)

$$u = \frac{IJ}{k}(v - a(\mathbf{x})) = \beta(\mathbf{x})v + \alpha(\mathbf{x}) \tag{10.3.43}$$

where

$$a(x) := \frac{MgL}{I}\sin(x_1)(x_2^2 + \frac{MgL}{I}\cos(x_1) + \frac{k}{I}) \qquad (10.3.44)$$
$$+ \frac{k}{I}(x_1 - x_3)(\frac{k}{I} + \frac{k}{J} + \frac{MgL}{I}\cos(x_1))$$

Therefore in the coordinates $y_1, \ldots, y_4$ with the control law (10.3.43) the system becomes

$$\dot{y}_1 = y_2 \qquad (10.3.45)$$
$$\dot{y}_2 = y_3$$
$$\dot{y}_3 = y_4$$
$$\dot{y}_4 = v$$

or, in matrix form,

$$\dot{\mathbf{y}} = A\mathbf{y} + \mathbf{b}v \qquad (10.3.46)$$

where

$$A = \begin{bmatrix} 0 & 1 & 0 & 0 \\ 0 & 0 & 1 & 0 \\ 0 & 0 & 0 & 1 \\ 0 & 0 & 0 & 0 \end{bmatrix} \quad \mathbf{b} = \begin{bmatrix} 0 \\ 0 \\ 0 \\ 1 \end{bmatrix} \qquad (10.3.47)$$

(xii) *Remark 10.3.5*

1. The above feedback linearization is actually global. In order to see this we need only compute the inverse of the change of variables (10.3.38)–(10.3.41). Inspecting (10.3.38)–(10.3.41) we see that

$$x_1 = y_1 \qquad (10.3.48)$$
$$x_2 = y_2$$
$$x_3 = y_1 + \frac{I}{k}(y_3 + \frac{MgL}{I}\sin(y_1))$$
$$x_4 = y_2 + \frac{I}{k}(y_4 + \frac{MgL}{I}\cos(y_1)\cdot y_2)$$

The inverse transformation is well defined and differentiable everywhere and, hence, the feedback linearization for the system (10.3.24) holds globally.

2. The transformed variables $y_1, \ldots, y_4$ are themselves physically meaningful. We see that

$$y_1 = x_1 = \text{link position} \qquad (10.3.49)$$
$$y_2 = x_2 = \text{link velocity}$$

$$y_3 = \dot{y}_2 = \text{ link acceleration}$$

$$y_4 = \dot{y}_3 = \text{ link jerk}$$

Since the motion trajectory of the link is typically specified in terms of these quantities they are natural variables to use for feedback.

(xiii) *Example 10.3.6*

One way to execute a step change in the link position while keeping the manipulator motion smooth would be to require a constant jerk during the motion. This can be accomplished by a cubic polynomial trajectory using the methods of Chapter Seven. Therefore, let us specify a trajectory

$$\theta_\ell^d(t) = y_1^d = a_1 + a_2 t + a_3 t^2 + a_4 t^3 \tag{10.3.50}$$

so that

$$y_2^d = \dot{y}_1^d = a_2 + 2a_3 t + 3a_4 t^2$$
$$y_3^d = \dot{y}_2^d = 2a_3 + 6a_4 t$$
$$y_4^d = \dot{y}_3^d = 6a_4$$

Then a linear control law that tracks this trajectory and that is essentially equivalent to the feedforward/feedback scheme of Chapter Eight is given by

$$v = \dot{y}_4^d - k_1(y_1 - y_1^d) - k_2(y_2 - y_2^d) - k_3(y_3 - y_3^d) - k_4(y_4 - y_4^d) \tag{10.3.51}$$

Applying this control law to the fourth order linear system (10.3.43) we see that the tracking error $e(t) = y_1 - y_1^d$ satisfies the fourth order linear equation

$$\frac{d^4 e}{dt^4} + k_4 \frac{d^3 e}{dt^3} + k_3 \frac{d^2 e}{dt^2} + k_2 \frac{de}{dt} + k_1 e = 0 \tag{10.3.52}$$

and, hence, the error dynamics are completely determined by the choice of gains $k_1, \ldots, k_4$.

Notice that the feedback control law (10.3.51) is stated in terms of the variables $y_1, \ldots, y_4$. Thus, it is important to consider how these variables are to be determined so that they may be used for feedback in case they cannot be measured directly. There are several approaches, most of which at the present time are active areas of research.

Although the first two variables, representing the link position and velocity, are easy to measure, the remaining variables, representing link acceleration and jerk, are difficult to measure with any degree of accuracy using present technology. One could measure the original variables $x_1, \ldots, x_4$ which represent the motor and link positions and velocities, and compute $y_1, \ldots, y_4$ using the transformation equations

(10.3.38)–(10.3.41). In this case the parameters appearing in the transformation equations would have to be known precisely. Another, and perhaps more promising, approach is to construct a dynamic observer to estimate the state variables $y_1, \dots, y_4$. The robust observer design for such systems is, at present, an open research problem.

10.4 FEEDBACK LINEARIZATION FOR *N*-LINK ROBOTS

In the general case of an n-link manipulator the dynamic equations represent a multi-input nonlinear system. The conditions for feedback linearization of multi-input systems are more difficult to state, but the conceptual idea is the same as the single-input case. That is, one seeks a coordinate systems in which the nonlinearities can be exactly canceled by one or more of the inputs. In the multi-input system we can also decouple the system, that is, linearize the system in such a way that the resulting linear system is composed of subsystems, each of which is affected by only a single one of the outer loop control inputs. Since we are concerned only with the application of these ideas to manipulator control we will not need the most general results in multi-input feedback linearization. Instead, we will use the physical insight gained by our detailed derivation of this result in the single-link case to derive a feedback linearizing control both for n-link rigid manipulators and for n-link manipulators with elastic joints directly.

(xiv) *Example 10.4.1*

We will first verify what we have stated previously, namely that for an n-link rigid manipulator the feedback linearizing control is identical to the inverse dynamics control of Chapter Eight. To see this, consider the rigid equations of motion (8.5.1), which we write in state space as

$$\begin{aligned}\dot{\mathbf{x}}_1 &= \mathbf{x}_2 \\ \dot{\mathbf{x}}_2 &= -M(\mathbf{x}_1)^{-1}\mathbf{h}(\mathbf{x}_1, \mathbf{x}_2) + M(\mathbf{x}_1)^{-1}\mathbf{u}\end{aligned} \tag{10.4.1}$$

with $\mathbf{x}_1 = \mathbf{q}$; $\mathbf{x}_2 = \dot{\mathbf{q}}$. In this case a feedback linearizing control is found by simply inspecting (10.4.1) as

$$\mathbf{u} = M(\mathbf{x}_1)\mathbf{v} + \mathbf{h}(\mathbf{x}_1, \mathbf{x}_2) \tag{10.4.2}$$

Substituting (10.4.2) into (10.4.1) yields

$$\begin{aligned}\dot{\mathbf{x}}_1 &= \mathbf{x}_2 \\ \dot{\mathbf{x}}_2 &= \mathbf{v}\end{aligned} \tag{10.4.3}$$

Equation 10.4.13 represents a set of n-second order systems of the form

$$\begin{aligned}\dot{x}_{1i} &= x_{2i} \\ \dot{x}_{2i} &= v_i \quad i = 1, \dots, n\end{aligned} \tag{10.4.4}$$

Comparing (10.4.2) with (8.3.4) we see indeed that the feedback linearizing control for a rigid manipulator is precisely the inverse dynamics control of Chapter Eight.

(xv) *Example 10.4.2 (Spong [7])*

If the joint flexibility is included in the dynamic description of an n-link robot the equations of motion as derived in section 8.2 are (ignoring damping for simplicity)

$$D(\mathbf{q}_1)\ddot{\mathbf{q}}_1 + \mathbf{h}(\mathbf{q}_1, \dot{\mathbf{q}}_1) + K(\mathbf{q}_1 - \mathbf{q}_2) = 0 \tag{10.4.5}$$

$$J\ddot{\mathbf{q}}_2 - K(\mathbf{q}_1 - \mathbf{q}_2) = \mathbf{u} \tag{10.4.6}$$

In state space, which is now $\mathbb{R}^{4n}$, we define state variables in block form

$$\mathbf{x}_1 = \mathbf{q}_1 \qquad \mathbf{x}_2 = \dot{\mathbf{q}}_1 \tag{10.4.7}$$

$$\mathbf{x}_3 = \mathbf{q}_2 \qquad \mathbf{x}_4 = \dot{\mathbf{q}}_2$$

Then from (10.4.5)–(10.4.6) we have :

$$\dot{\mathbf{x}}_1 = \mathbf{x}_2 \tag{10.4.8}$$

$$\dot{\mathbf{x}}_2 = -D(\mathbf{x}_1)^{-1}\{\mathbf{h}(\mathbf{x}_1, \mathbf{x}_2) + K(\mathbf{x}_1 - \mathbf{x}_3)\} \tag{10.4.9}$$

$$\dot{\mathbf{x}}_3 = \mathbf{x}_4 \tag{10.4.10}$$

$$\dot{\mathbf{x}}_4 = J^{-1}K(\mathbf{x}_1 - \mathbf{x}_3) + J^{-1}\mathbf{u} \tag{10.4.11}$$

This system is then of the form

$$\dot{\mathbf{x}} = \mathrm{f}(\mathbf{x}) + G(\mathbf{x})\mathbf{u} \tag{10.4.12}$$

In the single-link case we saw that the appropriate state variables with which to define the system so that it could be linearized by nonlinear feedback were the link position, velocity, acceleration, and jerk. Following the single-input example, then, we can attempt to do the same thing in the multi-link case and derive a feedback linearizing transformation blockwise as follows: Set

$$\mathbf{y}_1 = \mathrm{T}_1(\mathbf{x}_1) := \mathbf{x}_1 \tag{10.4.13}$$

$$\mathbf{y}_2 = \mathrm{T}_2(\mathbf{x}) := \dot{\mathbf{y}}_1 = \mathbf{x}_2 \tag{10.4.14}$$

$$\mathbf{y}_3 = \mathrm{T}_3(\mathbf{x}) := \dot{\mathbf{y}}_2 = \dot{\mathbf{x}}_2 \tag{10.4.15}$$

$$= -D(\mathbf{x}_1)^{-1}\{\mathbf{h}(\mathbf{x}_1, \mathbf{x}_2) + K(\mathbf{x}_1 - \mathbf{x}_3)\}$$

$$\mathbf{y}_4 = \mathrm{T}_4(\mathbf{x}) := \dot{\mathbf{y}}_3 \tag{10.4.16}$$

$$= -\frac{d}{dt}[D(\mathbf{x}_1)^{-1}]\{\mathbf{h}(\mathbf{x}_1, \mathbf{x}_2) + K(\mathbf{x}_1 - \mathbf{x}_3)\} - D(\mathbf{x}_1)^{-1}\{\frac{\partial \mathbf{h}}{\partial \mathbf{x}_1}\mathbf{x}_2$$

$$+ \frac{\partial \mathbf{h}}{\partial \mathbf{x}_2}[-D(\mathbf{x}_1)^{-1}(\mathbf{h}(\mathbf{x}_1, \mathbf{x}_2) + K(\mathbf{x}_1 - \mathbf{x}_3))] + K(\mathbf{x}_2 - \mathbf{x}_4)\}$$

$$:= \boldsymbol{a}_4(\mathbf{x}_1, \mathbf{x}_2, \mathbf{x}_3) + D(\mathbf{x}_1)^{-1}K\mathbf{x}_4$$

where for simplicity we define the function $\boldsymbol{a}_4$ to be everything in the definition of $\mathbf{y}_4$ except the last term, which is $D^{-1}K\mathbf{x}_4$. Note that $\mathbf{x}_4$ appears only in this last term so that $\boldsymbol{a}_4$ depends only on $\mathbf{x}_1, \mathbf{x}_2, \mathbf{x}_3$.

As in the single-link case, the above mapping is a global diffeomorphism. Its inverse can be found by inspection to be

$$\mathbf{x}_1 = \mathbf{y}_1 \tag{10.4.17}$$

$$\mathbf{x}_2 = \mathbf{y}_2 \tag{10.4.18}$$

$$\mathbf{x}_3 = \mathbf{y}_1 + K^{-1}(D(\mathbf{y}_1)\mathbf{y}_3 + \mathbf{h}(\mathbf{y}_1, \mathbf{y}_2)) \tag{10.4.19}$$

$$\mathbf{x}_4 = K^{-1}D(\mathbf{y}_1)(\mathbf{y}_4 - \boldsymbol{a}_4(\mathbf{y}_1, \mathbf{y}_2, \mathbf{y}_3)) \tag{10.4.20}$$

The linearizing control law can now be found from the condition

$$\dot{\mathbf{y}}_4 = \mathbf{v} \tag{10.4.21}$$

where $\mathbf{v}$ is a new control input. Computing $\dot{\mathbf{y}}_4$ from (10.4.16) and suppressing function arguments for brevity yields

$$\begin{aligned} \mathbf{v} = {} & \frac{\partial a_4}{\partial \mathbf{x}_1}\mathbf{x}_2 - \frac{\partial a_4}{\partial \mathbf{x}_2} D^{-1}(\mathbf{h} + K\,(\mathbf{x}_1 - \mathbf{x}_3)) \\ & + \frac{\partial a_4}{\partial \mathbf{x}_3}\mathbf{x}_4 + \frac{d}{dt}[D^{-1}]K\,\mathbf{x}_4 + D^{-1}K(J^{-1}\,K(\mathbf{x}_1 - \mathbf{x}_3) + J^{-1}\,\mathbf{u}) \\ =: {} & \boldsymbol{a}(\mathbf{x}) + b(\mathbf{x})\mathbf{u} \end{aligned} \tag{10.4.22}$$

where $\boldsymbol{a}(\mathbf{x})$ denotes all the terms in (10.4.22) but the last term, which involves the input $\mathbf{u}$, and $b(\mathbf{x}) := D^{-1}(\mathbf{x})KJ^{-1}$.

Solving the above expression for $\mathbf{u}$ yields

$$\begin{aligned} \mathbf{u} &= b(\mathbf{x})^{-1}(\mathbf{v} - \mathbf{a}(\mathbf{x})) \\ &=: \boldsymbol{\alpha}(\mathbf{x}) + \boldsymbol{\beta}(\mathbf{x})\mathbf{v} \end{aligned} \tag{10.4.23}$$

where $\boldsymbol{\beta}(\mathbf{x}) = JK^{-1}D(\mathbf{x})$ and $\boldsymbol{\alpha}(\mathbf{x}) = -b(\mathbf{x})^{-1}\boldsymbol{a}(\mathbf{x})$.

With the nonlinear change of coordinates (10.4.13)–(10.4.16) and nonlinear feedback (10.4.23) the transformed system now has the linear block form

$$\begin{aligned} \dot{\mathbf{y}} &= \begin{bmatrix} 0 & I & 0 & 0 \\ 0 & 0 & I & 0 \\ 0 & 0 & 0 & I \\ 0 & 0 & 0 & 0 \end{bmatrix} \mathbf{y} + \begin{bmatrix} 0 \\ 0 \\ 0 \\ I \end{bmatrix} \mathbf{v} \\ &=: A\,\mathbf{y} + B\,\mathbf{v} \end{aligned} \tag{10.4.24}$$

where $I = n \times n$ identity matrix, $0 = n \times n$ zero matrix, $\mathbf{y}^T = (\mathbf{y}_1^T, \mathbf{y}_2^T, \mathbf{y}_3^T, \mathbf{y}_4^T) \in \mathbb{R}^{4n}$, and $\mathbf{v} \in \mathbb{R}^n$. The system (10.4.24) represents a set of n decoupled quadruple integrators. The outer loop design can now proceed as before, because not only is the system linearized, but it consists of n subsystems each identical to the fourth order system (10.3.45).

10.5 INTRODUCTION TO OUTER LOOP DESIGN

The technique of feedback linearization is important in that it leads to a control design methodology for nonlinear systems. In the context of control theory, however, one should be highly suspicious of techniques that rely on exact mathematical cancellation of terms, linear or nonlinear, from the equations defining the system. One would never, for example, attempt to cancel an unstable pole in a linear system by the insertion of a zero. The feedback linearization result then should be viewed as a structural property of a class of nonlinear systems and not as a complete solution to the control design problem.

In this section, we investigate the effect of parameter uncertainty, computational error, model simplification, etc., and show that the most important property of feedback linearizable systems is not necessarily that the nonlinearities can be exactly canceled by nonlinear feedback, but rather that, once an appropriate coordinate system is found in which the system can be linearized, the nonlinearities are in the range space of the input. This property is highly significant and is exploited by so-called robust control techniques to guarantee performance in the realistic case that the nonlinearities in the system are not known exactly.

Consider first a single-input feedback linearizable system as in Section 10.3. After the appropriate coordinate transformation, the system can be written in the ideal case as

$$\dot{y}_1 = y_2 \tag{10.5.1}$$

$$\vdots$$

$$\dot{y}_n = v = \beta^{-1}(\mathbf{x})[u - \alpha(\mathbf{x})]$$

provided that u is given by (10.3.2) in order to cancel the nonlinear terms $\alpha(\mathbf{x})$ and $\beta(\mathbf{x})$. In practice such exact cancellation is not achievable and it is more realistic to suppose that the control law u in (10.3.2) is of the form

$$u = \hat{\alpha}(\mathbf{x}) + \hat{\beta}(\mathbf{x})v \tag{10.5.2}$$

where $\hat{\alpha}(\mathbf{x})$, $\hat{\beta}(\mathbf{x})$ represent the computed versions of $\alpha(\mathbf{x})$, $\beta(\mathbf{x})$, respectively. These functions may differ from the true $\alpha(\mathbf{x})$, $\beta(\mathbf{x})$ for several reasons. Because the inner loop control u is implemented digitally, there will be an error due to computational round-off and delay. Also, since the terms $\alpha(\mathbf{x})$, $\beta(\mathbf{x})$ are functions of the system parameters such as masses, and moments of inertia, any uncertainty in knowledge of these parameters will be reflected in $\hat{\alpha}(\mathbf{x})$, $\hat{\beta}(\mathbf{x})$. In addition, one may choose intentionally to simplify the control u by dropping various terms in the equations in order to facilitate on-line computation. If we now substitute the control law (10.5.2) into (10.5.1) we obtain

$$\dot{y}_1 = y_2 \qquad (10.5.3)$$

$$\vdots$$

$$\begin{aligned}\dot{y}_n &= \beta^{-1}(\mathbf{x})[\hat{\alpha}(\mathbf{x}) + \hat{\beta}(\mathbf{x})v - \alpha(\mathbf{x})] \\ &= v + \eta(y_1, \cdots, y_n, v)\end{aligned}$$

where the uncertainty η is given as

$$\eta(y_1, \cdots, y_n, v) = \{(\beta^{-1}\hat{\beta} - 1)v + \beta^{-1}(\hat{\alpha} - \alpha)\}|_{y = T^{-1}(x)} \qquad (10.5.4)$$

The system (10.5.3) can be written in matrix form as

$$\dot{\mathbf{y}} = A\mathbf{y} + \mathbf{b}\{v + \eta(\mathbf{y}, v)\} \qquad (10.5.5)$$

where A and $\mathbf{b}$ are given by (10.3.5).

Note that the system (10.5.5) is still nonlinear whenever $\eta \neq 0$. The practical implication of this is that if the outer loop control v is designed as a linear state feedback $v = -\mathbf{k}^T\mathbf{y}$ to place the poles of the system matrix A, various properties such as stability, etc., are no longer guaranteed by the location of the eigenvalues of $A - \mathbf{b}\mathbf{k}^T$. However, the nonlinearity in the system is of the form $\mathbf{b}\eta$, that is, it lies in the range space of the input. This fact, which we refer to as the **matching condition** after [5], is highly significant as it forms the basis for various (possibly nonlinear) techniques that can be used to design $\mathbf{v}$ in (10.5.5) to guarantee a desired performance of the system in spite of the uncertainty.

In the multi-input case a similar analysis shows that the application of an inner loop approximate feedback linearizing control to a feedback linearizable system results in a system of the form

$$\dot{\mathbf{y}} = A\mathbf{y} + B\{\mathbf{v} + \eta(\mathbf{y}, \mathbf{v})\} \qquad (10.5.6)$$

where B is $n \times m$, $\mathbf{v} \in \mathbb{R}^m$, and $\eta : \mathbb{R}^n \times \mathbb{R}^m \rightarrow \mathbb{R}^m$.

The system (10.5.6) can be represented by the block diagram of Figure 10-4. The application of the nonlinear inner loop control law results in a system which is "approximately linear." A common approach at this point is to decompose the control input $\mathbf{v}$ in (10.5.6) into two parts, the first to stabilize the "nominal linear system," represented by (10.5.6) with $\eta = 0$. In this case $\mathbf{v}$ can be taken as a linear state feedback control law designed to stabilize the nominal system and/or for tracking a desired trajectory. A second stage control Δv is then designed for robustness, that is, to guarantee the performance of the nominal design in the case that $\eta \neq 0$. The form of the outer loop control law is thus

$$\mathbf{v} = -K\mathbf{y} + r + \Delta\mathbf{v} \qquad (10.5.7)$$

where $K\mathbf{y}$ is a linear feedback designed to place the eigenvalues of A in a desired location, r is a reference input, which can be chosen as a

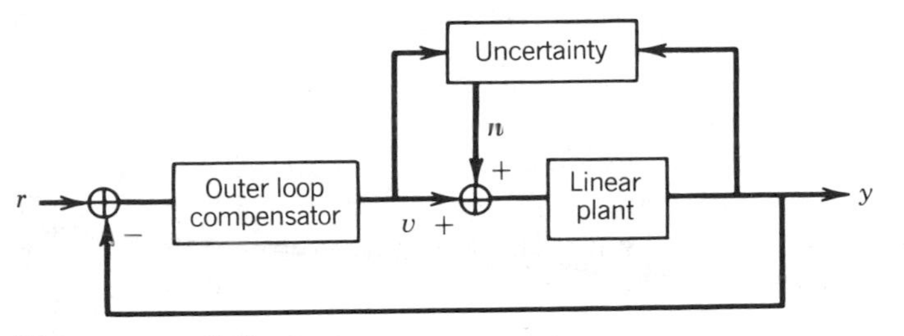

FIGURE 10-4
Block diagram for robust outer loop design.

feedforward signal for tracking a desired trajectory, and $\Delta\mathbf{v}$ represents an additional feedback loop to maintain the nominal performance despite the presence of the nonlinear term η.

10.6 OUTER LOOP DESIGN BY LYAPUNOV'S SECOND METHOD

In this section we discuss the outer loop design of feedback linearizable systems using the **Second Method of Lyapunov**, which generalizes the results previously derived for rigid robots in Chapter Eight.

Recall that the robust feedback linearization problem is to stabilize the nonlinear system

$$\dot{\mathbf{y}} = A\,\mathbf{y} + B\{\mathbf{v} + \eta(\mathbf{y},\mathbf{v})\} \tag{10.6.1}$$

by suitable choice of the outer loop control $\mathbf{v}$. If $\mathbf{y}$ represents the tracking error, then the stabilization of (10.6.1) is equivalent to the robust tracking problem. The approach that we use in this section is based on the theory of guaranteed stability of uncertain systems. The basic idea of this theory is that, although the uncertainty η in (10.6.1) is unknown, it may be possible to find a known function $\rho(\mathbf{y},t)$ satisfying

$$\|\,\eta(\mathbf{y},\mathbf{v})\,\| \leq \rho(\mathbf{y},t) \tag{10.6.2}$$

The point to notice in (10.6.2) is that ρ is independent of $\mathbf{v}$. If such a ρ can be found, then $\mathbf{v}$ can be chosen to "guarantee" stability of (10.6.1) so long as η remains bounded by the "worst case" (10.6.2).

The outer loop control law is chosen as

$$\mathbf{v} = -K\mathbf{y} + \Delta\mathbf{v} \tag{10.6.3}$$

which results in the system

$$\dot{\mathbf{y}} = \bar{A}\,\mathbf{y} + B\{\Delta\mathbf{v} + \bar{\eta}(\mathbf{y},\Delta\mathbf{v})\} \tag{10.6.4}$$

where $\bar{A}$ is Hurwitz, that is, has all of its eigenvalues in the left-half plane and

$$\bar{\eta}(\mathbf{y}, \Delta\mathbf{v}) = \eta(\mathbf{y}, -K\mathbf{y} + \Delta\mathbf{v}) \tag{10.6.5}$$

The design problem now is to choose $\Delta\mathbf{v}$ in (10.6.4) to guarantee stability of the overall system. Note that since the function η depends on the outer loop control $\mathbf{v}$ (or $\Delta\mathbf{v}$) it is not obvious that a function $\rho(\mathbf{y}, t)$ satisfying (10.6.2) can be found. In practice, additional assumptions on the uncertainty in the system must be made to guarantee the existence of such a function ρ. For notational simplicity, we will drop the overbars from $\bar{A}$ and $\bar{\eta}$ in (10.6.4) and call $\Delta\mathbf{v}$ simply $\mathbf{v}$. In other words, we consider (10.6.1) assuming now that A is Hurwitz.

The outer loop design now proceeds as in Chapter Eight.

Step 1: Given the system (10.6.1) with A Hurwitz, suppose we can find a continuous function $\rho(\mathbf{y}, t)$, bounded in t, satisfying the inequality

$$\|\eta\| \leq \rho(\mathbf{y}, t) \tag{10.6.6}$$

Step 2: Since A is Hurwitz, choose a $n \times n$ symmetric, positive definite matrix Q and let P be the unique positive definite symmetric solution to the Lyapunov equation

$$A^T P + PA + Q = 0 \tag{10.6.7}$$

Step 3: Choose the outer loop control $\mathbf{v}$ according to

$$\mathbf{v} = \begin{cases} -\rho(\mathbf{y}, t)\dfrac{B^T P\mathbf{y}}{\|B^T P\mathbf{y}\|} & \text{if } \|B^T P\mathbf{y}\| \neq 0 \\ 0 & \text{if } \|B^T P\mathbf{y}\| = 0 \end{cases} \tag{10.6.8}$$

Then we can prove the following:

(*xvi*) *Theorem 10.6.1*

Set $V(\mathbf{y}) = \mathbf{y}^T P\mathbf{y}$. Then with the outer loop control law (10.6.8) $\dot{V}(\mathbf{y}) < 0$ along solution trajectories of the system (10.6.1).

Proof: Identical to the proof of Theorem 8.5.1.

As before, we may also approximate (10.6.8) by a continuous control and prove uniform ultimate boundedness of solutions. Choose $\varepsilon > 0$ and set

$$\mathbf{v} = \begin{cases} -\rho(\mathbf{y}, t)\dfrac{B^T P\mathbf{y}}{\|B^T P\mathbf{y}\|} & \text{if } \|B^T P\mathbf{y}\| \geq \varepsilon \\ -\dfrac{\rho(\mathbf{y}, t)}{\varepsilon} B^T P\mathbf{y} & \text{if } \|B^T P\mathbf{y}\| < \varepsilon \end{cases} \tag{10.6.9}$$

(xvii) *Theorem 10.6.2*

The solution $\mathbf{y}(t)$ of the system (10.6.1) with initial condition $\mathbf{y}(t_0) = \mathbf{y}_0$ is uniformly ultimately bounded using the control law (10.6.9).

Proof: Identical to Theorem 8.5.3.

REFERENCES AND SUGGESTED READING

[1] BOOTHBY, W. M., *An Introduction to Differentiable Manifolds and Riemannian Geometry*, Academic Press, New York, 1975.

[2] FORREST-BARLACH, M.G., and BABCOCK, S.M., "Inverse Dynamics Position Control of a Compliant Manipulator," *Proc. IEEE Conference on Robotics and Automation*, San Francisco, pp. 196–205, Apr 1986.

[3] ISIDORI, A., *Nonlinear Control Systems: An Introduction*, Lecture Notes in Control and Information Sciences, Vol. 72, Springer–Verlag, Berlin, 1985.

[4] KAILATH, T., *Linear Systems*, Prentice-Hall, Englewood Cliffs, NJ, 1980.

[5] LEITMANN, G., "On the Efficacy of Nonlinear Control in Uncertain Linear Systems," *Trans. ASME, J of Dyn. Sys., Meas., and Cont.*, Vol. 102, pp. 95–102, June 1981.

[6] MARINO, R.W., and SPONG, M.W., "Nonlinear Control Techniques for Flexible Joint Manipulators: A Single Link Case Study," *Proc. 1986 IEEE Conference on Robotics and Automation*, San Francisco, pp. 1030–1026, April 1986.

[7] SPONG, M.W., "Modeling and Control of Elastic Joint Robots," *Trans. ASME, J of Dyn. Sys, Meas., and Cont.*, Vol. 109, pp. 310–319, Dec. 1987.

[8] SU, R., "On the Linear Equivalents of Nonlinear Systems," *Systems and Control Letters*, Vol. 2, 1981.

[9] VIDYASAGAR, M., *Nonlinear Systems Analysis*, Prentice-Hall, Englewood Cliffs, NJ, 1978.

PROBLEMS

10-1 Complete the proof of Lemma 10.2.4 by direct calculation.

10-2 Show that the function $h = z - \phi(x,y)$ satisfies the system (10.2.13) if ϕ is a solution of (10.2.8) and $\mathbf{X}_1, \mathbf{X}_2$ are defined by (10.2.11).

10-3 Show that if $h(x,y,z)$ satisfies (10.2.13), then, if $\frac{\partial h}{\partial z} \neq 0$, Equation 10.2.14 can be solved for z as $z = \phi(x,y)$ where ϕ satisfies (10.2.8). Also show that $\frac{\partial h}{\partial z} = 0$ can occur only in the case of the trivial solution $h = 0$ of (10.2.13).

10-4 Verify the expressions (10.3.18) and (10.3.19).

10-5 Derive the equations of motion (10.3.24) for the single-link manipulator with joint elasticity of Figure 10-2 using Lagrange's equations.

10-6 Repeat problem 10-5 where there is assumed to be viscous friction both on the link side and on the motor side of the spring in Figure 10-2.

10-7 Perform the calculations necessary to verify (10.3.30).

10-8 Derive the system of partial differential equations (10.3.36) from the conditions (10.3.32)–(10.3.35). Also verify (10.3.37).

10-9 Compute the change of coordinates (10.3.39)–(10.3.41).

10-10 Verify Equations 10.3.43–10.3.44.

10-11 Verify Equations 10.3.48.

10-12 Design and simulate a linear outer loop control law v for the system (10.3.24) so that the link angle $y_1(t)$ follows a desired trajectory $y_1^d(t) = \theta_\ell^d(t) = \sin 8t$. Use various techniques such as pole placement, linear quadratic optimal control, etc. (See reference [2] for some ideas.)

10-13 Consider again a single-link manipulator (either rigid or elastic joint). Add to your equations of motion the dynamics of a permanent magnet DC-motor. What can you say now about feedback linearizability of the system?

10-14 What happens to the inverse coordinate transformation (10.3.48) as the joint stiffness $k \to \infty$? Give a physical interpretation. Use this to show that the system (10.3.24) reduces to the equation governing the rigid joint manipulator in the limit as $k \to \infty$.

10-15 Consider the single-link manipulator with elastic joint of Figure 10-2 but suppose that the spring characteristic is nonlinear, that is, suppose that the spring force F is given by $F = \phi(q_1 - q_2)$, where ϕ is a diffeomorphism. Derive the equations of motion for the system and show that it is still feedback linearizable. Carry out the feedback linearizing transformation. Specialize the result to the case of a cubic characteristic, i.e., $\phi = k(q_1 - q_2)^3$. The cubic spring characteristic is a more accurate description for many manipulators than is the linear spring, especially for elasticity arising from gear flexibility.

10-16 Carry out the design of a robust feedback linearizing control law for a single link robot with joint flexibility using Lyapunov's Second Method. Simulate the controller for various reference trajectories. Plot the response of the link angle and the motor angle as a function of time. Discuss problems with real-time implementation of this control.

CHAPTER ELEVEN

VARIABLE STRUCTURE AND ADAPTIVE CONTROL

11.1 INTRODUCTION

In this chapter we introduce two additional techniques for robust control of manipulators, namely, **Variable Structure Control and Adaptive Control**. The intent here is not to give an exhaustive treatment of either subject, as both are extensive and are evolving areas of research. Our intent is merely to introduce our readers to alternate techniques that can be used for robust control of manipulators and other feedback linearizable systems so that they will then be prepared to tackle the research literature on the subject.

11.1.1 VARIABLE STRUCTURE SYSTEMS

We first discuss the so-called **Theory of Sliding Modes**. This approach is a special case of Variable Structure Systems and is actually quite similar in spirit to the Lyapunov approach that we discussed in Chapters Eight and Ten. (See [4] for a discussion of the relationship between the two theories.)

A variable structure system is one whose structure can be changed or switched abruptly according to a certain **switching logic** whose aim is to produce a desired overall behavior of the system. The simplest examples of variable structure systems are **relay** or **on-off** systems, in which the control input can have only two values, on or off. For example, home heating and cooling systems are of this type where the thermostat regulates the temperature by alternatively switching the heating or cooling system on and off. We can give a simple example to illustrate some of the unexpected properties of variable structure systems.

(i) *Example 11.1.1*

Consider the second order system shown in Figure 11-1. The control scheme is a simple proportional control with gains α or $\alpha+\beta$ depending on whether the switch is open or closed, respectively. We assume that $\alpha<1$ and $\alpha+\beta>1$. With the switch open the equations of motion of the system are given by

$$\dot{x}_1 = x_2 \qquad x_1 = x - r \tag{11.1.1}$$
$$\dot{x}_2 = -\alpha x_1$$

The system is thus oscillatory with phase plane trajectories shown in Figure 11-2.

With the switch closed the system is described by

$$\dot{x}_1 = x_2 \tag{11.1.2}$$
$$\dot{x}_2 = -(\alpha+\beta)x_1$$

and the solution trajectories are shown in Figure 11-3.

Equations 11.1.1 and 11.1.2 can be compactly expressed as

$$\dot{x}_1 = x_2 \tag{11.1.3}$$
$$\dot{x}_2 = -(\alpha + d\beta)x_1 \qquad d=0, 1$$

If we now implement a switching logic of the form

$$d = 0 \quad \text{if } x_1x_2 < 0$$
$$d = 1 \quad \text{if } x_1x_2 > 0 \tag{11.1.4}$$

the system will be asymptotically stable. To see this, we superimpose both sets of trajectories on a single phase plane as in Figure 11-4. The switching logic causes the system to follow the trajectories of (11.1.1) in the first and fourth quadrants and the trajectories of (11.1.2) in the

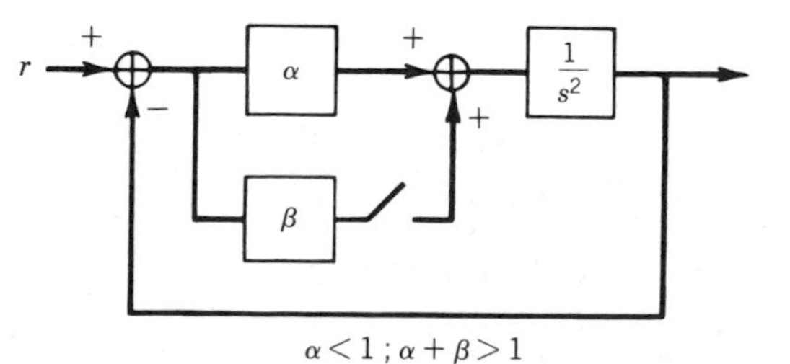

FIGURE 11-1
Switching controller.

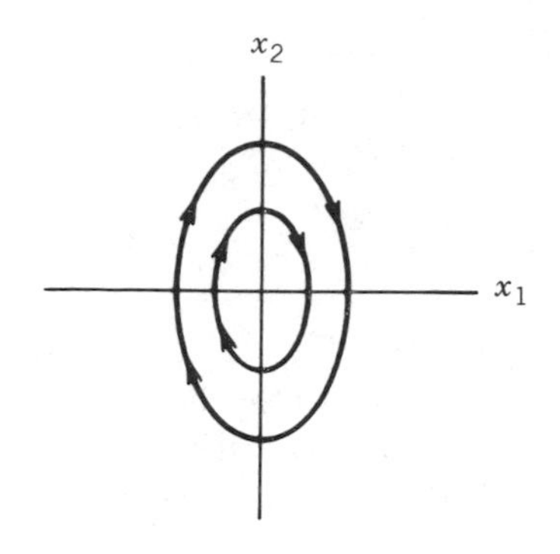

FIGURE 11-2
Phase portrait for gain α.

second and third quadrants. The resulting trajectory spirals into the origin. Notice that not only are the individual systems (11.1.1) and (11.1.2) important but also the chosen switching logic is vital. For example, in the above system if we choose a different switching logic

$$\begin{aligned} d &= 0 \quad \text{if } x_1x_2 > 0 \\ d &= 1 \quad \text{if } x_1x_2 < 0 \end{aligned} \tag{11.1.5}$$

then the resulting system is unstable with trajectories as shown in Figure 11-5.

11.1.2 PARAMETER ADAPTIVE CONTROL

The second approach to robot control that we consider in this chapter is **Adaptive Control**. All of the other robust control techniques that we consider in this text, including the variable structure control to follow, are based on worst case estimates of the uncertainty or mismatch between the plant and the inner loop control. In these approaches the inner loop control law is fixed and the gains in the outer loop are set according to the estimate of the uncertainty. The basic idea of adaptive control, on the other hand, is to change the values of the gains or other

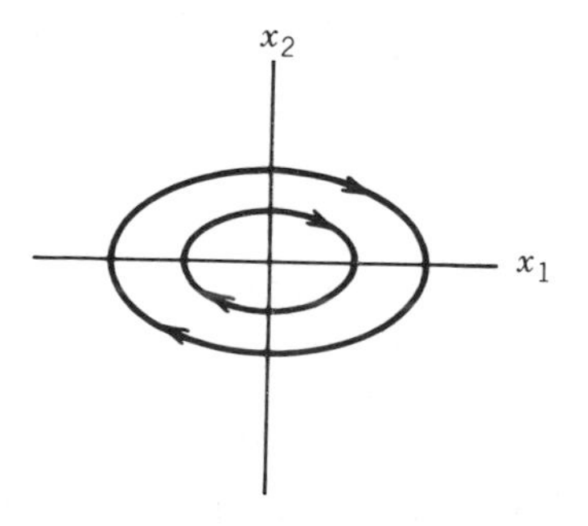

FIGURE 11-3
Phase portrait for gain $\alpha+\beta$.

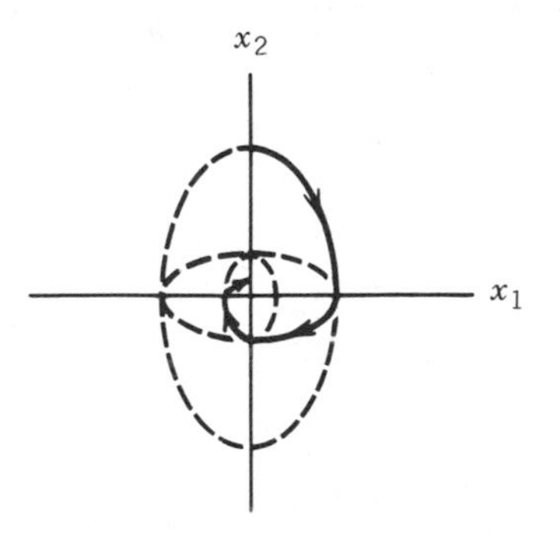

FIGURE 11-4
Phase portrait using the switching logic (11.1.4).

parameters in the control law according to some on-line algorithm. In this way the controller can "learn" an appropriate set of parameters during the course of its operation. This idea is especially useful for manipulators that are performing repetitive tasks. Without adaptation the tracking errors are also repetitive. With adaptation, the tracking performance can be improved through successive repetition.

The basic structure of an adaptive controller is shown in Figure 11-6. The adaptation algorithm accepts as its inputs the reference input and the plant output and updates the control law in some fashion with the goal of improving tracking performance.

In this chapter we limit our discussion to two techniques for adaptive control of rigid manipulators. The first technique is **adaptive inverse dynamics**. This means that the structure of the controller is fixed as an inverse dynamics control, and that some or all of the parameters defining the inner loop control are up-dated on-line. The second method that we discuss is a recent adaptive algorithm of Slotine and Li [13]. This method is interesting in that it makes fundamental use of the passive structure of the Lagrangian dynamics of rigid robots rather than attempting to exactly linearize the equations of motion.

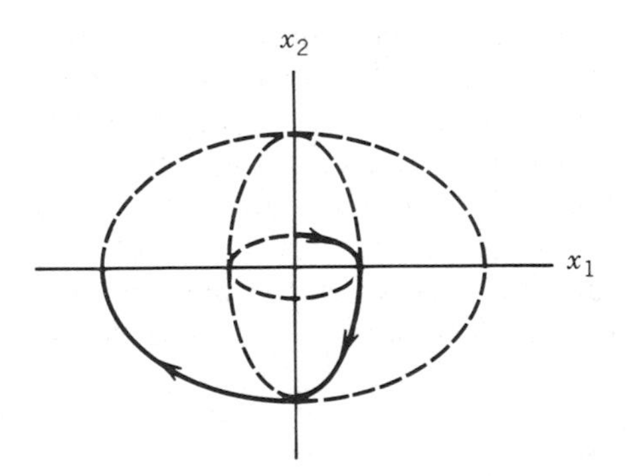

FIGURE 11-5
Phase portrait using the switching logic (11.1.5).

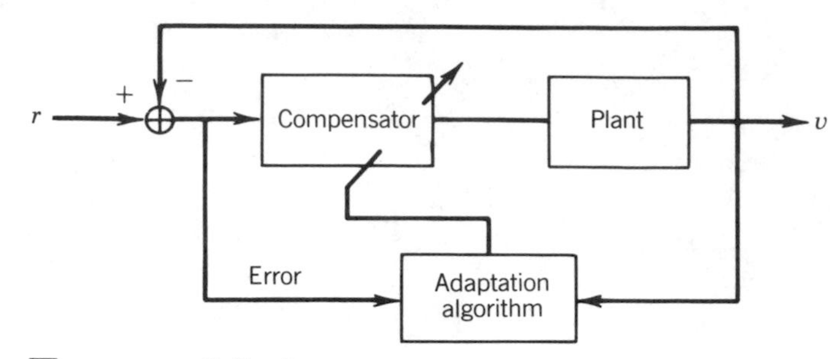

FIGURE 11-6
Basic structure of an adaptive controller.

11.2 THE METHOD OF SLIDING MODES[1]

In this section we introduce the theory of **Sliding Modes**, which is a particular approach to the design of variable structure systems. Developed in the Soviet Union more than 30 years ago, sliding mode controllers differ from simpler relay controllers in that they rely on extremely high speed switching among the control values. Recent advances in power electronics have made high speed switching practically implementable for many classes of systems such as electric motors. One should not be surprised therefore to find sliding mode theory receiving increasing attention in the robotics literature.

11.2.1 SLIDING MODE DESIGN FOR SINGLE-INPUT SYSTEMS

We will first discuss the design of sliding mode controllers for single-input systems. The idea behind sliding mode control is to choose a suitable surface in state space, typically a linear hypersurface, called the **switching surface**, and switch the control input on this surface. The control input is then chosen to guarantee that the trajectories near the sliding surface are directed toward the surface. (Refer to Figure 11-7.) Ideally then, *any* control input will suffice so long as the resulting trajectories are pointing toward the surface. Once the system is trapped on the surface, the closed loop dynamics are completely governed by the equations that define the surface. In this way, since the parameters defining the surface are chosen by the designer, the closed loop

[1]This section was greatly improved by Hebert Sira-Ramirez, who provided us with lecture notes and many helpful discussions.

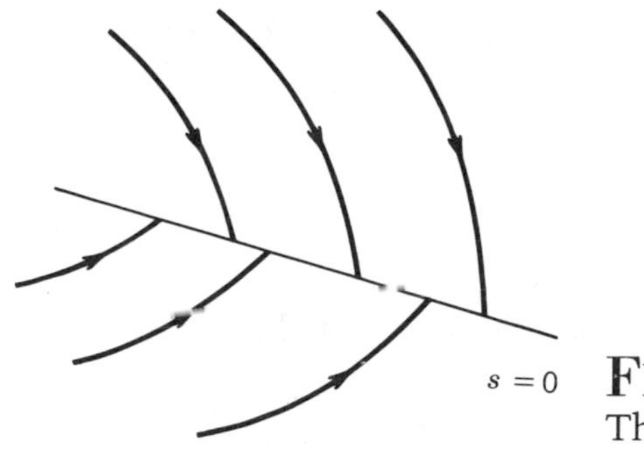

FIGURE 11-7
The notion of a sliding surface.

dynamics of the system will be independent of perturbations in the parameters of the system and robustness is achieved.

The design of a sliding mode control law can be broken down into three steps:

1. Define a suitable sliding surface, $s = 0$.
2. Choose a control law so that trajectories near the surface point toward the surface.
3. Determine the system dynamics on the surface.

The basic problems then are the specification of the gains in the controller so that the motion trajectory reaches the surface in finite time, the determination of the switching logic of the controller so that the motion on the surface is constrained to the surface, and the definition of the equations defining the surface that dictate the dynamic behavior of the system on the surface.

11.2.2 CONDITIONS FOR THE EXISTENCE OF SLIDING MOTION

Intuitively, if all nearby trajectories point toward the sliding surface then locally the system will become trapped on the surface. The behavior of the system constrained to the surface is called a **sliding regime**. Let us see how we might define such a sliding regime. Suppose that we have a variable structure system

$$\dot{\mathbf{x}} = \mathbf{f}(\mathbf{x}) = \begin{cases} \mathbf{f}^{+}(\mathbf{x}) & \text{if } s(\mathbf{x}) > 0 \\ \mathbf{f}^{-}(\mathbf{x}) & \text{if } s(\mathbf{x}) < 0 \end{cases} \tag{11.2.1}$$

We see that the system is not defined when $s = 0$, but we assume that

$$\lim_{s \to 0^-} \mathbf{f}(\mathbf{x}) = \mathbf{f}^{-}(\mathbf{x}) \tag{11.2.2}$$

$$\lim_{s \to 0^+} \mathbf{f}(\mathbf{x}) = \mathbf{f}^{+}(\mathbf{x}) \tag{11.2.3}$$

where the vector fields $\mathbf{f}^+(\mathbf{x})$ and $\mathbf{f}^-(\mathbf{x})$ are well-defined. Computing the derivative of s with respect to time yields

$$\frac{ds}{dt} = <\mathbf{ds}, \dot{\mathbf{x}}> = <\mathbf{ds}, \mathbf{f}(\mathbf{x})> \tag{11.2.4}$$

Since it is desired that the trajectory of the system in the vicinity of the sliding surface point toward the surface we see from (11.2.4) and Figure 11-8 that a sufficient condition for the existence of a sliding regime is given by

$$\lim_{s \to 0^-} \frac{ds}{dt} > 0 \tag{11.2.5}$$

$$\lim_{s \to 0^+} \frac{ds}{dt} < 0 \tag{11.2.6}$$

These two conditions are equivalent to

$$\lim_{s \to 0} s\frac{ds}{dt} < 0 \tag{11.2.7}$$

or,

$$\lim_{s \to 0} \tfrac{1}{2} \frac{d}{dt}(s^2) < 0 \tag{11.2.8}$$

which says that the function $V = {}^1\!/\!{}_2 s^2$ plays the role of a Lyapunov function candidate for the system relative to the surface $s = 0$.

Since the differential equation defining the system is discontinuous on the surface $s = 0$ we cannot define a solution in the usual sense on the sliding surface. A solution can be defined in a more general set-theoretic sense due to Fillipov[5], although a rigorous derivation of this result is beyond the scope of this text. Fillipov's solution on the surface $s = 0$ can be explained intuitively as follows. Since the switching logic will ideally switch infinitely fast between the two vector fields $\mathbf{f}^+(\mathbf{x})$ and $\mathbf{f}^-(\mathbf{x})$ on the surface, the system should respond as though governed by a suitable "average value" of $\mathbf{f}^+(\mathbf{x})$ and $\mathbf{f}^-(\mathbf{x})$, which we express as a convex combination

$$\dot{\mathbf{x}} = \mathbf{f}^0(\mathbf{x}) := \mu\mathbf{f}^+(\mathbf{x}) + (1-\mu)\mathbf{f}^-(\mathbf{x}) \quad 0 \le \mu \le 1 \tag{11.2.9}$$

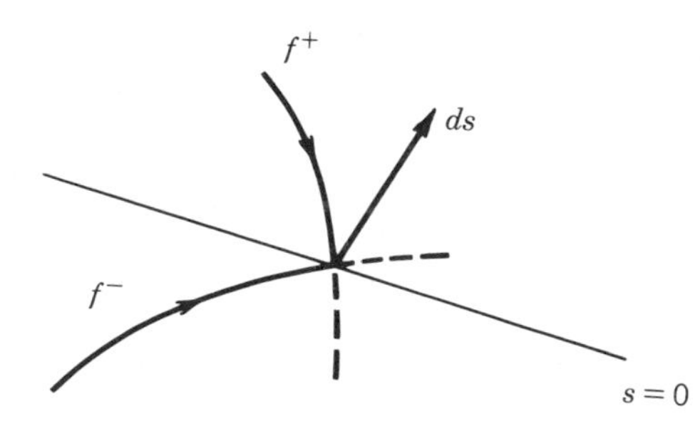

FIGURE 11-8
Defining the conditions for a sliding regime.

In order for the system to remain on the sliding surface the vector field $\mathbf{f}^0$ in (11.2.9) should be tangent to the surface at each point. This leads to the following specification of μ:

(i) *Lemma 11.2.1*

Defining $\mathbf{f}_0^+(\mathbf{x})$ and $\mathbf{f}_0^-(\mathbf{x})$ as

$$\mathbf{f}_0^+(\mathbf{x}) = \lim_{s \to 0^+} < \mathbf{ds}, \mathbf{f}^+(\mathbf{x}) > \tag{11.2.10}$$

$$\mathbf{f}_0^-(\mathbf{x}) = \lim_{s \to 0^-} < \mathbf{ds}, \mathbf{f}^-(\mathbf{x}) > \tag{11.2.11}$$

then, if a sliding motion exists on $s = 0$, we have

$$\mathbf{f}_0^-(\mathbf{x}) - \mathbf{f}_0^+(\mathbf{x}) > 0 \tag{11.2.12}$$

and

$$\mu(\mathbf{x}) = \frac{\mathbf{f}_0^-(\mathbf{x})}{\mathbf{f}_0^-(\mathbf{x}) - \mathbf{f}_0^+(\mathbf{x})} \tag{11.2.13}$$

Proof: The condition (11.2.12) holds if the trajectories near the surface are pointing toward the surface. Referring to Figure 11-9, we want the vector field $\mathbf{f}^0$ orthogonal to $\mathbf{ds}$, that is, we want $< \mathbf{ds}, \mathbf{f}^0 > = 0$, so that the system remains on the sliding surface. Using (11.2.9) we have

$$< \mathbf{ds}, \mathbf{f}^0 > \; = \mu < \mathbf{ds}, \mathbf{f}^+(\mathbf{x}) > \; + \; (1-\mu) < \mathbf{ds}, \mathbf{f}^-(\mathbf{x}) > \tag{11.2.14}$$

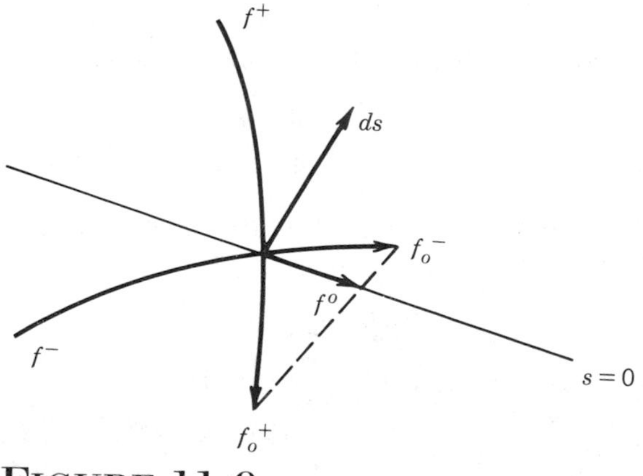

FIGURE 11-9
Fillipov's definition of equivalent system.

Defining μ by (11.2.13) and taking the limit as $s \to 0$ in (11.2.14) gives

$$< \mathbf{ds}, \mathbf{f}^0 > = \frac{\mathbf{f}_0^-(\mathbf{x})}{\mathbf{f}_0^-(\mathbf{x}) - \mathbf{f}_0^+(\mathbf{x})}\mathbf{f}_0^+(\mathbf{x}) + \left(1 - \frac{\mathbf{f}_0^-(\mathbf{x})}{\mathbf{f}_0^-(\mathbf{x}) - \mathbf{f}_0^+(\mathbf{x})}\right)\mathbf{f}_0^-(\mathbf{x}) = 0 \tag{11.2.15}$$

and the result follows.

11.2.3 EQUIVALENT CONTROL METHOD

Fillipov's notion of solution on $s = 0$ is rigorous but difficult to use for control design. A better approach for design is to introduce the so-called **equivalent control method** for defining the system behavior on the sliding surface. In the equivalent control method we do not average the vector fields directly, but in a sense we average the control input instead and define the response of the system on the surface as the response to this ideal averaged control.

The equivalent control is defined by the conditions

$$s = 0 \tag{11.2.16}$$

$$\dot{s} = 0 \tag{11.2.17}$$

The first condition (11.2.16) says that the system is on the sliding surface and the second condition (11.2.17) says that the system does not leave the surface. Consider a single-input nonlinear system that is linear in the control

$$\dot{\mathbf{x}} = \mathbf{f}(\mathbf{x}) + \mathbf{g}(\mathbf{x})u \tag{11.2.18}$$

The condition (11.2.17) becomes

$$\frac{\mathbf{ds}}{dt} = < \mathbf{ds}\ ,\ \mathbf{f} + \mathbf{g}u > = < \mathbf{ds}\ ,\ \mathbf{f} > + < \mathbf{ds}\ , \mathbf{g} > u = 0 \tag{11.2.19}$$

If $< \mathbf{ds}\ , \mathbf{g} > \neq 0$ then the equivalent control, denoted u_{eq}, is given as

$$u_{eq} = -\frac{< \mathbf{ds}\ , \mathbf{f} >}{< \mathbf{ds}\ , \mathbf{g} >} \tag{11.2.20}$$

The ideal sliding dynamics, that is, the ideal dynamics of the system on the sliding surface are then

$$\begin{aligned} \dot{\mathbf{x}} &= \mathbf{f}(\mathbf{x}) + \mathbf{g}(\mathbf{x})u_{eq} \\ &= \mathbf{f}(\mathbf{x}) - \mathbf{g}(\mathbf{x})\frac{< \mathbf{ds}, \mathbf{f} >}{< \mathbf{ds}, \mathbf{g} >} \\ &= \left(I - \frac{\mathbf{g}\,\mathbf{ds}}{< \mathbf{ds}, \mathbf{f} >}\right)\mathbf{f}(\mathbf{x}) \end{aligned} \tag{11.2.21}$$

11.2.4 INVARIANCE

We have mentioned before that the feedback linearization property is important because it allows satisfaction of the so-called matching conditions wherein all of the parametric uncertainty lies in the range space of the input. We now show why this is important. We show that a sliding motion, if it exists, is invariant under arbitrary nonlinear perturbations provided that the perturbations are in the range space of the input. For simplicity consider the single input system after application of an approximate feedback linearizing inner loop control

$$\dot{\mathbf{y}} = A\mathbf{y} + \mathbf{b}\{v + \eta(v, \mathbf{y})\} \tag{11.2.22}$$

where A and $\mathbf{b}$ are given by (10.3.5) and

$$\eta = \psi v + \phi \tag{11.2.23}$$

(i) *Lemma 11.2.2*

Let $s = \sum_{i=1}^{n-1} c_i y_i + y_n = 0$ define a sliding surface. Then

$$< \mathbf{ds}, \mathbf{b} > = 1 \tag{11.2.24}$$

Proof: Since $\mathbf{b}$ is given by (10.3.5) we have

$$< \mathbf{ds}, \mathbf{b} > = [\frac{\partial s}{\partial y_1}, \ldots, \frac{\partial s}{\partial y_n}] \begin{bmatrix} 0 \\ \cdot \\ \cdot \\ \cdot \\ 1 \end{bmatrix} = \frac{\partial s}{\partial y_n} = 1 \tag{11.2.25}$$

(ii) *Theorem 11.2.3*

Let ψ and ϕ be arbitrary perturbations defining η in (11.2.23). Then the ideal sliding motion of (11.2.22), provided it exists, is invariant with respect to ψ and ϕ.

Proof: Rewrite (11.2.22) as

$$\dot{\mathbf{y}} = A\mathbf{y} + \mathbf{b}(1 + \psi)v + \mathbf{b}\phi \tag{11.2.26}$$

Then, under ideal sliding conditions we have

$$\begin{aligned} 0 = \dot{s} &= < \mathbf{ds}, \dot{\mathbf{y}} > \\ &= < \mathbf{ds}, A\mathbf{y} + \mathbf{b}(1 + \psi)v_{eq} + \mathbf{b}\phi > \\ &= < \mathbf{ds}, A\mathbf{y} > + < \mathbf{ds}, \mathbf{b} > (1 + \psi)v_{eq} + < \mathbf{ds}, \mathbf{b} > \phi \\ &= < \mathbf{ds}, A\mathbf{y} > + (1 + \psi)v_{eq} + \phi \end{aligned} \tag{11.2.27}$$

where the last equality follows from Lemma 11.2.2. Therefore we have

$$(1 + \psi)v_{eq} = - < \mathbf{ds}, A\mathbf{y} > - \phi \tag{11.2.28}$$

Using the expression (11.2.28) the ideal sliding dynamics are governed by

$$\begin{aligned} \dot{\mathbf{y}} &= A\,\mathbf{y} + \mathbf{b}(1+\psi)v_{eq} + \mathbf{b}\phi \\ &= A\,\mathbf{y} + \mathbf{b}(-<\mathbf{ds}, A\,\mathbf{y}> - \phi) + \mathbf{b}\phi \\ &= (I - \mathbf{b}\,\mathbf{ds})A\,\mathbf{y} \end{aligned} \tag{11.2.29}$$

which is independent of ψ and ϕ.

11.2.5 MULTI-INPUT SYSTEMS

The design of sliding mode control laws for multi-input systems is considerably more complicated than for single-input systems. The easiest approach is to choose a sliding surface corresponding to each control input and then choose the variable structure control law so that a sliding motion exists *on the intersection of the sliding surfaces.* In the case of feedback linearizable systems, the design is simplified in that ideally the linearized system is decoupled into single-input systems. However, in the case of uncertain systems of the form

$$\begin{aligned} \dot{\mathbf{y}} &= A\,\mathbf{y} + B\{\mathbf{v} + \eta(\mathbf{v}, \mathbf{y})\} \\ &= A\,\mathbf{y} + B\{(I + \Psi)\mathbf{v} + \boldsymbol{\phi}\} \end{aligned} \tag{11.2.30}$$

the system remains coupled to some extent due to the presence of the uncertainty. Hence some care must be taken in specifying the control gains in the variable structure law.

In the multi-input case, we define m linear sliding surfaces $s_1, \ldots, s_m$, one for each of the inputs $v_1, \ldots, v_m$. This set of linear surfaces can be compactly expressed as a matrix equation

$$S = M\mathbf{y} = 0 \tag{11.2.31}$$

where M is an $m \times n$ matrix. As before we define an equivalent control by computing $\dot{S} = 0$, that is,

$$\dot{S} = M\dot{\mathbf{y}} = M(A\,\mathbf{y} + B(I + \Psi)\mathbf{v} + \Phi) \tag{11.2.32}$$

If $\det MB \neq 0$ and assuming that $I + \Psi$ is invertible, the equivalent control $\mathbf{v}_{eq}$ is found by setting (11.2.32) equal to zero as

$$\mathbf{v}_{eq} = -(MB)^{-1}(I + \Psi)^{-1}(-MA\,\mathbf{y} + MB\,\Phi) \tag{11.2.33}$$

11.2.6 VARIABLE STRUCTURE IMPLEMENTATION OF THE CONTROL LAW

It is easy to see that the equivalent control u_{eq} depends on knowledge of the parameters defining the system and is therefore sensitive to parametric uncertainty. The whole point of the sliding mode philosophy is to realize u_{eq} "on the average" by a variable structure control.

Returning to the single-input system (11.2.18) let us therefore construct a variable structure control law of the form

$$u = \begin{cases} -u^+ & \text{if } s > 0 \\ -u^- & \text{if } s < 0 \end{cases} \tag{11.2.34}$$

Without loss of generality[2] suppose $< \mathbf{ds}, \mathbf{g} > > 0$ and note from (11.2.20) that we may write

$$\begin{aligned} \frac{\mathbf{ds}}{dt} &= < \mathbf{ds}, \mathbf{f} > + < \mathbf{ds}, \mathbf{g} > \mathbf{u} \\ &= < \mathbf{ds}, \mathbf{g} > (u - u_{eq}) \end{aligned} \tag{11.2.35}$$

Therefore, from the sliding mode conditions (11.2.5)–(11.2.6), a sliding mode exists using the control law (11.2.34) if and only if u^+ and u^- satisfy

$$u^- < u_{eq} < u^+ \tag{11.2.36}$$

Equation 11.2.36 allows considerable freedom in the choice of the control law while maintaining a sliding regime. The value of u_{eq} need only be known with sufficient precision to be able to satisfy (11.2.36). The difference $u^+ - u^-$ is called the **control discontinuity**. Obviously the larger the uncertainty in u_{eq} the larger the control discontinuity must be in order to be sure that the inequalities in (11.2.36) remain satisfied.

One attractive way to implement the variable structure control law is as

$$.sp-5pu = -\lambda \, | \hat{u}_{eq} | \, sgn(s) \tag{11.2.37}$$

where $\hat{u}_{eq}$ is an estimate of the equivalent control, λ is a gain designed to guarantee (11.2.36) and $sgn(s)$ is the signum function

$$sgn(s) = \begin{cases} +1 & \text{if } s > 0 \\ -1 & \text{if } s < 0 \end{cases} \tag{11.2.38}$$

(i) *Example 11.2.4*

Consider the single-link manipulator

$$I\ddot{\theta} + MgL\sin(\theta) = u \tag{11.2.39}$$

and choose the inner loop control law

$$u = \hat{I}v + \widehat{MgL}\sin(\theta) \tag{11.2.40}$$

Substituting (11.2.40) into (11.2.39) yields

$$\ddot{\theta} = \frac{\hat{I}}{I}v + \frac{\Delta MgL}{I}\sin(\theta) \tag{11.2.41}$$

[2] If not, replace $\mathbf{u}$ by $-\mathbf{u}$.

For simplicity, suppose θ^d is constant and define the sliding surface

$$s = c\,(\theta - \theta^d) + \dot{\theta} \tag{11.2.42}$$

Then $\dot{s}$ is given by

$$\begin{aligned} \dot{s} &= c\dot{\theta} + \ddot{\theta} \\ &= c\,(s - c\,(\theta - \theta^d)) + \frac{\hat{I}}{I}v + \frac{\Delta MgL}{I}\sin(\theta) \\ &= cs - c^2(\theta - \theta^d) + \frac{\hat{I}}{I}v + \frac{\Delta MgL}{I}\sin(\theta) \end{aligned} \tag{11.2.43}$$

The equivalent control v_{eq} is then given as

$$v_{eq} = \frac{I}{\hat{I}}c^2(\theta - \theta^d) - \frac{\Delta MgL}{\hat{I}}\sin(\theta) \tag{11.2.44}$$

As before assume that $5 \leq I$, $MgL \leq 10$, and we take $\hat{I} = \widehat{MgL} = 5$. With these values we have the estimate

$$|\, v_{eq} \,| \leq 2c^2 \,|\, \theta - \theta^d \,| + 2/5 \tag{11.2.45}$$

Therefore we choose

$$|\, \hat{v}_{eq} \,| = 2c^2 \,|\, \theta - \theta^d \,| + 2/5 \tag{11.2.46}$$

and

$$v = -\lambda \,|\, \hat{v}_{eq} \,|\, sgn(s) \tag{11.2.47}$$

The step response is shown in Figure 11-10 and the control signal v is shown in Figure 11-11. The phase plane $(\theta - \theta^d, \dot{\theta})$ trajectory is shown in Figure 11-12. Note that the trajectory converges to the sliding surface given by (11.2.42) and follows the surface into the origin.

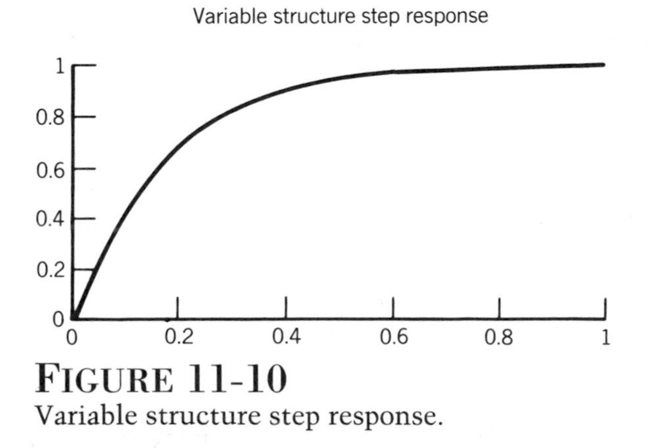

FIGURE 11-10
Variable structure step response.

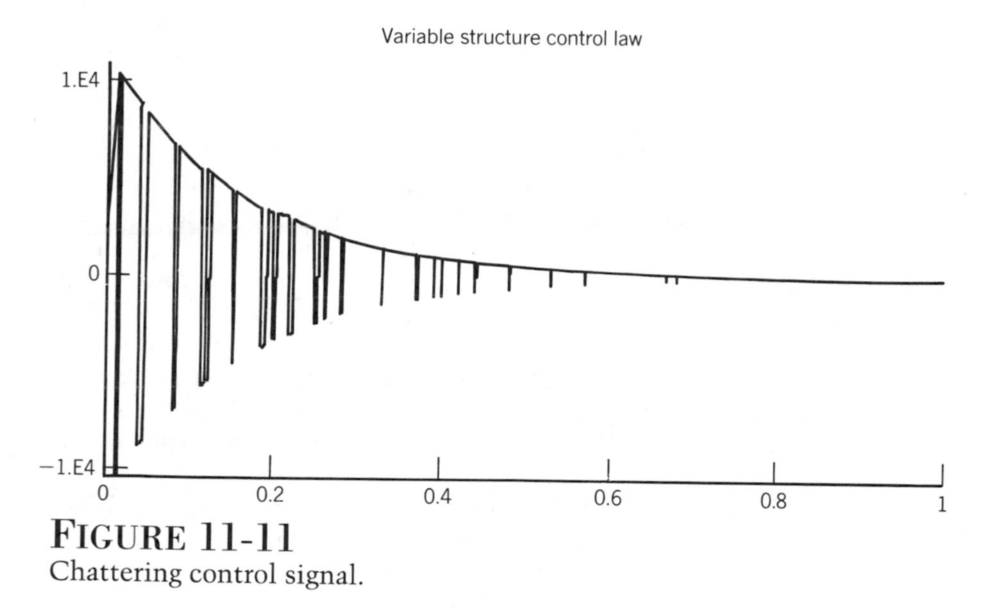

FIGURE 11-11
Chattering control signal.

11.2.7 CONTROL CHATTERING

In practice the implementation of variable structure controllers results in **control chattering**, as seen above. The ideal behavior of sliding mode controllers is achieved in the theoretical limit as the switching frequency becomes infinite. In practice the small, but nonzero delay in control switching will cause the trajectory to slightly overshoot the switching surface each time the control is switched. In practice, the larger the control discontinuity the more severe the control chatter will be. Thus there is a clear trade-off between the uncertainty over the true value of u_{eq}, and the amount of control effort.

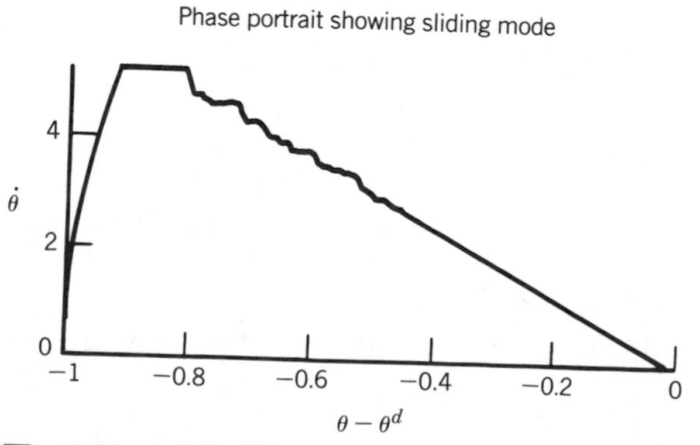

FIGURE 11-12
Phase portrait showing sliding surface.

Chattering is usually, but not always undesirable. In some cases chattering can excite high frequency, unmodeled dynamics in the system and lead to degradation in performance. However, in some cases, the control can be made to switch at a frequency well beyond the resonant frequencies of the system and the chattering will be acceptable. Such is the case with modern power electronic devices that can switch input voltages to the motors in the megahertz range. In other cases, such as when the joint torque is taken as the input, the high frequency switching is not achievable. Also, in hydraulic actuators, high frequency opening and closing of valves can lead to rapid wear which is also undesirable.

One approach to overcoming the undesirable effects of control chattering is to introduce what is known as a **boundary layer** around the sliding surface and approximate the switching control by a continuous control inside this boundary layer[14].

Thus, suppose in the above example the discontinuous control law (11.2.47) is replaced by the continuous approximation

$$v = -\lambda \, | \hat{v}_{eq} | \, sat(s/\varepsilon) \tag{11.2.48}$$

where the function $sat(\cdot)$ is defined by

$$sat(x) = \begin{cases} sgn(x) & \text{if } |x| > 1 \\ x & \text{if } |x| \leq 1 \end{cases} \tag{11.2.49}$$

The response with this control law is shown in Figure 11-13 with $\varepsilon = 0.2$. The control signal is shown in Figure 11-14, and the phase portrait is shown in Figure 11-15.

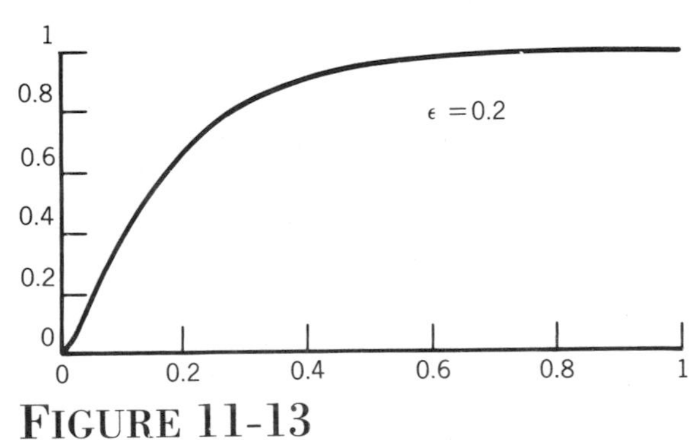

FIGURE 11-13
Step response—continuous approximation to variable structure control law.

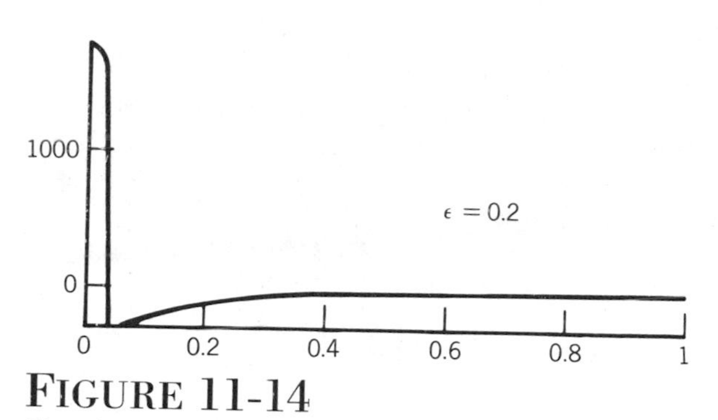

FIGURE 11-14
Continuous approximation to discontinuous control.

11.2.8 REMARK

So far, our results have been local in nature. In other words, we must assume that the initial conditions are "sufficiently close" to the sliding surface. For example, Theorem 11.2.3 holds independent of the magnitude of the perturbation η so long as the system starts on or very near the surface. In practice, however, the magnitude of η does matter in the sense that large perturbations may prevent the trajectory from ever reaching the surface in the first place. Therefore we need to address the property of reachability of the sliding surface.

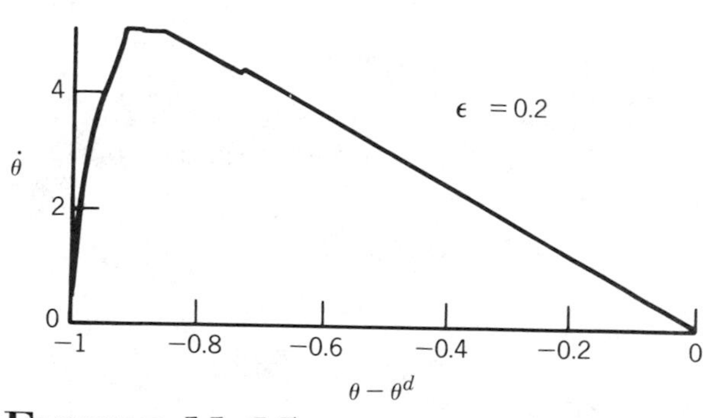

FIGURE 11-15
Phase portrait for continuous control.

11.2.9 REACHABILITY OF THE SLIDING SURFACE

In the case of a single-input feedback linearizable system we can guarantee attractivity of the sliding surface, that is, reachability, by choosing a Lyapunov function candidate

$$V = \frac{1}{2}s^2 \tag{11.2.50}$$

and showing that along trajectories of the system

$$\dot{V} = s\dot{s} < -\varepsilon \mid s \mid \tag{11.2.51}$$

for some $\varepsilon > 0$.

11.2.10 VARIABLE STRUCTURE CONTROL OF MANIPULATORS

We can now see how a sliding control scheme can be used in the outer loop design for robust trajectory tracking. Recalling the equations of motion (8.5.1) for a rigid manipulator with approximate feedback linearizing control (8.5.2) we see that the error dynamics are governed by

$$\begin{aligned} \dot{\mathbf{e}}_1 &= \mathbf{e}_2 \\ \dot{\mathbf{e}}_2 &= \mathbf{v} + \eta \end{aligned} \tag{11.2.52}$$

where

$$\eta = (M^{-1}\hat{M} - I)(\mathbf{v} + \ddot{\mathbf{q}}^d) + M^{-1}\mathbf{h} \tag{11.2.53}$$

If we choose a linear sliding surface for each degree-of-freedom of the manipulator

$$s_k\colon\ c_k e_{1k} + e_{2k} = 0 \tag{11.2.54}$$

where e_{1k}, e_{2k} are the position and velocity tracking errors, respectively for the k-th joint, then a variable structure outer loop control law can be designed as

$$\mathbf{v} = \ddot{\mathbf{q}}^d - K_1\dot{\mathbf{e}} - K_2\mathbf{e} + \Delta\mathbf{v} \tag{11.2.55}$$

where $\Delta\mathbf{v}$ is chosen as

$$\Delta\mathbf{v} = -\lambda \| \Delta\mathbf{v}_{eq} \| sgn(\mathbf{s}) \tag{11.2.56}$$

where $\mathbf{s}$ is the vector of sliding surface terms defined by (11.2.54).

We have given in this section only the simplest possible sliding mode designs in order to illustrate the theory in its most basic form. In reality, more complicated designs can be derived using this approach that explicitly trade off parameter uncertainty with control chattering and tracking performance. The reader is directed to the references at the end of this chapter for more details of this approach.

11.3 ADAPTIVE CONTROL

In this section[3] we discuss the application of adaptive control to the manipulator control problem. We first discuss the key feature of rigid robot dynamics that is exploited to derive adaptive control laws, which is **linearity in the parameters**. By this we mean that, although the equations of motion are nonlinear, the parameters of interest such as link masses, moments of inertia, and so forth appear as coefficients of known functions of the joint variables. By defining each coefficient as a separate parameter, a linear relationship results, so that we may write the dynamic equations of an n-link rigid robot as

$$M(\mathbf{q})\ddot{\mathbf{q}} + C(\mathbf{q},\dot{\mathbf{q}})\dot{\mathbf{q}} + \mathbf{g}(\mathbf{q}) = Y(\mathbf{q},\dot{\mathbf{q}},\ddot{\mathbf{q}})\mathbf{p} = \mathbf{u} \tag{11.3.1}$$

where $Y(\mathbf{q},\dot{\mathbf{q}},\ddot{\mathbf{q}})$ is an $n \times r$ matrix of known functions and $\mathbf{p}$ is an r-dimensional vector of parameters.

(i) *Example 11.3.1*

Consider the two-link planar arm of Example 6.4.2

$$\begin{bmatrix} m_1 l_{c_1}^2 + m_2(l_1^2 + l_{c_2}^2 + 2 l_1 l_{c_2}\cos q_2) + I_1 + I_2 & m_2(l_{c_2}^2 + l_1 l_{c_2}\cos q_2) + I_2 \\ m_2(l_{c_2}^2 + l_1 l_{c_2}\cos q_2) + I_2 & m_2 l_{c_2}^2 + I_2 \end{bmatrix} \begin{bmatrix} \ddot{q}_1 \\ \ddot{q}_2 \end{bmatrix}$$

$$+ \begin{bmatrix} -m_2 l_1 l_{c_2}\sin q_2(\dot{q}_1 \dot{q}_2 + \dot{q}_1^{\,2}) \\ m_2 l_1 l_{c_2}\sin q_2 \end{bmatrix} \tag{11.3.2}$$

$$+ \begin{bmatrix} (m_1 l_{c_1} + m_2 l_1) g\cos q_1 + m_2 l_{c_2} g\cos(q_1 + q_2) \\ m_2 l_{c_2}\cos(q_1 + q_2) \end{bmatrix} = \begin{bmatrix} u_1 \\ u_2 \end{bmatrix}$$

Defining parameters $p_1, \cdots, p_9$ as

$$\begin{array}{lll} p_1 = m_1 l_{c_1}^2 & p_4 = m_2 l_1 l_{c_2} & p_7 = m_1 l_{c_1} g \\ p2 = m_2 l_1^2 & p_5 = I_1 & p_8 = m_2 l_1 g \\ p_3 = m_2 l_{c_2}^2 & p_6 = I_2 & p_9 = m_2 l_{c_2} g \end{array} \tag{11.3.3}$$

we can write (11.3.2) as

$$Y(\mathbf{q}, \dot{\mathbf{q}}, \ddot{\mathbf{q}})\mathbf{p} = \mathbf{u} \tag{11.3.4}$$

[3] This section has benefited from helpful discussions with Romeo Ortega.

where $\mathbf{p} = [p_1, \cdots, p_9]^T$, and Y is given by (Problem 11-4)

$$Y(\mathbf{q}, \dot{\mathbf{q}}, \ddot{\mathbf{q}}) =$$

$$\begin{bmatrix} \ddot{q}_1 & \ddot{q}_2 & \ddot{q}_1+\ddot{q}_2 & 2\cos q_2\ddot{q}_1+\cos q_2\ddot{q}_2-2\sin q_2\dot{q}_1\dot{q}_2-\sin q_2\dot{q}_2^2 & \ddot{q}_1 & \ddot{q}_1+\ddot{q}_2 & \cos q_1 & \cos q_1 & \cos(q_1+q_2) \\ 0 & 0 & \ddot{q}_1+\ddot{q}_2 & \cos q_2\ddot{q}_1+\sin q_2 & \ddot{q}_2 & \ddot{q}_2 & 0 & 0 & \cos(q_1+q_2) \end{bmatrix} \quad (11.3.5)$$

Note that the choice of parameters in the above representation is not unique and that the dimension of the parameter space may depend on the particular choice of parameters. Also, some of the parameters in the above representation may be known, or at least, better known than others, and we may wish to update only those parameters that are poorly known. For example, we may know the parameters in the manipulator when it is not carrying any load. If the manipulator picks up an unknown load we may treat the load as part of link two, in which case only the parameters m_2, l_{c_2}, and I_2 would be unknown. A parameter estimation algorithm would then be needed only to estimate these parameters.

The above property of linearity in the parameters is crucial to the derivation of adaptive control results for manipulators. We will illustrate this first by a discussion of adaptive inverse dynamics.

11.3.1 ADAPTIVE INVERSE DYNAMICS

The inverse dynamics control law for the system

$$M(\mathbf{q})\ddot{\mathbf{q}} + \mathbf{h}(\mathbf{q}, \dot{\mathbf{q}}) = \mathbf{u} \quad (11.3.6)$$

is given as

$$\mathbf{u} = \hat{M}(\mathbf{q})(\ddot{\mathbf{q}}^d - K_0\mathbf{e} - K_1\dot{\mathbf{e}}) + \hat{\mathbf{h}}(\mathbf{q}, \dot{\mathbf{q}}) \quad (11.3.7)$$

where $\mathbf{q}^d$ is a desired trajectory, and $\mathbf{e} = \mathbf{q} - \mathbf{q}^d$ is the position tracking error. We assume that $\hat{M}$ and $\hat{\mathbf{h}}$ have the same functional form as M and $\mathbf{h}$ with estimated parameters $\hat{p}_1, \ldots, \hat{p}_r$. Thus

$$\hat{M}\ddot{\mathbf{q}} + \hat{\mathbf{h}} = Y(\mathbf{q}, \dot{\mathbf{q}}, \ddot{\mathbf{q}})\hat{\mathbf{p}} \quad (11.3.8)$$

where $\hat{\mathbf{p}}$ is the vector of estimated parameters.

We have seen that the mismatch $\hat{M} - M$, $\hat{\mathbf{h}} - \mathbf{h}$ results in poor tracking performance, and have also detailed approaches for modifying the control $\mathbf{v}$ to overcome the effects of this mismatch. The adaptive inverse dynamics approach, on the other hand, leaves the structure of the control law (11.3.7) fixed but updates the terms $\hat{M}$ and $\hat{\mathbf{h}}$. To see how this is accomplished we substitute (11.3.7) into (11.3.6) to obtain

$$M\ddot{\mathbf{q}} + \mathbf{h} = \hat{M}(\ddot{\mathbf{q}}^d - K_0\mathbf{e} - K_1\dot{\mathbf{e}}) + \hat{\mathbf{h}} \quad (11.3.9)$$

Adding and subtracting $\hat{M}\ddot{\mathbf{q}}$ on the left hand side of (11.3.9) and using (11.3.1) and (11.3.8) we can write

$$\hat{M}(\ddot{\mathbf{e}} + K_1\dot{\mathbf{e}} + K_0\mathbf{e}) = \tilde{M}\ddot{\mathbf{q}} + \tilde{\mathbf{h}} \tag{11.3.10}$$
$$= Y(\mathbf{q}, \dot{\mathbf{q}}, \ddot{\mathbf{q}})\tilde{\mathbf{p}}$$

where $\tilde{M} = \hat{M} - M$, $\tilde{\mathbf{h}} = \hat{\mathbf{h}} - \mathbf{h}$, and $\tilde{\mathbf{p}} = \hat{\mathbf{p}} - \mathbf{p}$. Finally, assuming $\hat{M}$ is invertible, the error dynamics may be written as

$$\ddot{\mathbf{e}} + K_1\dot{\mathbf{e}} + K_0\mathbf{e} = \hat{M}^{-1}Y\,\tilde{\mathbf{p}} := \Phi\,\tilde{\mathbf{p}} \tag{11.3.11}$$

where $\Phi = \hat{M}^{-1}Y$.

With gain matrices K_0, K_1 chosen as before to stabilize the double integrator system, (11.3.11) may be written in state space as

$$\dot{\mathbf{x}} = A\mathbf{x} + B\Phi\tilde{\mathbf{p}} \tag{11.3.12}$$

where

$$A = \begin{bmatrix} 0 & I \\ -K_0 & -K_1 \end{bmatrix} \quad B = \begin{bmatrix} 0 \\ I \end{bmatrix} \quad \mathbf{x} = \begin{bmatrix} \mathbf{e} \\ \dot{\mathbf{e}} \end{bmatrix} \tag{11.3.13}$$

We need to derive an algorithm for updating the parameter vector $\hat{\mathbf{p}}$ and to prove that the overall system is stable. To this end, choose a symmetric, positive definite matrix Q and solve the Lyapunov equation

$$A^TP + PA + Q = 0 \tag{11.3.14}$$

Next, choose the Lyapunov function candidate

$$V = \mathbf{x}^TP\mathbf{x} + \tilde{\mathbf{p}}^T\Gamma\tilde{\mathbf{p}} \tag{11.3.15}$$

where P is the unique symmetric positive definite solution of (11.3.14) and Γ is symmetric, positive definite. The time derivative of V along trajectories of (11.3.12) is computed to be

$$\dot{V} = -\mathbf{x}^TQ\mathbf{x} + 2\tilde{\mathbf{p}}^T[\Phi^TB^TP\mathbf{x} + \Gamma\dot{\tilde{\mathbf{p}}}\ ] \tag{11.3.16}$$

This suggests that we should choose a parameter update law to make the second term above zero, namely

$$\dot{\tilde{\mathbf{p}}} = -\Gamma^{-1}\Phi^TB^TP\mathbf{x} \tag{11.3.17}$$

Since the parameter vector $\mathbf{p}$ is constant, $\dot{\tilde{\mathbf{p}}} = \dot{\hat{\mathbf{p}}}$, and so we update the parameter vector $\hat{\mathbf{p}}$ according to the formula

$$\dot{\hat{\mathbf{p}}} = -\Gamma^{-1}\Phi^TB^TP\mathbf{x} \tag{11.3.18}$$

which results in

$$\dot{V} = -\mathbf{x}^TQ\mathbf{x} \le 0 \tag{11.3.19}$$

We have shown that the system is stable in the sense of Lyapunov. Equations 11.3.15–11.3.19 imply that V is a positive, nonincreasing function that is bounded below (by zero). Thus $\mathbf{x}(t)$ and $\tilde{\mathbf{p}}(t)$ are bounded, and $\mathbf{x}(t)$ is a so-called square integrable or L_2 function. Practically speaking, this is sufficient for our purposes since, under some additional continuity restrictions, an L_2 function must converge to zero as $t \to \infty$, which says that the tracking error converges to zero. However, in order to prove rigorously that the tracking error does indeed converge to zero several additional issues must be considered, most of which have occupied considerable space in the adaptive control literature. We must guarantee that the function Φ appearing in (11.3.12) remains bounded. First, since Φ contains the term $\hat{M}^{-1}$ we must guarantee that the estimate $\hat{M}$ never becomes singular. One way to do this is to restrict the estimated parameters $\hat{\mathbf{p}}$ to lie in a fixed (compact) region about the true parameters. This again requires some worst case estimates of the uncertainty. If the parameter estimate should ever leave this fixed region, the estimation algorithm should be designed to "reset" $\hat{\mathbf{p}}$ to the boundary of the region. See [2] for details of this approach.

Second, notice that Φ depends on the acceleration $\ddot{\mathbf{q}}$ of the manipulator. Since $\mathbf{x}$ and $\tilde{\mathbf{p}}$ are now bounded, the acceleration will be bounded as can be seen by solving for $\ddot{\mathbf{q}}$ from the equations of motion. Thus $\dot{\mathbf{x}}$ is bounded from (11.3.12). This means that $\mathbf{x}$ is uniformly continuous, and hence $\mathbf{x} \to 0$ as $t \to \infty$.

However, since Φ depends on the manipulator acceleration, in order to realize the parameter update law the acceleration needs to be measured (or estimated). Several recent papers have been devoted to the implementation of adaptive inverse dynamics without the need to measure the joint acceleration. Most of these results amount, in one guise or another, to estimating the acceleration from the velocity using a "first order filter" of the form

$$\ddot{\mathbf{q}}_r = \frac{s}{\tau s + 1}\dot{\mathbf{q}} \;, \quad s = \frac{d}{dt} \tag{11.3.20}$$

Note that if the time constant τ in (11.3.20) is small, then

$$\ddot{\mathbf{q}}_r \approx s\dot{\mathbf{q}} \approx \ddot{\mathbf{q}} \tag{11.3.21}$$

In practice this approach should be expected to work fairly well.

(i) *Example 11.3.2*

Consider the single-link manipulator

$$I\ddot{\theta} + MgL\sin(\theta) = u \tag{11.3.22}$$

with the inverse dynamics control

$$u = \hat{I}v + \widehat{MgL}\sin(\theta) \tag{11.3.23}$$

$$v = \ddot{\theta}^d - 2\omega(\dot{\theta} - \dot{\theta}^d) - \omega^2(\theta - \theta^d) \tag{11.3.24}$$

As before we will take $I=10=MgL$ and choose (in this case as our initial guess) $\hat{I}=5=\widehat{MgL}$. For these values of $\hat{I}$ and $\widehat{MgL}$ the response with $\omega=10$ has been shown in Figure 8-4. In this case, with $\omega=10$ the various quantities above are given as

$$\mathbf{p} = \begin{bmatrix} p_1 \\ p_2 \end{bmatrix} = \begin{bmatrix} I \\ MgL \end{bmatrix}, \quad \Phi^T = \frac{1}{\hat{p}_1}\begin{bmatrix} \ddot{\theta} \\ sin(\theta) \end{bmatrix} \tag{11.3.25}$$

$$A = \begin{bmatrix} 0 & 1 \\ -100 & -20 \end{bmatrix} \quad \mathbf{b} = \begin{bmatrix} 0 \\ 1 \end{bmatrix} \tag{11.3.26}$$

The solution of the Lyapunov equation for $Q = I$ is

$$P = \begin{bmatrix} 2.625 & 0.005 \\ 0.005 & 0.02525 \end{bmatrix} \tag{11.3.27}$$

The response of the system for a given desired trajectory is shown in Figure 11-16. The time evolution of the estimated parameters is shown in Figures 11-17 and 11-18. Note that the estimated parameters do not converge to their true values. However, our primary concern is with the tracking error, and we see that the tracking is now very good as a result of the parameter update law.

11.3.2 CONVERGENCE OF PARAMETERS

We note that, in the above example, tracking was achieved without the estimated parameters converging to their true values. It is well known in adaptive control circles that in order for the parameter estimates to converge to their true values, the reference trajectory must be "sufficiently rich", that is, the reference trajectory must sufficiently excite the dynamic response of the system so that the effects of the

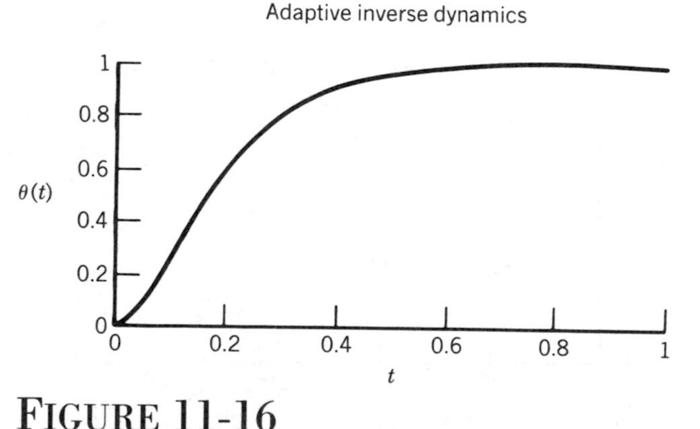

FIGURE 11-16
Tracking response—adaptive inverse dynamics.

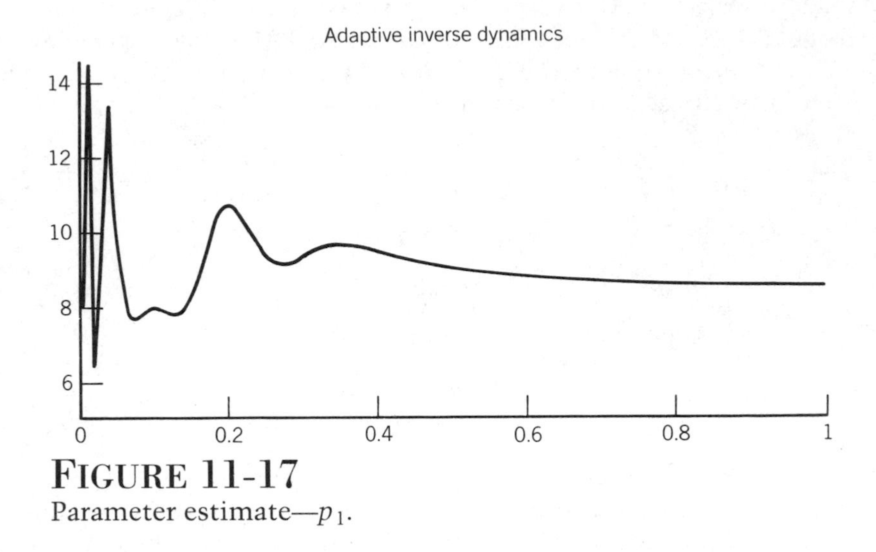

FIGURE 11-17
Parameter estimate—p_1.

various parameters can be distinguished. The conditions for parameter convergence are that the transfer function $b^T P(sI - A)^{-1} b$ satisfies a condition known as strict positive realness and that $\mathbf{q}^d$ satisfies

$$\alpha I \leq \int_{t_0}^{t_0+\rho} Y^T(\mathbf{q}^d, \dot{\mathbf{q}}^d, \ddot{\mathbf{q}}^d) Y(\mathbf{q}^d, \dot{\mathbf{q}}^d, \ddot{\mathbf{q}}^d) dt \leq \beta I \tag{11.3.28}$$

for all t_0 where α, β, ρ are positive numbers. (See [2].) Such a trajectory can be used as part of a "learning algorithm" to identify the system parameters, which can then be used in an inverse dynamics control law for tracking of any trajectory.

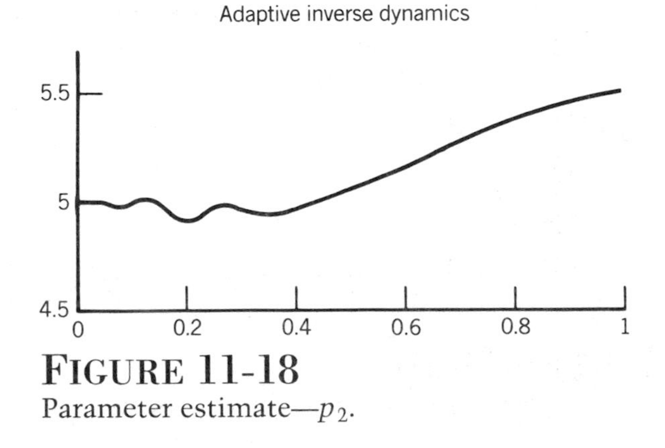

FIGURE 11-18
Parameter estimate—p_2.

11.3.3 THE ALGORITHM OF SLOTINE AND LI

We will next discuss a very recent result in adaptive control of manipulators due to Slotine and Li[13]. This approach is different from the inverse dynamics approach in that, even with exact knowledge of the parameters, the control law does not linearize the equations of motion of the robot. However, this approach overcomes the main drawbacks of the adaptive inverse dynamics control in that it does not require measurement or estimation of the manipulator acceleration nor does it require inversion of the estimated inertia matrix. Let us first illustrate the general idea of this approach in the known parameter case. Thus, given the manipulator dynamic equations (11.3.1), set

$$\mathbf{u} = M(\mathbf{q})\dot{\mathbf{v}} + C(\mathbf{q}, \dot{\mathbf{q}})\mathbf{v} + \mathbf{g}(\mathbf{q}) - K_D(\dot{\mathbf{q}} - \mathbf{v}) \tag{11.3.29}$$

and define $\mathbf{v}$ according to

$$\begin{aligned}\mathbf{v} &= \dot{\mathbf{q}}^d - \Lambda(\mathbf{q} - \mathbf{q}^d) \\ &= \dot{\mathbf{q}}^d - \Lambda\mathbf{e}\end{aligned}$$

where Λ is a (diagonal) matrix of positive gains. Define $\mathbf{r} = \dot{\mathbf{q}} - \mathbf{v} = \dot{\mathbf{e}} + \Lambda\mathbf{e}$ and substitute (11.3.29) into (11.3.1). The result is

$$M(\mathbf{q})\dot{\mathbf{r}} + C(\mathbf{q}, \dot{\mathbf{q}})\mathbf{r} + K_D\mathbf{r} = 0 \tag{11.3.30}$$

Note that the nonlinear control law (11.3.29) is distinct from the inverse dynamics control and does not lead to a linear system in the closed loop. Thus the stability analysis of (11.3.30) is more difficult than for the exact inverse dynamics control law. However, we will see that the above control law becomes useful when the problems of parameter uncertainty and parameter estimation are considered.

In order to show stability of the system (11.3.30), consider the function

$$V = {}^1\!/_2\,\mathbf{r}^T M(\mathbf{q})\mathbf{r} \tag{11.3.31}$$

Computing the time derivative of V along solutions of (11.3.31) yields

$$\dot{V} = \mathbf{r}^T M(\mathbf{q})\dot{\mathbf{r}} + {}^1\!/_2\mathbf{r}^T \dot{M}(\mathbf{q})\mathbf{r} \tag{11.3.32}$$

Substituting for $M(\mathbf{q})\dot{\mathbf{r}}$ from (11.3.30) and using the fact that $\dot{M} - 2C$ is skew symmetric we have

$$\dot{V} = -\mathbf{r}^T K_D \mathbf{r} \leq 0 \tag{11.3.33}$$

To show that (11.3.33) implies convergence of the tracking error to zero we must argue as follows[13]: First note that (11.3.31)–(11.3.33) imply that $\mathbf{r}$ is a (bounded) L_2 function. From the definition of $\mathbf{r}$ it follows that $\mathbf{e}$ and $\dot{\mathbf{e}}$ are bounded, and, in fact $\mathbf{e}$ goes to zero as t goes to infinity. (We can think of $\mathbf{e}$ as the output of a strictly proper, BIBO stable first order system with L_2 input $\mathbf{r}$.) From the system dynamics (11.3.30) $\dot{\mathbf{r}}$ is then bounded, from which it follows that $\mathbf{r}$, and hence $\dot{V}$, are uniformly continuous. Therefore $\mathbf{r}$ converges to zero, which further implies that $\dot{\mathbf{e}}$ converges to zero.

11.3.4 ADAPTIVE VERSION

The adaptive version of the above result proceeds as follows. Given the system dynamics (11.3.1) we choose the control law

$$\mathbf{u} = \hat{M}(\mathbf{q})\dot{\mathbf{v}} + \hat{C}(\mathbf{q}, \dot{\mathbf{q}})\mathbf{v} + \hat{\mathbf{g}}(\mathbf{q}) - K_D\mathbf{r} \tag{11.3.34}$$

Substituting this into the system (11.3.1) gives

$$M\ddot{\mathbf{q}} + C\dot{\mathbf{q}} + \mathbf{g} = \hat{M}\dot{\mathbf{v}} + \hat{C}\mathbf{v} + \hat{\mathbf{g}} - K_D\mathbf{r} \tag{11.3.35}$$

Now, since $\ddot{\mathbf{q}} = \dot{\mathbf{r}} + \dot{\mathbf{v}}$ and $\dot{\mathbf{q}} = \mathbf{r} + \mathbf{v}$ we can write (11.3.35) as

$$\begin{aligned} M\dot{\mathbf{r}} + C\mathbf{r} + K_D\mathbf{r} &= \tilde{M}\dot{\mathbf{v}} + \tilde{C}\mathbf{v} + \tilde{\mathbf{g}} \\ &= Y(\mathbf{q}, \dot{\mathbf{q}}, \mathbf{v}, \dot{\mathbf{v}})\tilde{\mathbf{p}} \end{aligned} \tag{11.3.36}$$

where $\tilde{M} = \hat{M} - M$, $\tilde{C} = \hat{C} - C$, $\tilde{\mathbf{g}} = \hat{\mathbf{g}} - \mathbf{g}$ and $\tilde{\mathbf{p}} = \hat{\mathbf{p}} - \mathbf{p}$. Note that the function Y does not depend on the manipulator acceleration, but only on $\dot{\mathbf{v}}$, which depends on the acceleration of the reference trajectory and the velocity tracking error.

Next, choose the Lyapunov function candidate

$$V = {}^1\!/\!_2\mathbf{r}^T M\mathbf{r} + {}^1\!/\!_2\tilde{\mathbf{p}}^T\Gamma\tilde{\mathbf{p}} \tag{11.3.37}$$

Then, along solution trajectories it follows that

$$\dot{V} = -\mathbf{r}^T K_D\mathbf{r} + \tilde{\mathbf{p}}^T\,(\Gamma\dot{\tilde{\mathbf{p}}} + Y^T\mathbf{r}) \tag{11.3.38}$$

If we choose the parameter adaptation law to make the second term zero, that is,

$$\dot{\hat{\mathbf{p}}} = \dot{\tilde{\mathbf{p}}} = -\Gamma^{-1}Y^T\mathbf{r} \tag{11.3.39}$$

then

$$\dot{V} = -\mathbf{r}^T K_D\mathbf{r} \leq 0 \tag{11.3.40}$$

and stability follows from an argument similar to that used previously.

(i) *Example 11.3.3*

Once again we consider the single-link manipulator

$$I\ddot{\theta} + MgL\,sin(\theta) = u \tag{11.3.41}$$

In this case the algorithm of Slotine and Li gives us

$$u = \hat{I}\dot{v} + \widehat{MgL}\sin(\theta) - K_D r \tag{11.3.42}$$

$$v = \dot{\theta}^d - \Gamma(\theta - \theta^d) \tag{11.3.43}$$

$$r = \dot{\tilde{\theta}} + \Gamma\tilde{\theta} \tag{11.3.44}$$

The closed loop system is then

$$I\dot{r} + K_D r = Y(\dot{v}, \theta)\tilde{\mathbf{p}} \tag{11.3.45}$$

where

$$Y(\dot{v}, \theta) = [\dot{v}, \sin(\theta)] \tag{11.3.46}$$

Note that Y does not depend on the manipulator acceleration. With $\hat{I} = \widehat{MgL} = 5$, the response of the system without parameter adaptation is shown in Figure 11-19. Note that the response is similar to that of the inverse dynamics control without adaptation. With the parameter update law (11.3.39) the response for the given desired trajectory is given in Figure 11-20. The time evolution of the estimated parameters is shown in Figures 11-21 and 11-22. Here also, the parameters do not converge to their true values but the tracking performance is excellent. As before, some sort of sufficient richness condition is necessary to insure convergence of the estimated parameters (See [15]).

11.3.5 The Problem of Unmodeled Dynamics

It is well known that dynamic uncertainty, that is, unmodeled dynamics such as joint flexibility, pose serious stability problems for adaptive control laws. We might ask, therefore, what the performance of the above adaptive control laws is when there is joint flexibility present in the manipulator? Here we reach the limits of knowledge in robot control theory. A stable adaptive control algorithm for the flexible joint model (8.2.22)–(8.2.23) does not exist at present. Even a rigorous analysis of the effects of joint flexibility on the above adaptive control laws awaits further work. Using our intuition gained so far we might expect the following behavior: When the motor variables are used in the adaptive feedback law, the flexible joint system will remain

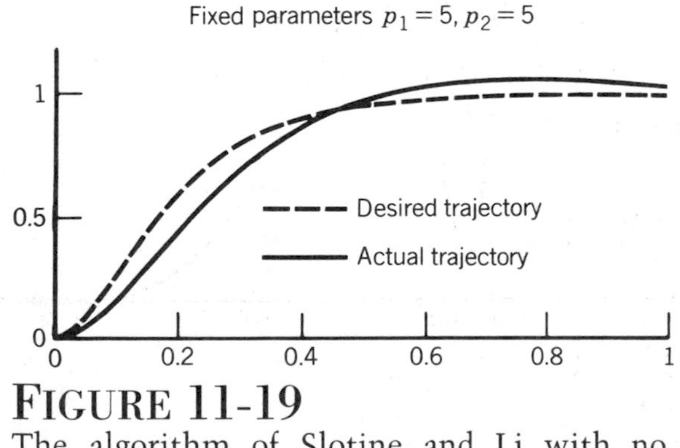

FIGURE 11-19
The algorithm of Slotine and Li with no parameter estimation.

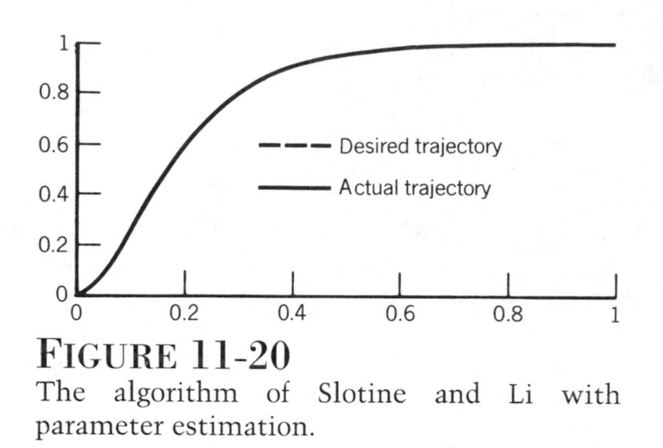

FIGURE 11-20
The algorithm of Slotine and Li with parameter estimation.

stable provided that the gains in the adaptation law are not too large and provided that there is sufficient damping present in the system. Tracking performance will be degraded since the link angles are not directly controlled. When the link variables are used in the feedback control law stability will be difficult if not impossible to maintain.

We now end our treatment of robot control with examples of adaptive control applied to a flexible joint robot. We offer these comments and examples in the hope of enticing and challenging the reader to seek rigorous solutions to these problems. Much work indeed remains to be done in robot control theory.

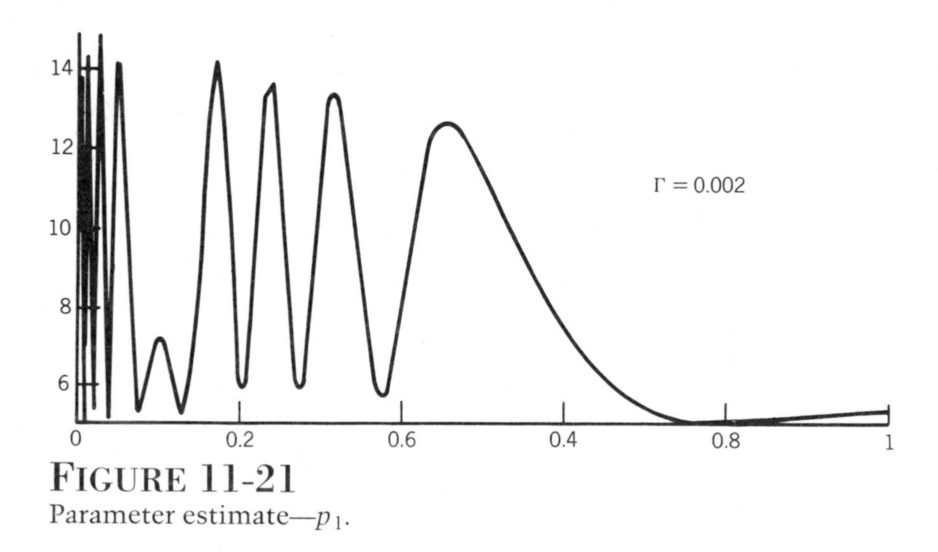

FIGURE 11-21
Parameter estimate—p_1.

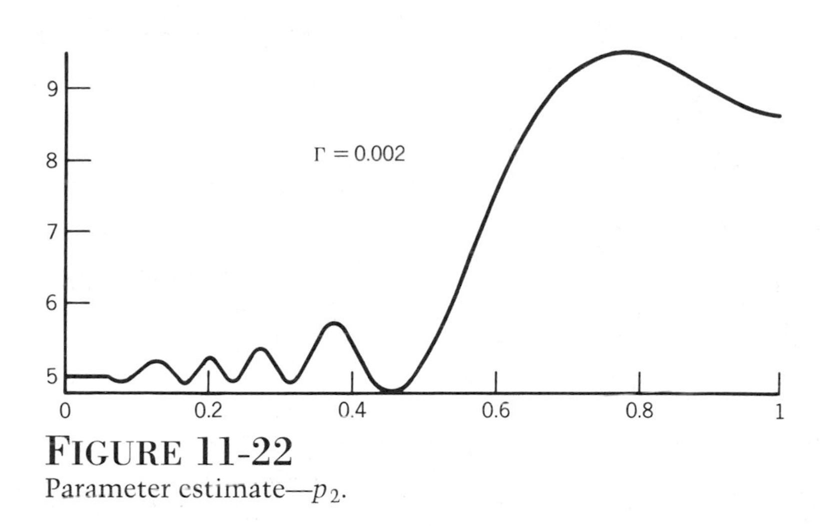

FIGURE 11-22
Parameter estimate—p_2.

(i) *Example 11.3.4*

Consider the single-link robot with joint flexibility and damping

$$I\ddot{\theta}_1 + B_1\dot{\theta}_1 + MgL\sin(\theta_1) + k(\theta_1 - \theta_2) = 0 \tag{11.3.47}$$

$$J\ddot{\theta}_2 + B_2\dot{\theta}_2 - k(\theta_1 - \theta_2) = u \tag{11.3.48}$$

Let us apply the algorithm of Slotine and Li[4] to the system (11.3.47)–(11.3.48), but ignore the joint flexibility. In other words, we assume $k = \infty$ and that the system is therefore governed by

$$(I + J)\ddot{\theta} + (B_1 + B_2)\dot{\theta} + MgL\, sin(\theta) = u \tag{11.3.49}$$

Note that there are now three parameters to estimate, namely,

$$p_1 = I + J \quad p_2 = B_1 + B_2 \quad p_3 = MgL \tag{11.3.50}$$

Applying the adaptive control law derived for (11.3.49) to the flexible joint system leads to the response shown in Figure 11-23 for stiffness $k = 1000$ provided that the motor variable θ_2 is used in the control law.

The corresponding parameter estimates are shown in Figure 11-24. For lower stiffness, $k = 100$, the response is shown in Figure 11-25. Note that the motor shaft angle θ_2 tracks the reference trajectory but that the link angle θ_1 is not controlled and hence oscillates.

[4]The adaptive inverse dynamics could be applied also, with similar behavior.

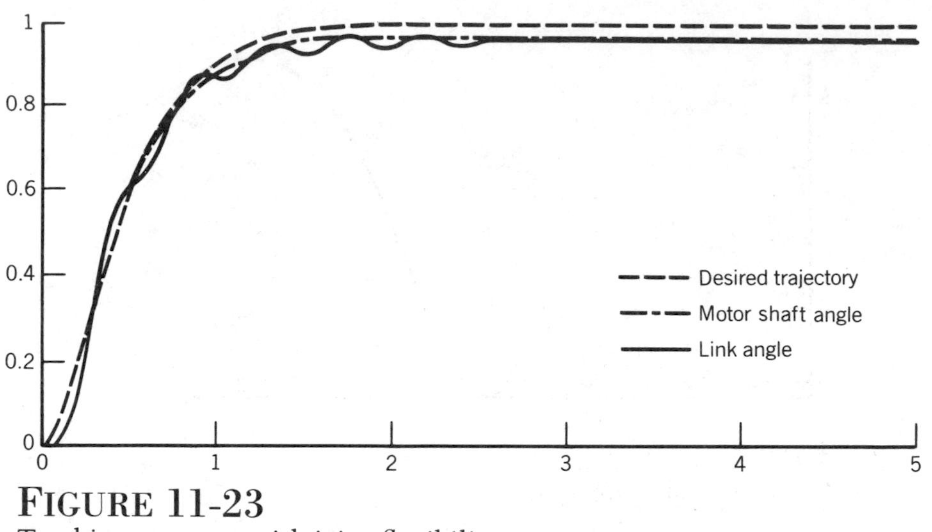

FIGURE 11-23
Tracking response with joint flexibility present.

If we attempt to remedy this by using the link angle θ_1 in the control law, we will drive the system unstable. The reader is invited to verify this statement by computer simulation. Adaptive control of the system (11.3.47)–(11.3.48), therefore, remains an open research problem.

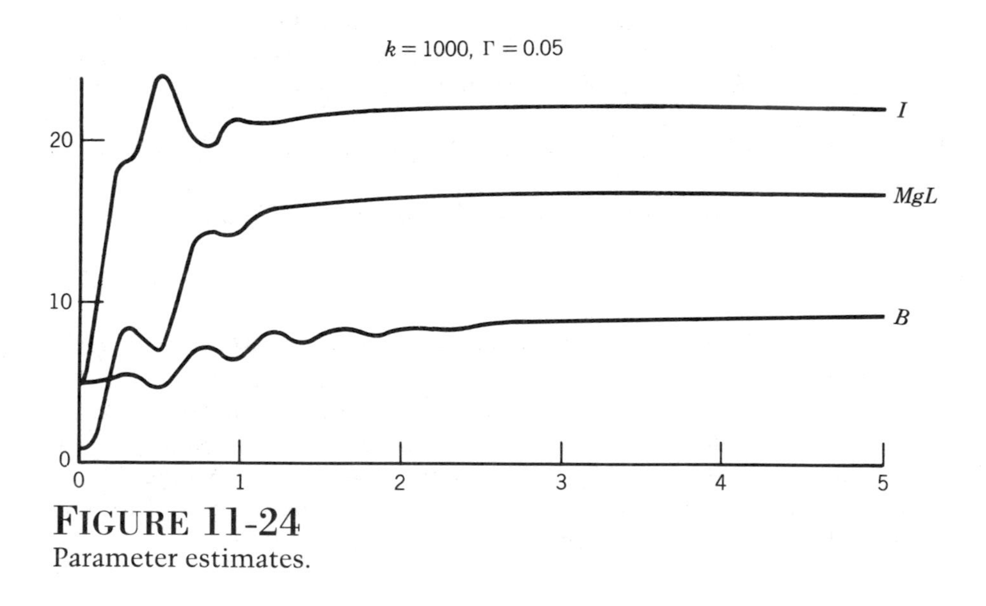

FIGURE 11-24
Parameter estimates.

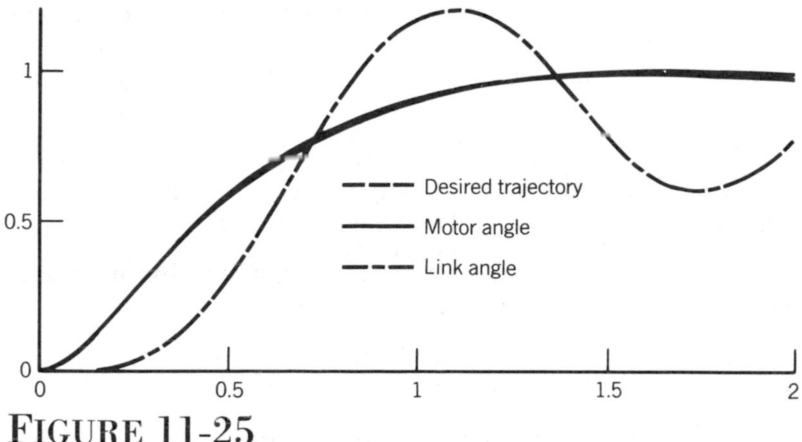

FIGURE 11-25
Tracking response with very flexible joint.

REFERENCES AND SUGGESTED READING

[1] ASTROM, K.J., "Theory and Applications of Adaptive Control: A Survey," *Automatica*, Vol. 19, p. 471, 1983.

[2] CRAIG, J., J., *Adaptive Control of Mechanical Manipulators*, Addison–Wesley, Reading, MA, 1988.

[3] CRAIG, J.J., HSU, P., and SASTRY, S., "Adaptive Control of Mechanical Manipulators," *IEEE Int. Conf. Robotics and Automation*, San Francisco, CA, Mar. 1986.

[4] DECARLO, R.A., ZAK, S.H., and MATHEWS, G.P., "Variable Structure Control of Nonlinear Multivariable Systems: A Tutorial," *Proc. of the IEEE*, pp. 212-232, Mar. 1988

[5] FILIPPOV, A.F., "Differential Equations with Discontinuous Right-Hand Sides," *Ann. Math Soc. Transl.*, Vol. 42, pp. 199-232, 1964.

[6] HSIA, T.C., "Adaptive Control of Robot Manipulators," *IEEE Int. Conf. Robotics and Automation*, San Francisco, CA, Mar. 1986.

[7] ITKIS, U., *Control Systems of Variable Structure*, Wiley, New York, 1976.

[8] KHOSLA, P., and KANADE, T., "Parameter Identification of Robot Dynamics," *IEEE Conf. on Decision and Control*, Ft. Lauderdale, FL, Dec. 1985.

[9] MORGAN, R.G., and OZGUNNER, U., "A Decentralized Variable Structure Control Algorithm for Robotic Manipulators," *IEEE Trans. Robotics and Automation*, Vol. RA-1, No.1, pp. 57–65, Mar. 1985.

[10] SIRA-RAMIREZ, H., and SPONG, M.W., "Variable Structure Control of Flexible Joint Robot Manipulators," *IASTED Journal of Robotics and Automation*, submitted Mar. 1987.

[11] SLOTINE, J.-J. E., "Sliding Controller Design for Nonlinear Systems," *Int. J. Control*, Vol. 40, No. 2, 1984.

[12] SLOTINE, J.-J. E., "The Robust Control of Robot Manipulators," *Int. J. Robotics Research*, Vol. 4, No. 2, 1985.

[13] SLOTINE, J.-J. E., and LI, W., "On the Adaptive Control of Robot Manipulators," *Int. J. Robotics Research*, Vol. 6, No. 3, pp. 49–59, Fall 1987.

[14] SLOTINE, J.-J. E., and SASTRY, S. S., "Tracking Control of Nonlinear Systems Using Sliding Surfaces with Applications to Robot Manipulators," *Int. J. Control*, Vol. 39, No. 2, 1983.

[15] LI, W., and SLOTINE, J.-J. E., "Parameter Estimation Strategies for Robotic Applications," *ASME Winter Annual Meeting*, DSC-Vol. 6, Boston, MA, Dec. 1987.

[16] SLOTINE, J.-J. E., and SPONG, M.W., "Robust Robot Control with Bounded Input Torques," *J. Robotic Systems*, Vol. 2, No. 4, 1985.

[17] UTKIN, V.I., *Sliding Modes and Their Application in Variable Structure Systems*, MIR Publishers, Moscow, 1978.

[18] UTKIN, V.I., "Variable Structure Systems with Slidings," *IEEE Trans. Automatic Control*, Vol. AC-22, No. 2, pp. 212–222, 1977.

[19] WALCOTT, B.L., and ZAK, S.H., "State Observation of Nonlinear Uncertain Dynamical Systems," *IEEE Trans. Automation Control*, Vol. AC-32, No. 2, pp. 166–170, 1987.

[20] YOUNG, K.K.D. "Controller Design for a Manipulator Using Theory of Variable Structure Systems,"*IEEE Trans. Sys. Man. Cyber.*, SMC-8, No.2, Feb. 1978.

PROBLEMS

11-1 Consider the single link rigid manipulator of Example 11.2.4. Assume that the various parameters given are exact but that there is an unknown load $0 < M < 10kg$ at the end-effector.

a) Design an inner loop approximate feedback linearizing control for the system. In this case you will have to choose a nominal value $\hat{M}$ for the unknown mass M. Try three cases, $\hat{M}=0$, $\hat{M}=5$, and $\hat{M}=10$.

b) Based on your inner loop control and the bounds on the actual value of M, design an outer loop control v using the method of sliding modes so that the link angle θ_ℓ tracks a sinusoidal trajectory $\theta_\ell^d = \sin 8t$.

c) Simulate your design for various values of M, say $M = 0$, $M = 2.5$, $M = 9$. Which nominal choice, if any, for $\hat{M}$ gives the best response?

11-2 Apply the control law designed in Problem 11-1, directly to the flexible joint system of Example 11.3.4 in order to investigate the effect of unmodeled dynamics. What happens for various values of the joint stiffness k?

11-3 Given the two-link planar manipulator of Example 6.4.1, carry out a complete analysis, controller design, and simulation using any of the methods of this chapter. Investigate various effects on your own. For example,

a) What happens if the input joint torques are constrained by $|u_i| \leq u_{i,max}$ for $i = 1, 2$? Simulate your design for various values of $u_{i,max}$. How can you modify your controller in case the torque constraints result in poor performance?

b) What happens if there is measurement noise in the system?

c) What happens if there is gear backlash or static friction present? Can you suggest ways to overcome these problems, either by software or hardware modification?

d) Suppose it is desired for the end-effector to track a straight line in Cartesian space. How fast can you track a given straight line with your design? Use realistic torque constraints for this one. What happens, if anything, in the neighborhood of singular configurations? Explain.

11-4 Verify Equation 11.3.5.

11-5 Suppose that in Example 11.3.1 all of the parameters of the manipulator are initially known but that the manipulator picks up an unknown mass m. Assume that m is just a point mass. Which parameters in (11.3.1) are now unknown? What would the dimension of the parameter space be to estimate the unknown parameters?

APPENDIX A

LINEAR ALGEBRA

In this book we assume that the reader has some familiarity with basic properties of vectors and matrices, such as matrix addition, subtraction, multiplication, matrix transpose, and determinants. These concepts will not be defined here. For additional background see [1].

We use lower case boldface letters $\mathbf{a}$, $\mathbf{b}$, $\mathbf{c}$, $\mathbf{x}$, $\mathbf{y}$, etc., to denote vectors in $\mathbb{R}^n$. Uppercase letters A, B, C, R, etc., denote matrices. The symbol $\mathbb{R}$ will denote the set of real numbers, and $\mathbb{R}^n$ will denote the usual vector space on n-tuples over $\mathbb{R}$.

Unless otherwise stated, vectors will be defined as column vectors. Thus, the statement $\mathbf{x} \in \mathbb{R}^n$ means that

$$\mathbf{x} = \begin{bmatrix} x_1 \\ \cdot \\ \cdot \\ \cdot \\ x_n \end{bmatrix}, \quad x_i \in \mathbb{R} \tag{A.1}$$

The vector $\mathbf{x}$ is thus an n-tuple, arranged in a column with components $x_1, \ldots, x_n$. We will frequently denote this as

$$\mathbf{x} = [x_1, \ldots, x_n]^T \tag{A.2}$$

where the superscript T denotes transpose. The length or **norm** of a vector $\mathbf{x} \in \mathbb{R}^n$ is

$$\|\mathbf{x}\| = (x_1^2 + \cdots + x_n^2)^{1/2} \tag{A.3}$$

The **scalar product**, $<\mathbf{x},\mathbf{y}>$, of two vectors $\mathbf{x}$ and $\mathbf{y}$ belonging to $\mathbb{R}^n$ is a real number defined by

$$<\mathbf{x},\mathbf{y}> = \mathbf{x}^T\mathbf{y} = x_1y_1 + \cdots + x_ny_n \tag{A.4}$$

Thus,

$$\|\mathbf{x}\| = <\mathbf{x},\mathbf{x}>^{1/2} \tag{A.5}$$

The scalar product of vectors is commutative, that is,

$$<\mathbf{x},\mathbf{y}> = <\mathbf{y},\mathbf{x}> \tag{A.6}$$

We also have the useful inequalities,

$$|<\mathbf{x},\mathbf{y}>| \leq \|\mathbf{x}\|\,\|\mathbf{y}\| \quad \text{(Cauchy–Schwartz)} \tag{A.7}$$

$$\|\mathbf{x}+\mathbf{y}\| \leq \|\mathbf{x}\| + \|\mathbf{y}\| \quad \text{(Triangle Inequality)} \tag{A.8}$$

For vectors in $\mathbb{R}^3$ the scalar product can be expressed as

$$<\mathbf{x},\mathbf{y}> = \|\mathbf{x}\|\,\|\mathbf{y}\|\cos(\theta) \tag{A.9}$$

where θ is the angle between the vectors $\mathbf{x}$ and $\mathbf{y}$.

The **outer product** of two vectors $\mathbf{x}$ and $\mathbf{y}$ belonging to $\mathbb{R}^n$ is an $n \times n$ matrix defined by

$$\mathbf{x}\mathbf{y}^T = \begin{bmatrix} x_1y_1 & \cdot & \cdot & x_1y_n \\ x_2y_1 & \cdot & \cdot & x_2y_n \\ \cdot & \cdot & \cdot & \cdot \\ x_ny_1 & \cdot & \cdot & x_ny_n \end{bmatrix} \tag{A.10}$$

From (A.10) we can see that the scalar product and the outer product are related by

$$<\mathbf{x},\mathbf{y}> = \mathbf{x}^T\mathbf{y} = Tr(\mathbf{x}\mathbf{y}^T) \tag{A.11}$$

where the function $Tr(\cdot)$ denotes the trace of a matrix, that is, the sum of the diagonal elements of the matrix.

We will use $\mathbf{i}$, $\mathbf{j}$ and $\mathbf{k}$ to denote the standard unit vectors in $\mathbb{R}^3$

$$\mathbf{i} = \begin{bmatrix} 1 \\ 0 \\ 0 \end{bmatrix}, \quad \mathbf{j} = \begin{bmatrix} 0 \\ 1 \\ 0 \end{bmatrix}, \quad \mathbf{k} = \begin{bmatrix} 0 \\ 0 \\ 1 \end{bmatrix} \tag{A.12}$$

Using this notation a vector $\mathbf{x} = [x_1, x_2, x_3]^T$ may be written as

$$\mathbf{x} = x_1\mathbf{i} + x_2\mathbf{j} + x_3\mathbf{k} \tag{A.13}$$

The **vector product** or **cross product** $\mathbf{x}\times\mathbf{y}$ of two vectors $\mathbf{x}$ and $\mathbf{y}$ belonging to $\mathbb{R}^3$ is a vector $\mathbf{c}$ defined by

$$\mathbf{c} = \mathbf{x} \times \mathbf{y} = \det \begin{bmatrix} \mathbf{i} & \mathbf{j} & \mathbf{k} \\ x_1 & x_2 & x_3 \\ y_1 & y_2 & y_3 \end{bmatrix} \tag{A.14}$$

$$= (x_2y_3 - x_3y_2)\mathbf{i} + (x_3y_1 - x_1y_3)\mathbf{j} + (x_1y_2 - x_2y_1)\mathbf{k}$$

The cross product is a vector whose magnitude is

$$\|\mathbf{c}\| = \|\mathbf{x}\| \ \|\mathbf{y}\| \sin(\theta) \tag{A.15}$$

where θ is the angle between $\mathbf{x}$ and $\mathbf{y}$ and whose direction is given by the right hand rule shown in Figure A-1.

A right-handed coordinate frame $x-y-z$ is a coordinate frame with axes mutually perpendicular and that also satisfies the right hand rule as shown in Figure A-2. We can remember the right hand rule as being the direction of advancement of a right-handed screw rotated from the positive x axis is rotated into the positive y axis through the smallest angle between the axes.

The cross product has the properties

$$\begin{aligned} \mathbf{x}\times\mathbf{y} &= -\mathbf{y}\times\mathbf{x} \\ \mathbf{x}\times(\mathbf{y}+\mathbf{z}) &= \mathbf{x}\times\mathbf{y}+\mathbf{x}\times\mathbf{z} \\ \alpha(\mathbf{x}\times\mathbf{y}) &= (\alpha\mathbf{x})\times\mathbf{y} = \mathbf{x}\times(\alpha\mathbf{y}) \end{aligned} \tag{A.16}$$

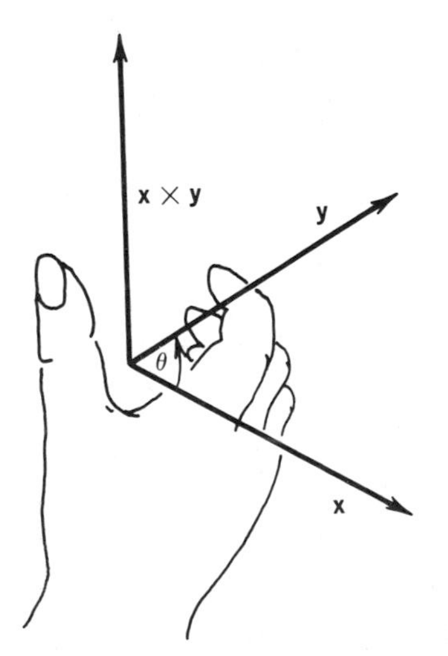

FIGURE A-1
The right hand rule.

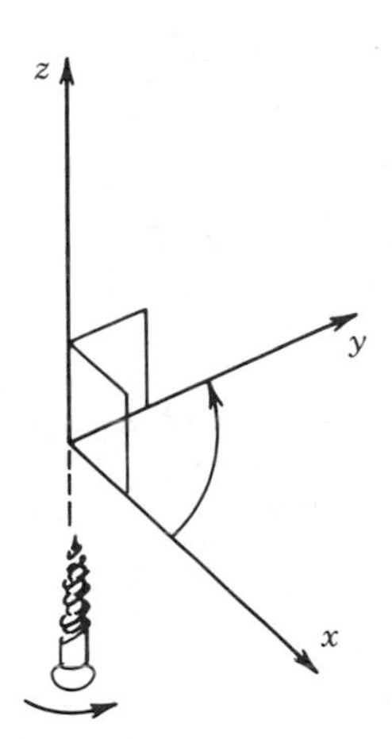

FIGURE A-2
A right-handed coordinate frame.

A.1.1 DIFFERENTIATION OF VECTORS

Suppose that the vector $\mathbf{x}(t) = (x_1(t), \ldots, x_n(t))^T$ is a function of time. Then the time derivative $\dot{\mathbf{x}}$ of $\mathbf{x}$ is just the vector

$$\dot{\mathbf{x}} = (\dot{x}_1(t), \ldots, \dot{x}_n(t))^T \tag{A.17}$$

that is, the vector can be differentiated coordinatewise. Likewise, the derivative dA/dt of a matrix $A = (a_{ij})$ is just the matrix $(\dot{a}_{ij})$. Similar statements hold for integration of vectors and matrices. The scalar and vector products satisfy the following product rules for differentiation similar to the product rule for differentiation of ordinary functions.

$$\frac{d}{dt} < \mathbf{x}, \mathbf{y} > = < \frac{d\mathbf{x}}{dt}, \mathbf{y} > + < \mathbf{x}, \frac{d\mathbf{y}}{dt} > \tag{A.18}$$

$$\frac{d}{dt}(\mathbf{x} \times \mathbf{y}) = \frac{d\mathbf{x}}{dt} \times \mathbf{y} + \mathbf{x} \times \frac{d\mathbf{y}}{dt} \tag{A.19}$$

A.1.2 CHANGE OF COORDINATES

A matrix can be thought of as representing a linear transformation from $\mathbb{R}^n$ to $\mathbb{R}^n$ in the sense that A takes a vector $\mathbf{x}$ to a new vector $\mathbf{y}$ according to

$$\mathbf{y} = A\mathbf{x} \tag{A.20}$$

$\mathbf{y}$ is called the **image** of $\mathbf{x}$ under the transformation A. If the vectors $\mathbf{x}$ and $\mathbf{y}$ are represented in terms of the standard unit vectors $\mathbf{i}$, $\mathbf{j}$, and $\mathbf{k}$, then the columns of A are themselves vectors which represent the images of the basis vectors $\mathbf{i}$, $\mathbf{j}$, $\mathbf{k}$. Often it is desired to represent vectors with respect to a second coordinate frame with basis vectors $\mathbf{e}$, $\mathbf{f}$, and $\mathbf{g}$. In this case the matrix representing the same linear transformation as A, but relative to this new basis, is given by

$$A' = T^{-1}AT \tag{A.21}$$

where T is a non-singular matrix with column vectors **e**, **f**, **g**. The transformation $T^{-1}AT$ is called a **similarity transformation** of the matrix A.

A.1.3 EIGENVALUES AND EIGENVECTORS

The **eigenvalues** of a matrix A are the solutions in s of the equation

$$\det(sI - A) = 0 \qquad \text{(A.22)}$$

The function, $\det(sI - A)$ is a polynomial in s called the **characteristic polynomial** of A. If s_e is an eigenvalue of A, an **eigenvector** of A corresponding to s_e is a nonzero vector $\mathbf{x}$ satisfying the system of linear equations

$$(s_e I - A)\mathbf{x} = 0 \qquad \text{(A.23)}$$

or, equivalently,

$$A\mathbf{x} = s_e \mathbf{x} \qquad \text{(A.24)}$$

If the eigenvalues $s_1, \ldots, s_n$ of A are distinct, then there exists a similarity transformation $A' = T^{-1}AT$, such that A' is a diagonal matrix with the eigenvalues $s_1, \ldots, s_n$ on the main diagonal, that is,

$$A' = diag[s_1, \ldots, s_n] \qquad \text{(A.25)}$$

A set of vectors $\{\mathbf{x}_1, \ldots, \mathbf{x}_n\}$ is said to be **linearly independent** if and only if

$$\sum_{i=1}^{n} \alpha_i \mathbf{x}_i = 0 \qquad \text{(A.26)}$$

implies

$$\mathbf{x}_i = 0 \quad \text{for all } i \qquad \text{(A.27)}$$

The **rank** of a matrix A is the the largest number of linearly independent rows (or columns) of A. Thus the rank of an $n \times m$ matrix can be no greater than the minimum of n and m.

REFERENCES AND SUGGESTED READING

[1] BARNETT, S., *Matrix Methods for Engineers and Scientists*, McGraw–Hill Book Company (UK) Limited, 1979.

APPENDIX B

STATE SPACE THEORY OF DYNAMICAL SYSTEMS

(i) *Definition B.1*

A vector field **f** is a continuous function $\mathbf{f}: \mathbb{R}^n \to \mathbb{R}^n$.

We can think of a differential equation

$$\dot{\mathbf{x}}(t) = \mathbf{f}(\mathbf{x}(t)) \tag{B.1}$$

as being defined by a vector field **f** on $\mathbb{R}^n$. A solution $t \to \mathbf{x}(t)$ of (B.1) with $\mathbf{x}(t_0) = \mathbf{x}_0$ is then a curve C in $\mathbb{R}^n$, beginning at $\mathbf{x}_0$ parametrized by t, such that at each point of C, the vector field $\mathbf{f}(\mathbf{x}(t))$ is tangent to C. $\mathbb{R}^n$ is then called the **state space** of the system (B.1). For two dimensional systems, we can represent

$$t \to \begin{bmatrix} x_1(t) \\ x_2(t) \end{bmatrix} \tag{B.2}$$

by a curve C in the plane.

(ii) *Example B.2*

Consider the two-dimensional system

$$\dot{x}_1 = x_2 \qquad x_1(0) = x_{10} \tag{B.3}$$

$$\dot{x}_2 = -x_1 \qquad x_2(0) = x_{20} \tag{B.4}$$

In the phase plane the solutions of this equation are circles of radius

$$r = x_{10}^2 + x_{20}^2 \tag{B.4}$$

To see this consider the equation

$$x_1^2(t) + x_2^2(t) = r \tag{B.5}$$

Clearly the initial conditions satisfy this equation. If we differentiate (B.5) in the direction of the vector field $\mathbf{f} = (x_2, -x_1)^T$ that defines (B.3)–(B.4) we obtain

$$2x_1\dot{x}_1 + 2x_2\dot{x}_2 = 2x_1x_2 - 2x_2x_1 = 0 \tag{B.6}$$

Thus $\mathbf{f}$ is tangent to the circle. The graph of such curves C in the $x_1 - x_2$ plane for different initial conditions are shown in Figure B-1. The x_1–x_2 plane is called the **phase plane** and the trajectories of the system (B.3)–(B.4) form what is called the **phase portrait**. For linear systems of the form

$$\dot{\mathbf{x}} = A\mathbf{x} \tag{B.7}$$

in $\mathbb{R}^2$ the phase portrait is determined by the eigenvalues and eigenvectors of A. For example, consider the system

$$\dot{x}_1 = x_2 \tag{B.8}$$

$$\dot{x}_2 = x_1 \tag{B.9}$$

In this case

$$A = \begin{bmatrix} 0 & 1 \\ 1 & 0 \end{bmatrix} \tag{B.10}$$

The phase portrait is shown in Figure B-2. The lines ℓ_1 and ℓ_2 are in the direction of the eigenvectors of A and are called **eigen-subspaces** of A.

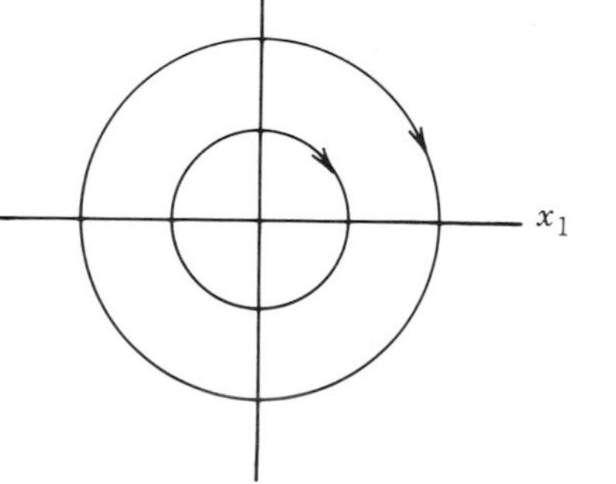

FIGURE B-1
Phase portrait for Example B.1.

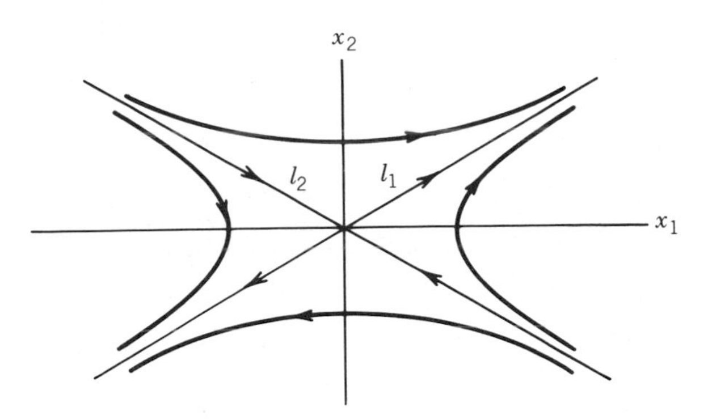

FIGURE B-2
Phase portrait for example.

B.1.1 STATE SPACE REPRESENTATION OF LINEAR SYSTEMS

Consider a single-input/single-output linear control system with input u and output y of the form

$$a_n \frac{d^n y}{dt^n} + a_{n-1} \frac{d^{n-1}y}{dt^{n-1}} + \cdots + a_1 \frac{dy}{dt} + a_0 y = u \tag{B.11}$$

The characteristic polynomial, whose roots are the open loop poles, is given as

$$p(s) = a_n s^n + a_{n-1}s^{n-1} + \cdots + a_0 \tag{B.12}$$

For simplicity we suppose that $p(s)$ is monic, that is, $a_n = 1$. The standard way of representing (B.11) in state space is to define n state variables $x_1, x_2, \ldots, x_n$ as

$$\begin{aligned} x_1 &= y \\ x_2 &= \dot{y} = \dot{x}_1 \\ x_3 &= \ddot{y} = \dot{x}_2 \\ &\cdot \\ &\cdot \\ &\cdot \\ x_n &= \frac{d^{n-1}y}{dt^{n-1}} = \dot{x}_{n-1} \end{aligned} \tag{B.13}$$

and express (B.11) as the system of first order differential equations

$$\begin{aligned}
\dot{x}_1 &= x_2 \\
\dot{x}_2 &= x_3 \\
\dot{x}_{n-1} &= x_n \\
\dot{x}_n &= \frac{d^n y}{dt^n} = -a_0 y - a_1 \frac{dy}{dt} - \cdots - a_{n-1} \frac{d^n y}{dt^n} + u \\
&= -a_0 x_1 - a_1 x_2 - \cdots - a_{n-1} x_n + u
\end{aligned} \tag{B.14}$$

In matrix form this system of equations is written as

$$\begin{bmatrix} \dot{x}_1 \\ \cdot \\ \cdot \\ \cdot \\ \cdot \\ \dot{x}_n \end{bmatrix} = \begin{bmatrix} 0 & 1 & . & . & 0 \\ 0 & 0 & 1 & . & 0 \\ & & & . \; . & \\ & & & & 1 \\ & & & & \\ -a_0 & . & . & . & -a_{n-1} \end{bmatrix} \begin{bmatrix} x_1 \\ \cdot \\ \cdot \\ \cdot \\ \cdot \\ x_n \end{bmatrix} + \begin{bmatrix} 0 \\ 0 \\ \\ \\ 0 \\ 1 \end{bmatrix} u \tag{B.15}$$

or

$$\dot{\mathbf{x}} = A\mathbf{x} + \mathbf{b}u \qquad x \in \mathbb{R}^n$$

The output y can be expressed as

$$\begin{aligned}
y &= [1, 0, \ldots, 0]\mathbf{x} \\
&= \mathbf{c}^T \mathbf{x}
\end{aligned} \tag{B.16}$$

It is easy to show that

$$\det(sI - A) = s^n + a_{n-1}s^{n-1} + \cdots + a_1 s + a_0 \tag{B.17}$$

and so the last row of the matrix A consists of precisely the coefficients of the characteristic polynomial of the system, and furthermore the eigenvalues of A are the open loop poles of the system.

In the Laplace domain, the transfer function $\frac{Y(s)}{U(s)}$ is equivalent to

$$\frac{Y(s)}{U(s)} = \mathbf{c}^T (sI - A)^{-1} \mathbf{b} \tag{B.18}$$

REFERENCES AND SUGGESTED READING

[1] KAILATH, T., *Linear Systems*, Prentice–Hall, Englewood Cliffs, NJ, 1980.

[2] VIDYASAGAR, M., *Nonlinear Systems Analysis*, Prentice–Hall, Englewood Cliffs, NJ, 1978.

APPENDIX C

LYAPUNOV STABILITY

We give here some basic definitions of stability and Lyapunov functions and present a sufficient condition for showing stability of a class of nonlinear systems. For simplicity we treat only time-invariant systems. For a more general treatment of the subject the reader is referred to [1].

(*i*) *Definition C.1*

Consider a nonlinear system on $\mathbb{R}^n$

$$\dot{\mathbf{x}} = \mathbf{f}(\mathbf{x}) \tag{C.1}$$

where $\mathbf{f}(\mathbf{x})$ is a vector field on $\mathbb{R}^n$ and suppose that $\mathbf{f}(\mathbf{0}) = \mathbf{0}$. Then the origin in $\mathbb{R}^n$ is said to be an **equilibrium point** for (C.1). If initially the system (C.1) satisfies $\mathbf{x}(t_0) = \mathbf{0}$ then the function $\mathbf{x}(t) \equiv \mathbf{0}$ for $t > t_0$ can be seen to be a solution of (C.1) called the **null** or **equilibrium** solution. In other words, if the system represented by (C.1) starts initially at the equilibrium, then it remains at the equilibrium thereafter. The question of stability deals with the solutions of (C.1) for initial conditions away from the equilibrium point. Intuitively, the null solution should be called stable if, for initial conditions close to the equilibrium, the solution remains close thereafter in some sense. We can formalize this notion into the following

(ii) *Definition C.2*

The null solution $\mathbf{x}(t) = \mathbf{0}$ is **stable** if and only if, for any $\varepsilon > 0$ there exist $\delta(\varepsilon) > 0$ such that

$$\|\mathbf{x}(t_0)\| < \delta \text{ implies } \|\mathbf{x}(t)\| < \varepsilon \text{ for all } t > t_0 \tag{C.2}$$

This situation is illustrated by Figure C-1 and says that the system is stable if the solution remains within a ball of radius ε around the equilibrium, so long as the initial condition lies in a ball of radius δ around the equilibrium. Notice that the required δ will depend on the given ε. To put it another way, a system is stable if "small" perturbations in the initial conditions, results in "small" perturbations from the null solution.

(iii) *Definition C.3*

The null solution $\mathbf{x}(t) = \mathbf{0}$ is **asymptotically stable** if and only if there exists $\delta > 0$ **such that**

$$\|\mathbf{x}(t_0)\| < \delta \text{ implies } \|\mathbf{x}(t)\| \to 0 \text{ as } t \to \infty \tag{C.3}$$

In other words, asymptotic stability means that if the system is perturbed away from the equilibrium it will return asymptotically to the equilibrium. The above notions of stability are local in nature, that is, they may hold for initial conditions "sufficiently near" the equilibrium point but may fail for initial conditions farther away from the equilibrium. Stability (respectively, asymptotic stability) is said to be **global** if it holds for arbitrary initial conditions.

We know that a linear system

$$\dot{\mathbf{x}} = A\mathbf{x} \tag{C.4}$$

will be globally asymptotically stable provided that all eigenvalues of the matrix A lie in the open left half of the complex plane. For nonlinear systems stability cannot be so easily determined.

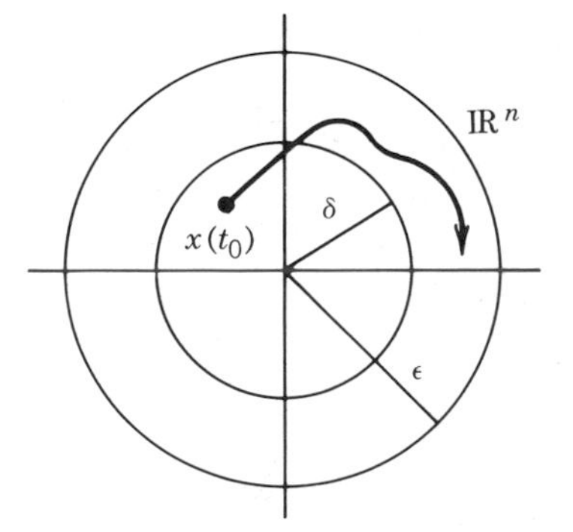

FIGURE C.1
Illustrating the definition of stability.

Another important notion related to stability is the notion of **uniform ultimate boundedness** of solutions.

(iv) Definition C.4

A solution $\mathbf{x}(t):[t_0, \infty] \rightarrow \mathbb{R}^n$ of (C.1) with initial condition $\mathbf{x}(t_0) = \mathbf{x}_0$ is said to **uniformly ultimately bounded** (u.u.b.) with respect to a set S if there is a nonnegative constant $T(\mathbf{x}_0, S)$ such that

$$\mathbf{x}(t) \in S \text{ for all } t \geq t_0 + T$$

Uniform ultimate boundedness says that the solution trajectory of (C.1) beginning at $\mathbf{x}_0$ at time t_0 will ultimately enter and remain within the set S. If the set S is a small region about the equilibrium, then uniform ultimate boundedness is a practical notion of stability, which is useful in control system design.

C.1.1 QUADRATIC FORMS AND LYAPUNPOV FUNCTIONS

(i) Definition C.5

Given a symmetric matrix $P = (p_{ij})$ the scalar function

$$V(\mathbf{x}) = \mathbf{x}^T P \mathbf{x} = \sum_{i,j=1}^{n} p_{ij} x_i x_j \tag{C.5}$$

is said to be a **quadratic form**. $V(\mathbf{x})$, equivalently the quadratic form, is said to be **positive definite** if and only if

$$V(\mathbf{x}) > 0 \tag{C.6}$$

for $\mathbf{x} \neq 0$. Note that $V(0) = 0$. $V(\mathbf{x})$ will be positive definite if and only if the matrix P is a positive definite matrix, that is, has all eigenvalues positive.

The level surfaces of V, given as solutions of $V(\mathbf{x}) = constant$ are ellipsoids in $\mathbb{R}^n$. A positive definite quadratic form is like a norm. In fact, given the usual norm $||\mathbf{x}||$ on R^n, the function V given as

$$V(\mathbf{x}) = \mathbf{x}^T \mathbf{x} = ||\mathbf{x}||^2 \tag{C.7}$$

is a positive definite quadratic form.

(ii) Definition C.6 (Lyapunov Function Candidate):

Let $V(\mathbf{x}) : \mathbb{R}^n \rightarrow \mathbb{R}$ be a continuous function with continuous first partial derivatives in a neighborhood of the origin in $\mathbb{R}^n$. Further suppose that V is positive definite, that is, $V(0) = 0$ and $V > 0$ for $\mathbf{x} \neq 0$. Then V is called a **Lyapunov Function Candidate** (for the system (C.1).

The positive definite function V is also like a norm. For the most part we will be utilizing Lyapunov function candidates that are quadratic forms, but the power of Lyapunov stability theory comes from the fact that any function may be used in an attempt to show stability of a given system provided it is a Lyapunov function candidate according to the above definition.

By the derivative of V along trajectories of (C.1), or the derivative of V in the direction of the vector field defining (C.1), we mean

$$\dot{V}(t) = <\mathbf{d}V, \mathbf{f}> = \frac{\partial V}{\partial x_1} f_1(\mathbf{x}) + \cdots + \frac{\partial V}{\partial x_n} f_n(\mathbf{x}) \tag{C.8}$$

Suppose that we evaluate the Lyapunov function candidate V at points along a solution trajectory $\mathbf{x}(t)$ of (C.1) and find that $V(t)$ is decreasing for increasing t. Intuitively, since V acts like a norm, this must mean that the given solution trajectory must be converging toward the origin. This is the idea of Lyapunov stability theory.

C.1.2 Lyapunov Stability

(i) *Theorem C.7*

The null solution of (C.1) is stable if there exists a Lyapunov function candidate V such that $\dot{V}$ is negative semi-definite along solution trajectories of (C.1), that is, if

$$\dot{V} = < \mathbf{d}V, \mathbf{f}(\mathbf{x}) > = \mathbf{d}V^T \mathbf{f}(\mathbf{x}) \leq 0 \tag{C.9}$$

Equation (C.9) says that the derivative of V computed along solutions of (C.1) is nonpositive, which says that V itself is nonincreasing along solutions. Since V is a measure of how far the solution is from the origin, (C.9) says that the solution must remain near the origin. If a Lyapunov function candidate V can be found satisfying (C.9) then V is called a **Lyapunov Function** for the system (C.1). Note that Theorem C.7 gives only a sufficient condition for stability of (C.1). If one is unable to find a Lyapunov function satisfying (C.9) it does not mean that the system is unstable. However, an easy sufficient condition for instability of (C.1) is for there to exist a Lyapunov function candidate V such that $\dot{V} > 0$ along at least one solution of the system.

(ii) *Theorem C.8*

The null solution of (C.1) is asymptotically stable if there exists a Lyapunov function candidate V such that $\dot{V}$ is strictly negative definite along solutions of (C.1), that is,

$$\dot{V}(\mathbf{x}) < 0 \tag{C.10}$$

The strict inequality in (C.10) means that V is actually decreasing along solution trajectories of (C.1) and hence the trajectories must be converging to the equilibrium point.

(*iii*) *Corollary C.9*

Let V be a Lyapunov function candidate and let S be any level surface of V, that is,

$$S(c_0) = \{\mathbf{x} \in \mathbb{R}^n \mid V(\mathbf{x}) = c_0\} \tag{C.11}$$

for some constant $c_0 > 0$. Then a solution $\mathbf{x}(t)$ of (C.1) is uniformly ultimately bounded with respect to S if

$$\dot{V} = \langle \mathbf{d}V, \mathbf{f}(\mathbf{x}) \rangle < 0 \tag{C.12}$$

for $\mathbf{x}$ outside of S.

If $\dot{V}$ is negative outside of S then the solution trajectory outside of S must be pointing toward S as shown in Figure C-2. Once the trajectory reaches S we may or may not be able to draw further conclusions about the system, except that the trajectory is trapped inside S.

C.1.3 LYAPUNOV STABILITY FOR LINEAR SYSTEMS

Consider the linear system (C.4) and let

$$V(\mathbf{x}) = \mathbf{x}^T P \mathbf{x} \tag{C.13}$$

be a Lyapunov function candidate, where P is symmetric and positive definite. Computing $\dot{V}$ along solutions of (C.4) yields

$$\begin{aligned} \dot{V} &= \dot{\mathbf{x}}^T P \mathbf{x} + \mathbf{x}^T P \dot{\mathbf{x}} \\ &= \mathbf{x}^T (A^T P + PA) \mathbf{x} \\ &= -\mathbf{x}^T Q \mathbf{x} \end{aligned} \tag{C.14}$$

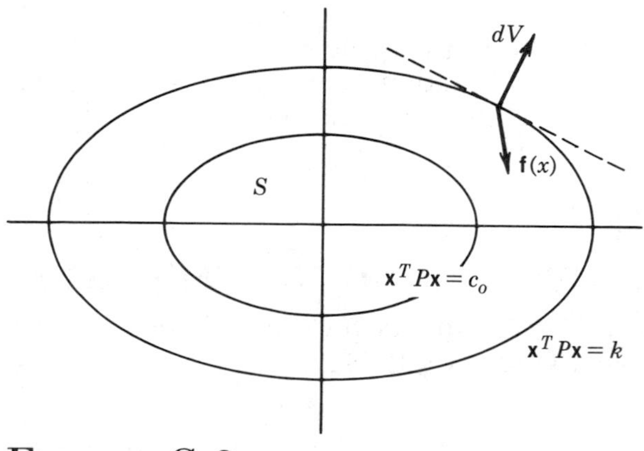

FIGURE C-2
Illustrating ultimate boundedness.

where we have defined Q as

$$A^T P + PA = -Q \tag{C.15}$$

Theorem (C.8) now says that if Q given by (C.15) is positive definite (it is automatically symmetric since P is) then the linear system (C.4) is stable. One approach that we can now take is to first fix Q to be symmetric, positive definite and solve (C.15), which is now called the **matrix Lyapunov equation**, for P. If a symmetric positive definite solution P can be found to this equation, then (C.4) is stable and $\mathbf{x}^T P \mathbf{x}$ is a Lyapunov function for the linear system (C.4). The converse to this statement also holds. In fact, we can summarize these statements as

(i) Theorem C.10

Given an $n \times n$ matrix A then all eigenvalues of A have negative real part if and only if for every symmetric positive definite $n \times n$ matrix Q, the Lyapunov equation (C.11) has a unique positive definite solution P.

Thus, we can reduce the determination of stability of a linear system to the solution of a system of linear equations, namely, (C.11), which is certainly easier than finding all the roots of the characteristic polynomial and, for large systems, is more efficient than, say, the Routh test.

The strict inequality in (C.7) may be difficult to obtain for a given system and Lyapunov function candidate. We therefore discuss **LaSalle's Theorem** which can be used to prove asympotic stability even when $\dot{V}$ is only negative semi-definite.

C.1.4 LASALLE'S TEOREM

Given the system (C.1) suppose a Lyapunov function candidate V is found such that, along solution trajectories

$$\dot{V} \leq 0 \tag{C.16}$$

Then (C.1) is asymptotically stable if $\dot{V}$ does not vanish identically along any solution of (C.1) other than the null solution, that is, (C.1) is asymptotically stable if the only solution of (C.1) satisfying

$$\dot{V} \equiv 0 \tag{C.17}$$

is the null solution.

REFERENCES AND SUGGESTED READING

[1] Vidyasagar, M., *Nonlinear Systems Analysis*, Prentice–Hall, Englewood Cliffs, N.J, 1978.

INDEX

G

H

I

J

K

L

M

N

O

P

Q

R